HORIZONS IN EARTH SCIENCE RESEARCH

VOLUME 11

Horizons in Earth Science Research

Additional books in this series can be found on Nova's website under the Series tab.

Additional e-books in this series can be found on Nova's website under the e-book tab.

HORIZONS IN EARTH SCIENCE RESEARCH

HORIZONS IN EARTH SCIENCE RESEARCH

VOLUME 11

BENJAMIN VERESS
AND
JOZSI SZIGETHY
EDITORS

nova publishers
New York

For permission to use material from this book please contact us:
Telephone 631-231-7269; Fax 631-231-8175
Web Site: http://www.novapublishers.com

Library of Congress Cataloging-in-Publication Data

ISBN: 978-1-63321-296-1

ISSN: 2159-1350

Published by Nova Science Publishers, Inc. † New York

CONTENTS

PREFACE

This book presents and discusses research in the field of earth science, with a focus on high resolution insight on the submarine mudflows of the gulf of Mexico; shale and its beneficial and dangerous characteristics for human beings; publicly available open access data and machine learning model-predictions applied with open source GIS for the entire Antarctic ocean: a first meta-analysis and synthesis from 53 charismatic species; a case study on ice sheets self-destructing and causing rapid climate change; conservation planning and management of limited water resources in arid and semi-arid areas as economic resources; laser Raman spectroscopy study on the growth zoning of feldspar, zircon and apatite; statistical prediction of winter rainfall in Patagonia (Argentina); and research of seismicity at an oceanic and sea bottoms.

Chapter 1 – The seafloor offshore of the Mississippi delta is characterized by numerous submarine mudflows, which have caused extensive damage to infrastructure in the area. These mudflows are thought to be triggered by storm-induced high waves. Hurricanes can cause much damage in the Gulf of Mexico and in this regard, the 2005 hurricane season was unprecedented. Post-hurricane surveys carried out by the industry determined that much of the damage was caused by mudflows that originated on the upper portions of the Mississippi birdsfoot delta. The authors have monitored those shallow mudflows in multi-year high resolution surveys (multibeam, side scan sonar, subbottom profiler techniques, cores).

The authors imaged the shallow portions of eight separate mudflows, including the source of one of the flows. The seafloor imagery shows spectacular detail of the head of the mudflows, and its connection with a dredge dump location. The five westernmost flows originate at a known dredge dump site and appear significantly wider than the three flows to the east. The eastern flows are more clearly delineated in the seafloor imagery and follow more tortuous downslope paths. This suggests a possible anthropogenic component to mudflow sourcing in this area. The subbottom profiler data show large blocks of material containing intact stratigraphy located within the mudflow gullies. High dips were observed in numerous blocks in a range of directions. The bases of these blocks were imaged in places, suggesting that the blocks possibly move downslope by sliding on top of more rigid material. Alternatively, the mudflows may be characterized by rigid blocks being transported along by a viscous / more fluid matrix. No high relief mudflows were observed in these data.

The morphologies on the seafloor and how they change seem to be directly related to the amount of big storms/hurricanes in the area. The surveyed seafloor did not show evidence of any change when the hurricanes season was weak. Surveying after a big hurricane season showed how the seafloor's morphologies can change and move downslope in the Gulf.

Chapter 2 – Shale is a laminated and fissile clastic (detrital) sedimentary rock, whose main ingredients include clays and silts. A multitude of accessory ingredients could also be present, depending on the chemistry of the provenance rocks, whose minerals were altered/weathered, and transported, by water and wind, to depositional basins for compaction and cementation. Because of the compositional variation, it is hard to find two shale outcrops chemically alike over the Earth surface. The major minerals, found in shales, are: kaolin, illite, smectite (montmorillonite), vermiculite, chlorite, feldspars, micas, and quartz. They were mostly identified by scanning electron microscopes and microprobe analyses because of their small dimensions, not exceeding 0.02mm. Many other minerals could not be identified due to their extremely fine sizes. Due to their complex compositions, shales cannot be chemically classified with precision. Rather, their types are summarized according to their colors (gray, black, brown, red, blue, green, etc.) which reflect the presence of dominant minerals, or organic materials. Also, the cement, which solidifies the loose ingredients, plays an important role in coloring the shale (red, for instance, would indicate the presence of the cementing hematite Fe_2O_3; and the dark color, the presence of dead organisms or oil compounds). In many ways, shales are useful to human beings. They form impervious layers over oil pools, preventing the petroleum from escaping as gases to the atmosphere. They also form impervious bases for perched and other aquifers. When metamorphosed to cleavable slates, the slaty sheets are used in roofing (commonly used in Europe), as tiles to pave the floors, and as small framed boards for pupils. Shales are hosts to oil compounds (oil shales), and to beneficial metallic mineral deposits (Kupferschiefer, Cu-deposit in Europe, and uranium). However, the shale presence, beneath earth-fill dams is dangerous to their stability. By getting wet, shales become muddy with a weak resistance, causing thus differential subsidence of , and small faults along the dam, leading to water seepage. This situation, observed at Toledo Bend dam, between Louisiana and Texas, forced the engineers to drop the lake water surface to the dam's base, and to ignore the electricity generation project, in order to save the dam for tourism. The shale wetting would increase the compactibility and would effect the stability of streets, highways, and buildings.

Chapter 3 – The Antarctic Treaty is of global importance and it is very clear in its statement and spirit that science is to be shared with the global community. However, huge amounts of data still remain to be made freely available to the public. Available data is often in data formats that are not easily accessible or limited to just certain software platforms, and the associated metadata is often poorly done or non existent. There is tremendous progress brought by publicly available open access 'presence only' (location only) data sets though, e.g. for scientific and conservation management questions. While this is fully in line with ongoing data initiatives such as the International Polar Year (IPY 2007), the International Council for Science (ww.icsu.org) and the Cyber-infrastructure with U.S. NSF and its relatives, such compiled and ready-to-use datasets in a good format do not exist or are still very difficult to come by. Here the authors make such data sets publicly available for free for 53 species and for further use by the global community. Secondly, they provide a first application of such datasets using free tools such as machine learning software in R, as well as open source geographic information systems (such as QGIS, R and SAGA GIS). While these resulting data can be used and improved further, here they present a first and robust meta-analysis and synthesis from these experiences and models.

Chapter 4 – Recent global warming has greatly reduced the area of perennial sea ice in the Arctic Ocean, caused mountain glaciers to retreat and allowed major ice streams draining the Antarctic and Greenland ice sheets to approach and even exceed surge velocities. This has brought new urgency to the concept that large ice sheets, past, present, and future, may have the capacity to self-destruct causing rapid and permanent changes in global climate and sea level. This possibility has focused attention on these ice streams, beginning with Jakobshavn Isbrae, long the world's fastest ice stream and discharging seven percent of Greenland ice, a discharge rate that has now doubled following catastrophic disintegration of its buttressing ice shelf in Jakobshavn Isfjord in 1997. A geometrical solution to the longitudinal force balance was combined with a simple mass balance to obtain two solutions to the resulting force-and-mass balance from 1986 to 2006. The two solutions are obtained using two approaches to quantify the loss of buttressing. The best match to changes in surface elevations and velocities along the ice stream, especially the rapidly lowering surface during this timespan, is provided by the solution that has the strongest reduction of ice-bed coupling up the ice stream, quantified by a rapid increase in the "floating fraction" of ice. This pattern seems to be repeating for other Greenland and Antarctic ice streams. If rapid surface lowering and ice discharge spreads to all major ice streams and lasts for centuries, even accelerating over time, self-destruction of these two remaining ice sheets seems inevitable. Without the high albedo of ice sheets, global warming would increase until a new global equilibrium with the new boundary conditions was established. The other possibility is the observed rapid ice thinning and discharge may be transient and lead to a new stable equilibrium that keeps these ice sheets largely intact with no significant change in global climate and sea level. The authors results from Jakobshavn Isbrae were unable to determine which scenario is most likely, but time will tell, probably over the next few decades as observational data accumulate and our modeling capability improves.

Chapter 5 – Thirty-four million people live in Sudan, and environmental pollution is a major concern throughout the country; therefore industry, communities, local authorities and central government, to deal with pollution issues, should adopt an integrated approach. Most polluters pay little or no attention to the control and proper management of polluting effluents. This may be reflect lack of enforceable legislation and/or the fear of spending money on the treatment of their effluent prior to discharge. Present resources must be strictly monitored and managed effectively if further deterioration is to be avoided. Community participation in water management programmes should fully involve women, who play a central role in the provision and safeguarding of water for domestic use in rural areas. The pricing system recognises the economic value of water, as stated in the principles of the policy.

Chapter 6 – Zircon, apatite and feldspar are widely distributed in the crust, and they are common in rocks of various genetic types. Feldspar occurs as the main mineral while zircon and apatite are important accessory minerals of igneous rocks. Chemical composition and crystal structure of these minerals offer temperature, pressure and other formation environmental parameters, and they record the origin, migration and evolution of magma as well.

Ordering degree of alkali feldspars, zoning structure of plagioclase feldspars, zoning structure and elements composition of zircon, zoning structure and geochemical features of apatite were researched in this experiment with LRM technology, and the variation of Raman spectra was summarized and qualitative and semi-quantitative relationships were established.

Combined these minerals' information with regional tectonic situation and thermal evolution history, the forming environment of magmatic crystallization and its geological significance were discussed.

Chapter 7 – This chapter describes rainfall evolution in southern Argentina (Patagonia) and faces with the possibility to predict winter rainfall using oceanic and atmospheric predictors. The Andes Mountain ranges along the west of Patagonia and a large plateau extends towards the east. A decrease of rainfall was observed in all seasons in central and east Patagonia meanwhile an increment is observed in southern part of Los Andes. The study area was regionalized using a principal component analysis and four regions could be defined. Almost all regions showed an annual cycle with maximum rainfall in autumn-winter. Therefore, several different statistical models: an autoregressive integrated moving average model (ARIMA), Holt Winter (HW), Climate Prediction Tool (CPT) and an ensemble of all, called multimodel were applied to predict MJJ (May to July) Patagonia rainfall in four regions defined in Patagonia. They were designed with MJJ rainfall for the period 1975-2007 in CPT model and for the period 1988-2007 in ARIMA and Holt Winter because this last two models use 1975-1987 values to generate the final scheme. All the models were proved for 2008-2012. The main result of this work is that the consideration of ARIMA and HW in the multimodel ensemble improves the efficiency obtained using CPT. This result was supported by the calculation of efficiency coefficients which represent the ability to detect over and sub normal rainfall.

Chapter 8 – The modern world seismological network consists mainly of stations on land while most earthquakes occur on seas and oceans. To create a single tectonic scheme of the Earth it is necessary to develop global network of bottom stations. However, there are many problems. In this paper the features of seismic signals in the bottom waters are discussed, which still has not been sufficiently studied. These problems are including: coupling a seismograph housing, resonance phenomena in recording signals on the bottom, noises from near bottom currents, and others. Describes broadband bottom seismographs and bottom observatory, developed in the Shirshov Institute of Oceanology Russian Academy of Science, taking into account the specific conditions recording at the sea bottom. There are present some results of studies with bottom seismographs. It is considered the possibility of using the bottom observatory for the study hydro-physic precursors of strong earthquakes in the water areas.

In: Horizons in Earth Science Research. Volume 11
Editors: Benjamin Veress and Jozsi Szigethy
ISBN: 978-1-63321-296-1

Chapter 1

HIGH RESOLUTION INSIGHT ON THE SUBMARINE MUDFLOWS OF THE GULF OF MEXICO

***A. García-García*[1,2] *and D. Orange*[1,3]**
[1]University of California in Santa Cruz, Santa Cruz, CA, US
[2]Monterey Peninsula College, Monterey, CA, US
[3]Oro Negro Exploration LLC, Jakarta, Indonesia

ABSTRACT

The seafloor offshore of the Mississippi delta is characterized by numerous submarine mudflows, which have caused extensive damage to infrastructure in the area. These mudflows are thought to be triggered by storm-induced high waves. Hurricanes can cause much damage in the Gulf of Mexico and in this regard, the 2005 hurricane season was unprecedented. Post-hurricane surveys carried out by the industry determined that much of the damage was caused by mudflows that originated on the upper portions of the Mississippi birdsfoot delta. We have monitored those shallow mudflows in multi-year high resolution surveys (multibeam, side scan sonar, subbottom profiler techniques, cores).

We imaged the shallow portions of eight separate mudflows, including the source of one of the flows. The seafloor imagery shows spectacular detail of the head of the mudflows, and its connection with a dredge dump location. The five westernmost flows originate at a known dredge dump site and appear significantly wider than the three flows to the east. The eastern flows are more clearly delineated in the seafloor imagery and follow more tortuous downslope paths. This suggests a possible anthropogenic component to mudflow sourcing in this area. The subbottom profiler data show large blocks of material containing intact stratigraphy located within the mudflow gullies. High dips were observed in numerous blocks in a range of directions. The bases of these blocks were imaged in places, suggesting that the blocks possibly move downslope by sliding on top of more rigid material. Alternatively, the mudflows may be characterized by rigid blocks being transported along by a viscous / more fluid matrix. No high relief mudflows were observed in these data.

The morphologies on the seafloor and how they change seem to be directly related to the amount of big storms/hurricanes in the area. The surveyed seafloor did not show evidence of any change when the hurricanes season was weak. Surveying after a big hurricane season showed how the seafloor's morphologies can change and move downslope in the Gulf.

Keywords: Submarine mudflows, hurricanes, seafloor, multibeam, subbotom profiler, Gulf of Mexico

INTRODUCTION

The seafloor off the coast of Louisiana has been studied in detail in the past due to two main factors: the vast infrastructure related to the oil and gas industry that has been built in the area, and the instability of the seafloor and subsurface sediments. The area has endured multiple seasonal hurricanes and tropical storms, one of the deadliest and costliest in the U.S. being the 2005 Atlantic hurricane season, with seven major hurricanes (categories 3, 4, or 5 on the Saffir/Simpson Hurricane Wind Scale), (Blake et al., 2011). Research into the late Holocene records on land shows that major hurricanes also affected the area in the past (Liu and Fearn, 2000; Wallace and Anderson, 2013).

Historically, pipelines have been the most common means used to transport crude oil produced in the Gulf of Mexico (GoM), (Cranswick, 2001). As of June 2001, there were approximately 44,218 km (27,569 mi) of pipeline on the seafloor of the GoM. Most of these pipelines connect to shelf and near-shelf facilities; a small percentage supports deepwater operations (Cranswick, 2001). These numbers have increased with time, along with platforms and supportive infrastructure constructed on the seafloor. Based on the existing inventory of exploratory prospects for the deepwater operators and continuing successes on the shelf, there will be strong exploration and production (E&P) activity in the GoM for decades to come (Kulkarni, 2013). All of these infrastructures have the potential of being affected by big storms and movement of the sediments underneath.

The Mississippi River delta has a high sediment capacity with 0.6-1.2 meters/year (23-47 inches/year) of sediment accumulation near the mouth, up to 3-4.6 m (10-15 feet) in a river flood, being initially accumulated in the upper edge of the delta (Handley, 1982). The sediments of the submerged delta consist of thick, weak, low permeability sediments that can be deposited rapidly making the area very unstable (Coleman et al., 1982; Handley, 1982; Hitchcock et al., 2010). Large hurricanes in the GoM can generate waves large enough to trigger submarine slope failures, commonly referred to as "mudslides" or "mudflows", moving rapidly downslope and affecting the seafloor structures. This became particularly evident in 1969 when the large (at least 21 m high) waves generated by hurricane Camille produced a mudslide in the South Pass Block 70 area of the GoM that destroyed one offshore oil platform and severely damaged at least one other (Sterling and Strobeck, 1973). Following this mudslide, considerable attention was devoted by the offshore oil industry to developing models that could be used to predict when mudslides might occur and to estimate movements and risks (Nodine et al., 2010; Hitchcock et al., 2010). Some of these mudflows have been recently studied and mapped in detail by different authors (García-García et al., 2008; Hitchcock et al., 2006, 2010; Levey et al., 2009; Nodine et al., 2010; Roberts, 1984).

REGIONAL SETTING

The Gulf of Mexico is a Mediterranean-type sea located at the southeastern corner of North America. It is bordered by the United States to the north (Florida, Alabama, Mississippi, Louisiana, Texas), five Mexican states to the west (Tamaulipas, Veracruz, Tabasco, Campeche, Yucatan), and the island of Cuba to the southeast (Gore, 1992; Darnell and Defenbaugh, 1990).

The Gulf measures approximately 1,600 kilometers from east to west, 900 km from north to south, and has a surface area of 1.5 million square km. The marine shoreline from Cape Sable, Florida to the tip of the Yucatan peninsula extends ~5,700 km, with another 380 km of shore on the northwest tip of Cuba. If bays and other inland waters are included, the total shoreline increases to over 27,000 km in the U.S. alone. Its small basin (Menard, 1967) resembles a large pit with a broad shallow rim. Approximately 38% of the Gulf is comprised by shallow and intertidal areas (< 20 m deep). The area of the continental shelf (< 180 m) and continental slope (180 - 3,000 m) represent 22% and 20% respectively, and abyssal areas deeper than 3,000 m comprise the final 20% (Gore, 1992).

The Northern Gulf of Mexico extends from just east of the Mississippi Delta near Biloxi to the eastern side of Apalachee Bay. It consists of a wide shelf, shelf break, slope and rise morphology. The present physiography of the slope is dominated by salt tectonics, in which hummocky bathymetry is dominated by irregularly shaped salt withdrawal basins and the intervening structural highs that are often underlain by diapers (Bouma et al., 1990; Winker and Booth, 2000). Terrigenous siliclastic deposition, dominated in the Pleistocene and Holocene by the Mississippi River, characterizes most of the basins and intra-basin settings (Winker and Booth, 2000).

Mudflows

Mudflows are the primary mechanism of sediment movement in the area and are the result of ongoing processes associated with historic and recent sedimentation by the Mississippi River (Bea et al., 1975, Coleman et al., 1982, Roberts, 1984). The seafloor slopes are very gentle (less than 1 degree) and the mudflows are triggered by cyclic loading associated with large, long wavelength waves generated during significant storm events (Bea et al., 1975, Coleman et al., 1982). Based upon surveys of infrastructure damaged during hurricane Ivan, Thompson et al. (2005) inferred that mudflows triggered by Ivan moved a 16.7 km (27 mile) pipeline section 580 m (1900 ft). Geotechnical work conducted in the mudflow region suggests that the mobile sediment in the mudflows can be 12.2 – 24.4 m (40 – 80 ft) thick (Hooper, 2005).

Atlantic Hurricane Seasons (2004-2008)

The area of interest in the Gulf of Mexico is affected by seasonal hurricanes and tropical storms (see Figure 1). The 2004 Atlantic hurricane season was well above average in activity, with fifteen named storms and one of the highest accumulated cyclone energy totals ever observed (Ivan, National Hurricane Center, NOAA). The 2005 season was the most active season in recorded history (National Hurricane Center, NOAA). The impact was widespread and ruinous with at least 2,280 deaths and record damages of over $128 billion USD. Of the storms that made landfall, five of the season's seven major hurricanes—Dennis, Emily, Katrina, Rita, and Wilma—were responsible for most of the destruction. The most catastrophic effects of the season were felt on the United States' Gulf Coast, where a 10 m storm surge from Hurricane Katrina caused devastating flooding that inundated New Orleans, Louisiana, and destroyed many structures on the Mississippi coastline. After that, the 2006 Atlantic hurricane season was much less active and was the first since 2001 in which no hurricanes made landfall in the United States. The 2007 season was an active season but 2008 was much more active, with 16 named storms formed, 8 of them hurricanes and 5 of those major hurricanes; the third most costly on record (Ike).

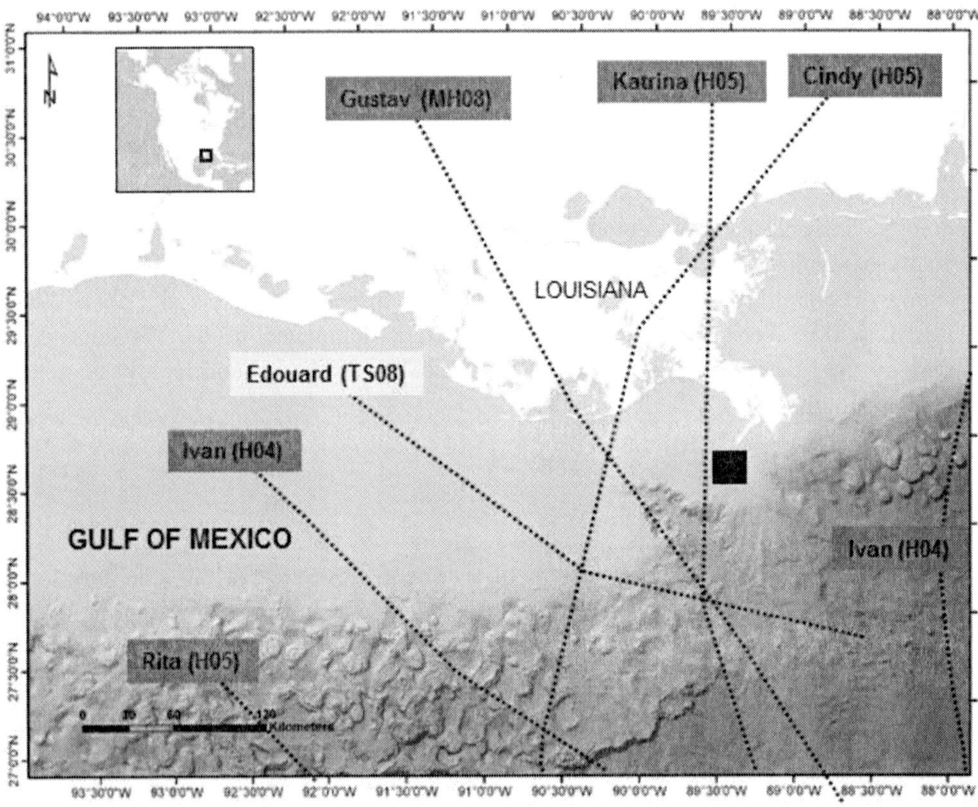

Figure 1. Area of interest (black square) south of the Mississippi river delta, offshore Louisiana, USA (see map in inset). The main paths for the major hurricanes (MH), hurricanes (H) and tropical storms (TS) for 2004, 2005 and 2008 that affected the area of interest are showed (track information taken from the National Hurricane Center NOAA).

Materials and Methods

In order to monitor and obtain high resolution data of the shallow seafloor mudflows (18 m to 50 m water depth) we run several surveys in 2007-2009 aboard the R/V Pelican, based at the Louisiana Universities Marine Consortium (LUMCON) facility in Cocodrie, LA (http://www.lumcon.edu/). Data were acquired using a pole-mounted EM3002 (300 kHz) multibeam system, and a towed dual frequency Benthos Datasonics SIS-1000 with a side scan sonar (90-110 kHz frequency) and sub-bottom profiler (2-7 kHz frequency) fish (see Figure 2), which were run over an 8 km by 2 km grid at a 100 m line spacing pattern. We surveyed in summer 2007 and in February 2009 (after the strong hurricane Atlantic season of 2008). We run additional coring programs (box cores) in 2008 to groundtruth the datasets.

Figure 2. (A) RV Pelican at LUMCON's facility; (B) towed Benthos (Datasonics) SIS-1000 with 90-110 kHz sidescan sonar and 2-7 kHz subbottom chirp profiler; (C) Pole-mounted 300 kHz Kongsberg Simrad EM3002 multibeam system, (130 degree swath width, 254 beams).

We used Triton Imaging Inc. software to acquire and interpret the side scan sonar and subbottom seismic lines and CARIS for the multibeam datasets.

Results

The high resolution seafloor data of the area shows two very different types of morphologies (Figure 3): smooth areas of seafloor that are relatively high and deeper broad gullies that show a rougher and more irregular topography (numbers 1-8). Farther downslope these gullies coalesce into broad mudflows whose surface may be irregular or smooth. Locally smooth areas that are within mudflows can be differentiated by mapping the lateral and downslope extent of allochthonous mud.

We imaged the shallow portions of eight separate mudflows, including the source of one of the flows (6) (see Figure 3). The seafloor imagery shows spectacular detail of the head of the mud flows, and its connection with a dredge dump location. The five westernmost flows originate at a known dredge dump site and appear significantly wider than the three flows to the east. The eastern flows are more clearly delineated in the seafloor imagery and follow more tortuous downslope paths. The subbottom profiler data show large blocks of material containing intact stratigraphy located within the mudflow gullies (Figure 4, see detail in Figure 7). High dips were observed in numerous blocks in both the strike and dip directions. The base of these blocks was imaged in places, suggesting that the blocks move downslope by sliding on top of more viscous material. No high relief mudflows were observed.

In a few sites, box corers were taken to groundtruth the indirect datasets (Figures 3 and 5). All cores showed a good recovery with 50-70 cm of very soft clay, sometimes with organic matter (and H_2S smell). We did not see any difference in the grain size from the sediment recovered from areas of gullies versus areas of smooth seafloor, although there were differences in the color.

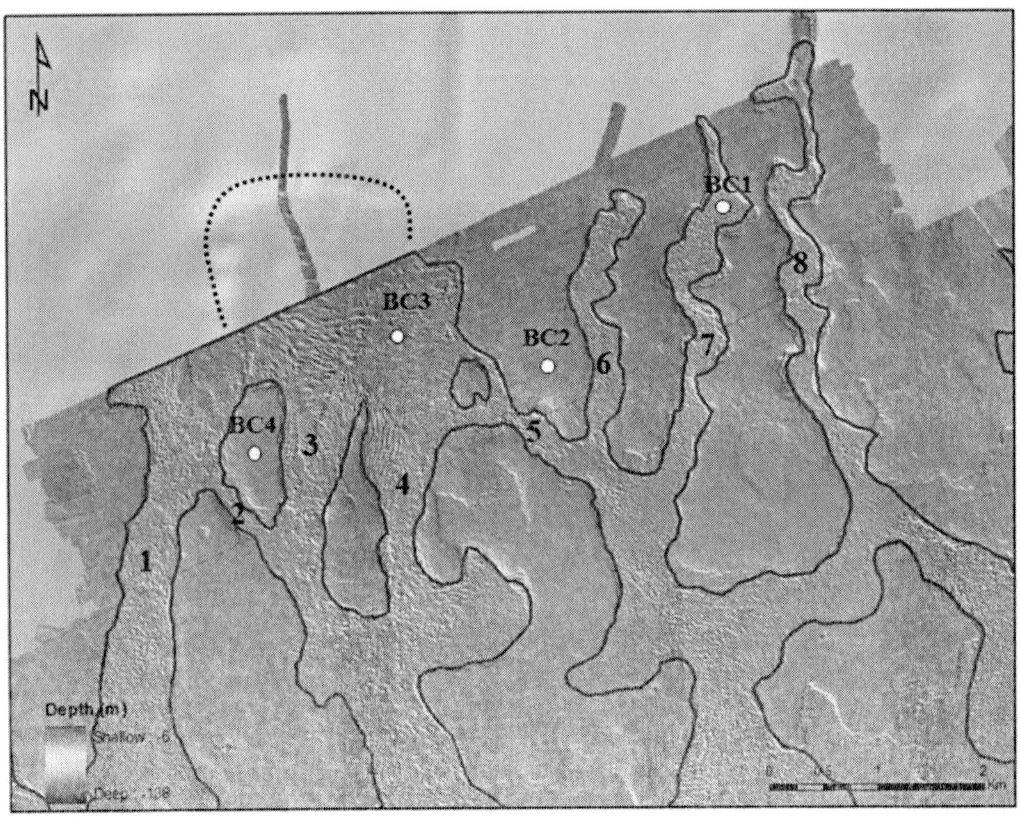

Figure 3. High resolution multibeam bathymetric image of the shallow seafloor off the Mississippi River Delta (meters), with a 5 m grid cell size. The identified gullies have been numbered 1 to 8.

Location of box cores shown as white dots (BC1 to BC4). Location of the Corps of Engineers dredge dump location outlined with dashed line.

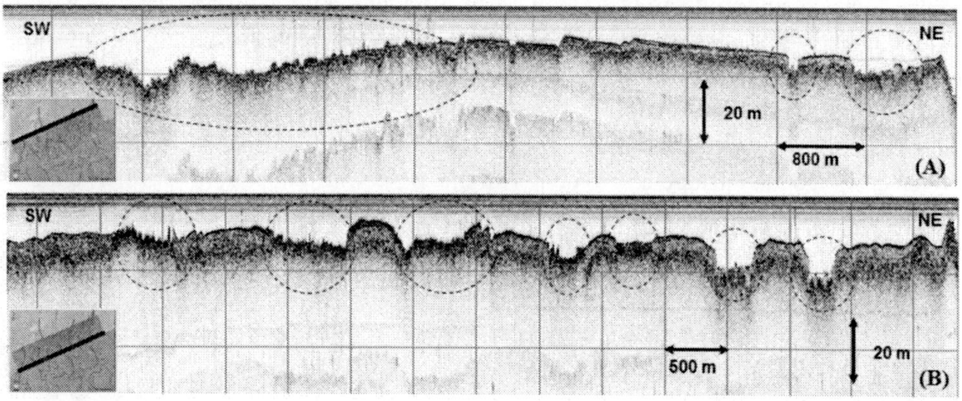

Figure 4. SBP SW-NE lines across the shallow mudflows. (A) Line indicating up to 4 m of visible sediment showing smooth areas and rough ones (gullies, indicated by dashed lines); (B) southernmost line showing up to 7 m of visible sediment, showing the different morphologies within the mudflows.

Latitude	Longitude	Core	Description
28°52'.713	-89°24'.412	BC1	Oily (shiny) olive grey soft mud with shells
28°52'.006	-89°25'.414	BC2	Oily (shiny) olive grey soft mud with shells
28°52'.167	-89°26'.527	BC3	5cm-soft grey homogenous clay above 10 cm of dark black mud with high organic matter (H2S smell), above 10 cm of soft grey clay
28°52'.513	-89°27'.063	BC4	2cm-soft brown mud with tube worms, above 10 cm sandy clay with shell fragments, above sandy mud layer with shell fragments

Figure 5. Photograph of box core BC3 and location and description of all boxcores.

DISCUSSION

The high resolution datasets show a very different seafloor in the east and west of the surveyed area (Figures 3 and 4). We interpret the smooth areas of seafloor to be unfailed, with the broad gullies of irregular topography representing mudflows. The eastern flow paths (Figure 6) have very narrow morphologies, indicating the mudflows are channelled through them. The seafloor in between gullies shows much smoother morphologies and a lobe shape correlated to the downslope movement (to the SE). The subbottom profiler data shows sediment thickness up to 10 m in the smooth regions between the gullies. Hooper (1980) called these 'crusts', which perfectly matches the morphologies.

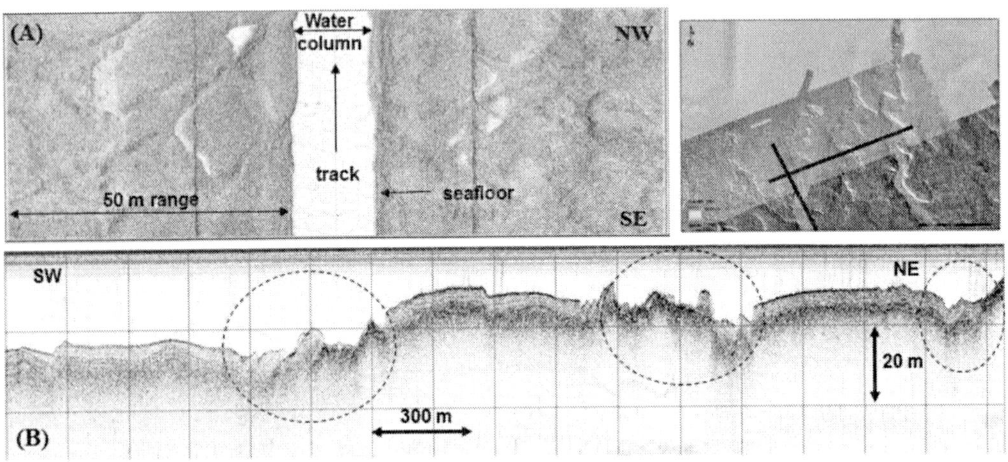

Figure 6. Detail of eastern seafloor showing the narrow gullies (multibeam image and side scan sonar (A); and the rough topography in a subbottom profiler line (B).

In the western seafloor mudflows appear significantly wider than the three flows to the east (Figure 7) and are located directly south of a known dredge dump site (EPA region 6). This U.S. Army Corps of Engineers site correlates well with the area of widest gullies. Numerous studies show that these dumping sites affect the seafloor morphology and rapid change of the seafloor (Wienberg and Bartholoma, 2005). In this case, we note a clear correlation that the dredge dump strongly impacts sediment distribution and seafloor morphology and leads to wider and larger mudflows. Another anthropogenic impact on this seafloor is trawling, which is common on the side scan sonar records during our surveys. It can modify the physical properties of the seafloor sediments and the sediment fluxes (Churchill, 1989), not only affecting the continental shelves seafloors but also the deep seafloor (Puig et al., 2012).

Interpreting the seismic data allows us to measure the sediment accumulated in the area. A map imaging the total amount of sediment, called an isopach map, shows thicknesses up to 5 meters (Figure 8). The areas with high sediment accumulation are located in the smooth areas of seafloor whilst the gullies have close to minimal accumulation. The Corps dump site is located above the western surveyed seafloor, which has the minimum interpreted sediment and the widest gullies. We interpret that that area serves as a passage of sediment downslope and that the morphology of the western seafloor has been strongly affected by that location of that dumping site. We suggest that the seafloor would look like the eastern seafloor if the site was to be removed in the future. It is likely all gullies accumulate the same amount of sediment, and then, move downslope. Their morphology thought seems to be affected by this anthropogenic process.

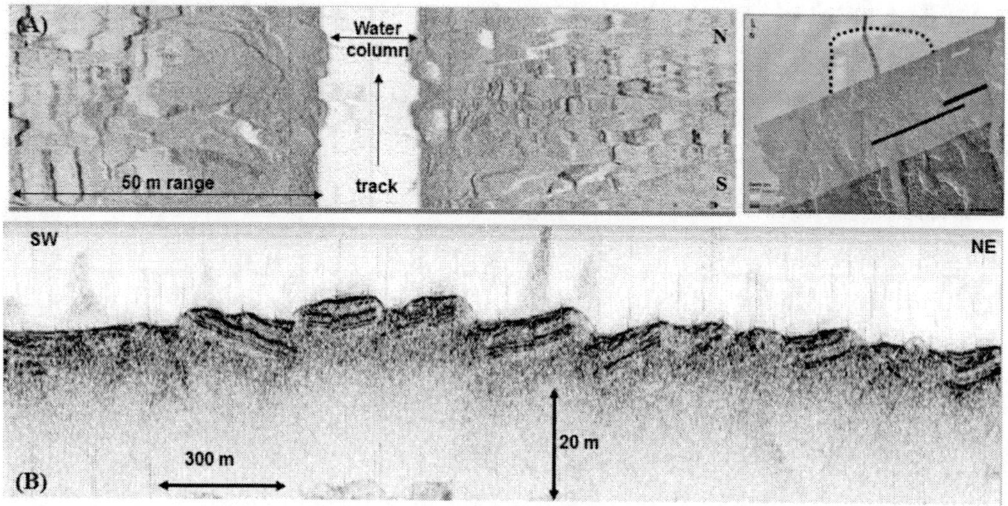

Figure 7. Detail of western seafloor showing the wide gullies (multibeam image and side scan sonar, A) and the U.S. Army Corps of Engineers dredge dump location (dashed line); (B) Subbottom profiler line (shorter line) shows examples of tilted blocks (intact stratigraphy) with visible thicknesses of 5-10 meters.

The fact that the box cores taken did not show differences in the grain size tells us all the sediment has the same source and is impacted by the same forces/flows. The main difference in this seafloor is the morphologies and that is directly related to where the gullies and the dumping site are located. What the box corers did show were differences in the color for the clay. The western area had olive color that did not occur in the western sites. Although only four cores were taken, it is possible the color difference can be explained by the different minerals the sediment sources have. This would correlate also with the source of material in the west area being the dredging areas off Louisiana, material that doesn't reach the easternmost gullies.

The occurrence of gas in sediments is well known in the Northern Gulf of Mexico (Anderson and Bryant, 1990; Hooper, 1980; Roberts and Carney, 1997). In particular, this already unstable area has been known for containing shallow gas and seeping sediments. A preliminary analysis of the box cores collected showed signs of gas, with the strong sulphide smell and the widespread evidence of bubbles in the box cores. The content of gas in the sediments would further weaken the sediment and decrease the potential for seafloor instability.

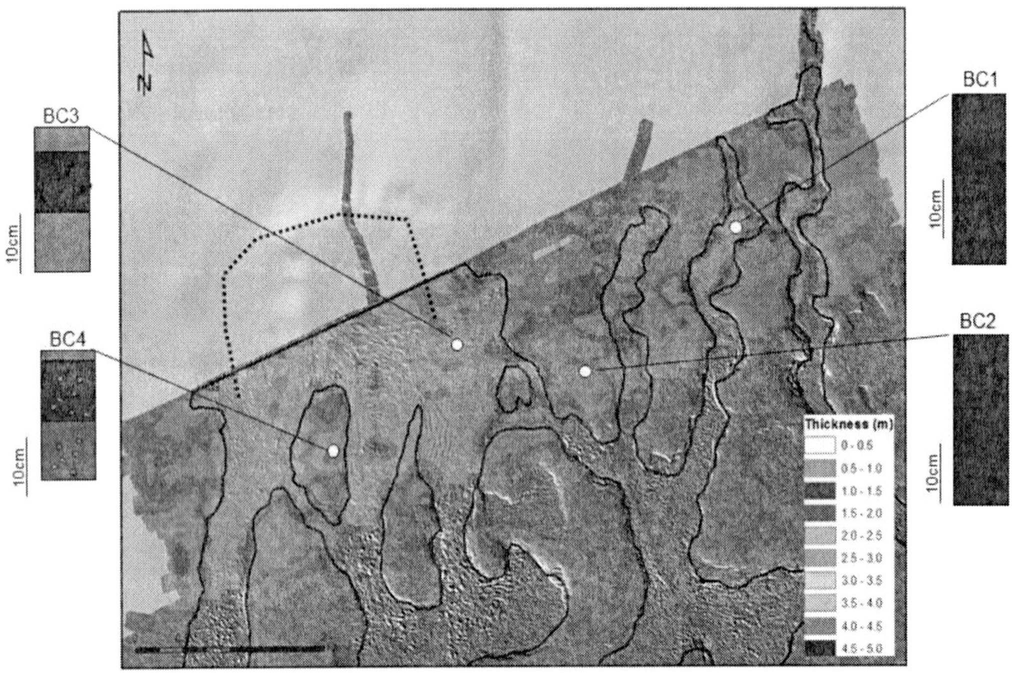

Figure 8. Isopach map showing the recent sedimentation in meters after interpretation of the subbottom profiler data. Note the lack of it along the gullies and in particular, in the western area, where the Corps Eng. Site is located. Reconstruction of box cores also shown (see description in figure 5).

Conclusion

We have monitored the GoM shallow mudflows in multi-year high resolution surveys, which allowed us to document the seafloor and subsurface sediments with high resolution datasets.

The morphology of the head of the western mudflows is interpreted to be affected by the proximity of a dredge dump location. The western mudflows appear significantly wider than the three mudflows to the east. The eastern flows are more clearly delineated in the seafloor imagery and follow more tortuous downslope paths.

The subbottom profiler data show large blocks of material containing intact stratigraphy located within the mudflow gullies. The isopach map shows minimal sediment accumulation at the gullies (interpreted as bypass areas for mudflows going downslope). The maximum thicknesses, up to 5 m, are located at the smooth seafloor, in between gullies. High dips were observed in numerous blocks in a range of directions. The bases of these blocks were imaged in places, suggesting that the blocks move downslope by sliding on top of more rigid material. Alternatively, the mudflows may be characterized by rigid blocks being transported along by a viscous / more fluid matrix, where sediment accumulation offsets the volumetric removal of material downslope.

The morphologies on the seafloor and how they change seem to be directly related to the amount of big storms/hurricanes in the area. The surveyed seafloor did not show evidence of any change when the hurricanes season was weak. Surveying after a big hurricane season showed how the seafloor's morphologies can change and move downslope in the gulf.

Acknowledgments

We want to thank all the crew from the R/V Pelican cruises as well as the AOA Geophysics Inc. colleagues that helped in earlier discussions, especially Dr. Marianne Mulrey, Ms. Jennifer Henderson, and Mr. Matt Levey. Acoustic data were acquired and processed by Seafloor Systems, Inc. We used the Triton Imaging Inc. software to interpret the side scan sonar and subbotom seismic lines. The research herein was funded by the U.S. Navy through the Office of Naval Research (contract numbers N00014-03-1-142, N00014-10-1-0583, and N00014-14-1-0172).

References

Anderson, A.L. and Bryant W.R. (1990) Gassy sediment occurrence and properties: Northern Gulf of Mexico. *Geo-Marine Letters*, Volume 10, Issue 4, pp 209-220.

Bea, R. G., H. A. Bernard, P. Arnold, and E. H. Doyle (1975), Soil Movements and forces developed by wave-induced slides in the Mississippi Delta: *Journal of Petroleum Technology*, v 27: 500-514.

Blake, E.S., Lansea, C.W., and Gibney, E.J. (2011) The deadliest, costliest, and most intense united states tropical cyclones from 1851 to 2010 (and other frequently requested hurricane facts). NOAA Technical Memorandum NWS NHC-6.

Bouma, A.H., Roberts, H.H., and Coleman, J.M. (1990) Acoustical and geological characteristics of near-surface sediments, upper continental slope of northern Gulf of Mexico. *Geo-Marine Letters*, Volume 10, Issue 4, pp 200-208.

Churchill, J.H. (1989) The effect of commercial trawling on sediment resuspension and transport over the Middle Atlantic Bight continental shelf. *Cont. Shelf Res*. 9, 841-865.

Coleman, J. M., D. B. Prior, and L.E. Garrison (1982). Subaqueous sediment instabilities in the offshore Mississippi River delta*: Bureau of Land Management Open-File Report* 80-02: 1-49.

Cranswick, D. (2001) Brief Overview of Gulf of Mexico OCS Oil and Gas Pipelines: Installation, Potential Impacts, and Mitigation Measures. MMS, UIS Department of the Interior, OCS Report MMS 2001-067.

Darnell, R.M. and R.E. Defenbaugh. 1990. Gulf of Mexico: Environmental Overview and History of Environmental Research. *American Zoologist* 30:3-6.

Handley, L.R. (1982) Mississippi River delta mudslides: techniques allow for continued development with reduced risk. *Geo-Marine Letters* 2, 179-184.

Hitchcock, C., Givler, R., Angell, M., and Hooper, J. (2006) A Pilot Study for Regionally-Consistent Hazard Susceptibility Mapping of Submarine udslides. *OTC Paper* 18323.

Hitchcock, C., Givler, R., Angell, M., and Hooper, J. (2010) GIS-based assessment of submarine mudflow hazard offshore of the Mississippi Delta, Gulf of Mexico. In:

Mosher, D.C. et al. (eds.), Submarine Mass Movements and Their Consequences, *Advances in Natural and Technological Hazards Research*, vol. 28: 353-364.

Hooper, JR (1980) Crustal Layers in Mississippi delta mudflows. OTC paper, 3770-MS, DOI 10.4043/3770-MS.

Hooper, J. R. (2005), Hurricane Ivan as a Geologic Force: Mississippi Delta Front Seafloor Failures, Proc. Offshore Technology Conference, Houston, Texas, OTC Paper No. 17737, 4 pp.

García-García, A., Henderson, J., Henning, A., Orange, D., Mulrey, M. and RV Pelican field party. (2008) High-Resolution Imagery of Mudflows in the Mississippi River Delta: Results from Multibeam, Sidescan Sonar and Subbottom Profiler Data. The 5th International Conference on High-resolution Surveys in Shallow Water, New Hampshire, October 2008.

Gore, R.H. 1992. The Gulf of Mexico. Pineapple Press, Inc. Sarasota Florida. 384 p.

Kulkarni, P. (2013) Regional Report Gulf of Mexico. *World Oil*, April 2013, vol. 234, No. 4.

Levey, M., García-García, A. and Orange, D. (2009) Evidence of hurricane-induced mudflows and their characteristics near the West Delta, Gulf of Mexico. EOS Trans., AGU 90 (52), Fall. Meet. Suppl., Abstract NH43B-1310.

Liu, K-b. and Fearn, ML (2000) Holocene History of Catastrophic Hurricane Landfalls along the Gulf of Mexico Coast Reconstructed from Coastal Lake and Marsh Sediments. Section 3.2. In: Ning et al. (eds.) Current Stresses and Potential Vulnerabilities: Implications of Global Change for the Gulf Coast Region of the United States, Franklin Press, Inc. and GCRCC.

Menard, HW (1967) Transitional Types of Crusts under Small Ocean Basins. *Journal of Geophysical Research,* vo9l. 72, No. 12: 3061-3076.

National Hurricane Center NOAA, http://www.nhc.noaa.gov/index.shtml

Nodine. MC, Gilbert, RB, Cheaon, JY, Wright, SG, and Ward, EG (2010) Risk Analysis for Hurricane-Induced Submarine Mudslides. In: Mosher, D.C. et al. (eds.), Submarine Mass Movements and Their Consequences, *Advances in Natural and Technological Hazards Research*, vol. 28: 353-364.

Roberts, H. H. (1984), Sedimentology of mudflow deposits in the Mississippi River delta-front environment. *AAPG Bulletin*, vol 68, n 4: 521-522.

Harry H. Roberts and Robert S. Carney (1997) Evidence of episodic fluid, gas, and sediment venting on the northern Gulf of Mexico continental slope Economic Geology, December 1997, v. 92, p. 863-879

Puig, P., Canals, M., Company, J.B., Martin, J., Amblas, D., Lastras, G., Palanques, A. (2012) Ploughing the deep sea floor. Nature, 489: 286-290.

Sterling, G. H., and E. E. Strobeck (1973), "The Failure of the South Pass 70 B Platform in Hurricane Camille," *Proc. Offshore Technology Conference*, Houston, Texas, OTC Paper No. 1898.

Thompson, J., M. Garrett, M. Taylor, and T. George (2005), "Sonar surveys for pipeline inspection show extent of pipeline displacement and seafloor instability following Hurricane Ivan," *Proc. Offshore Technology Conference,* Houston, Texas, OTC Paper No. 17738.

Wallace, D.J. and Anderson, JB (2013) Evidence of similar probability of intense hurricane strikes for the Gulf of Mexico over the late Holocene. *Geology*, v. 38, 6: 511-514.

Wienberg, C. and Bartholoma, A. (2005) Acoustic seabed classification in a coastal environment (outer Weser Estuary, German Bigth)-a new approach to monitor dredging and dredge spoil disposal. *Cont Shelf Res*. 25: 1143-1156.

Winker, C. D., and J. R. Booth (2000) Sedimentary dynamics of the salt-dominated continental slope, Gulf of Mexico: Integration of observations from the seafloor, near-surface, and deep subsurface, in P. Weimer, R. M. Slatt, J. Coleman, N. C. Rosen, H. Nelson, A. H. Bouma, M. J. Styzen, and D. T. Lawrence, eds., Deep-water Reservoirs of the World: SEPM Gulf Coast Section 20th Annual Research Conference, p. 1059-1086.

In: Horizons in Earth Science Research. Volume 11
Editors: Benjamin Veress and Jozsi Szigethy
ISBN: 978-1-63321-296-1

Chapter 2

SHALE: ITS BENEFICIAL AND DANGEROUS CHARACTERISTICS FOR HUMAN BEINGS

Robert F. Mahfoud*
McNeese State University, Lake Charles, Louisiana, US

ABSTRACT

Shale is a laminated and fissile clastic (detrital) sedimentary rock, whose main ingredients include clays and silts. A multitude of accessory ingredients could also be present, depending on the chemistry of the provenance rocks, whose minerals were altered/weathered, and transported, by water and wind, to depositional basins for compaction and cementation. Because of the compositional variation, it is hard to find two shale outcrops chemically alike over the Earth surface. The major minerals, found in shales, are: kaolin, illite, smectite (montmorillonite), vermiculite, chlorite, feldspars, micas, and quartz. They were mostly identified by scanning electron microscopes and microprobe analyses because of their small dimensions, not exceeding 0.02mm. Many other minerals could not be identified due to their extremely fine sizes. Due to their complex compositions, shales cannot be chemically classified with precision. Rather, their types are summarized according to their colors (gray, black, brown, red, blue, green, etc.) which reflect the presence of dominant minerals, or organic materials. Also, the cement, which solidifies the loose ingredients, plays an important role in coloring the shale (red, for instance, would indicate the presence of the cementing hematite Fe_2O_3; and the dark color, the presence of dead organisms or oil compounds). In many ways, shales are useful to human beings. They form impervious layers over oil pools, preventing the petroleum from escaping as gases to the atmosphere. They also form impervious bases for perched and other aquifers. When metamorphosed to cleavable slates, the slaty sheets are used in roofing (commonly used in Europe), as tiles to pave the floors, and as small framed boards for pupils. Shales are hosts to oil compounds (oil shales), and to beneficial metallic mineral deposits (Kupferschiefer, Cu-deposit in Europe, and uranium). However, the shale presence, beneath earth-fill dams is dangerous to their stability. By getting wet, shales become muddy with a weak resistance, causing thus differential subsidence of , and small faults along the dam, leading to water seepage. This situation, observed at

* To whom all correspondence should be addressed. Tel: (337) 475-5755; email: rmahfoud@mcneese.edu.

Toledo Bend dam, between Louisiana and Texas, forced the engineers to drop the lake water surface to the dam's base, and to ignore the electricity generation project, in order to save the dam for tourism. The shale wetting would increase the compactibility and would effect the stability of streets, highways, and buildings.

1. INTRODUCTION

Shale is a laminated and fissile clastic (detrital) siliceous sedimentary rock. Its thin laminae do not exceed 1mm in thickness (Bates and Jackson, 1980). The term fissility is characteristically applied to differentiate the sedimentary shales from the metamorphic slates with slaty cleavage formed under pressure and temperature exerted on shales. Fissility refers to the breakage of shale's laminae into small pieces. The regional metamorphic slate easily breaks in sheets along the slaty cleavage planes where the bonding is weak. The dominant ingredients in shales are clays and silts. Clays are the finest Si-bearing minerals in nature. Their sizes are measured in Angstroms (A) (1mm = 1000 microns; and 1micron = 10,000 angstroms). The criterion, used here by the author, is that the clay percentage in shales, or in mudstones, should be over 50 vol. per cent (the lamination differentiates shales from mudstones). The clays, brought into the depositional basins (seas, lakes, etc.) by rivers, winds, and glaciers, remain in suspension (because of their light weights) as long as the water is agitated by waves. They fall to the bottom by gravity when the basins become totally calm. At the bottom, the clays and silts form fluffy layers filled with water. Under the accumulation load, the layers would be compressed and cemented to form shales (wherever the accumulation is in a shallow water, mudstones will form). The clay deposit can take place at any depth; and can be mixed with whatever rock or organic materials are in the water or at the bottom. Because of their light weights, clays can remain in the air for years around the Earth until an appropriate calm prevails, which allows then the gravity to pull them down to the surface. The provenance of clays is difficult to pinpoint.

Silts are somewhat heavier and bigger than clays. They need a stronger wind to be lifted into the air. They go down to the surface when the wind strength diminishes to a weak velocity, and form hills of loess (loess is a silt-blown deposit). A hill of this kind was observed at the outskirt of Vicksburg, Mississippi; and a huge deposit is known to be present in China. Loess is a fertile soil for agriculture. When carried by rivers, silts would deposit soon after the river hits the depositional water, i.e. at the slope of the basins (clays would go in suspension until the water calms down as mentioned before). The settled clays and silts would mix to form the fluffy layers as already described.

2. COMPOSITION

2.1. Silicate Minerals

The clay minerals, which dominantly cooperate in the formation of shales, are numerous. However, only a few could be determined by scanning electron microscopes and microprobe analyses. The other contributors are very fine, and not yet identified. The determined minerals include:

Kaolin: $Al_2Si_4O_{10}(OH)_8$
Montmorillonite (Smectite): $(Na, Ca)_{0.33}(Al, Mg)_2Si_4O_{10}(OH)_2 \cdot nH_2O$
Illite: $(K, H_2O)(Al, Mg, Fe)_2(Si, Al)_4O_{10}[(OH)_2 \cdot H_2O]$
Vermiculite: $(Mg, Fe, Al)_3(Si, Al)_4O_{10}(OH)_2 \cdot 4H_2O$
Muscovite: $KAl_2(Al, Si_3O_{10})(OH)_2$
Biotite: $K(Mg, Fe)_3(Al, Fe)Si_3O_{10}(OH, F)_2$
Chlorite: $(Mg, Fe, Al)_6(Al, Si)_4O_{10}(OH)_8$
Feldspars
a) K feldspar: $KAlSi_3O_8$
b) Ca-Na-feldspars (plagioclase series)
Anorthite: $CaAl_2Si_2O_8$ (most calcic plagioclase)
Albite: $NaAlSi_3O_8$ (most sodic plagioclase)
Quartz: SiO_2

The chemical formulas were copied from Berry, Mason and Dietrich's mineralogy book (1983). All the listed minerals contain Si; therefore, they are call silicates. Except feldspars and quartz, all the others are classified as phyllosilicates, or sheet structures, where the Si/O ratio is 2/5 or 4/10. The feldspars and quartz are tektosilicates, or framework structures, where the Si/O ratio is 1/2. (In the feldspars, Al should be added to the numbers of Si to make the ratio Si/O = ½) (Berry et al., 1983; McLean and Gribble, 1988; Nesse, 2000; Perkins, 2002).

2.2. Phyllosilicate Structures

There are two basic layers which form the phyllosilicate structures. One layer contains the Si-O-tetrahedra (a tetrahedron is a structure with four equilateral triangles, where the Si-ion is at the center touched by the four oxygens (SiO_4). The three oxygens, at the base of the tetrahedron, are shared with three neighboring tetrahedral. This arrangement would give a Si/O ratio of 1 Si/2.5 Oxygens per one tetrahedron, or 2 Si/5 Oxygens per two tetrahedral, or 4/10 per 4 tetrahedra. The Si-O-tetrahedron layer is overlain by another different layer that contains the rest of the phyllosilicate formula (except K^+, Na^+, H^+, etc.). The ions here are arranged in octahedral with the cations in the center. Those two layers (the tetrahedral and octahedral) form one unit with 1:1 in some phyllosilicate minerals (like kaolin, serpentine, etc.). Others (montmorillonite, illite, vermiculite, etc.) have two tetrahedral layers: one at the bottom, and another reversed above the octahedral layer. Here, the unit would be 2:1. In chlorite, the unit is 2:2 (an octahedron is a structure with 8 faces in two pyramids with a square in between, one pyramid is reversed at the bottom, overlain by another right-side-up). The thickness of a kaolin unit, for instance, is between 7 and 8A (angstroms). The ions, inside the formula's parentheses, can substitute for each other because of their close ionic radii. The potassium (K^{+1}), Na^{+1}, H^{+1}, etc. weakly bond the units together by occupying the space between them.

Readers are referred to books on mineralogy to look at the above mentioned structures (Berry et al., 1983; McLean and Gribble, 1988; Nesse, 2000; Perkins, 2002).

2.3. Effects of the Diagenesis Process

It is important to mention that the above mentioned phyllosilicates and tektosilicates are not the only minerals that settle at the bottoms of the depositional basins, though they are the dominant ones found in shales. The fine materials that are suspended in water (seas, lakes...) are a mixture of mineral fragments, amorphous matters, dead organisms (plants, and animals), colloids, etc..., in addition to many ions and radicals (CO_3^{-2}, SO_4^{-2}, etc...). When the water becomes still, all of them would precipitate as fluffy layers (alternating with water). Under the sediment load, most of the water would be compressed out, and sedimentary layers would form, and would be lithified to form a laminated shale. During that time (i.e. from the fluffy deposit to lithification), but before any metamorphic mineral is formed, under higher pressure and temperature, the diagenetic process plays its role. By definition, the diagenesis includes all the chemical, organic, and structural changes which take place inside the sediment deposit from the moment of precipitation until the beginning of metamorphism (Bates and Jackson, 1980; Ehlers and Blatt, 1980; Raymond, 1995; Boggs, 2001). New minerals would form, and others would disappear because of the change in the chemistry, and in the pH/Eh conditions, which renders the environment favorable for new minerals, and unfavorable for others. As an example, if the montmorillonite clay adsorbed Al^{+3} and K^{+1}, present in the surrounding, it would change to illite. On the other hand, if the appropriate K^{+1} is not present, illite would not form. By adsorbing Na^{+1}, a Ca-feldspar would become a Na-rich plagioclase. Another example, if Fe^{+3} encounters a reducing Eh, it would gain 1 electron and would change to Fe^{+2}; and consequently a red hematite mineral would be converted to black magnetite Fe_3O_4, or to siderite if CO_2 is present ($FeCO_3$), or to pyrite or marcasite if S^{-1} is present (FeS_2). As a conclusion, the color of the sediments, and consequently the shale, would change during the diagenesis process, which is controlled by the chemistry and pH/Eh (Krumbine and Garret, 1952).

3. Types

The author suggests that the types would be differentiated here according to their colors (gray, green, red brown, dark, etc...)

3.1. Gray Shales

Studies and field observations showed that the gray shale was formed in a shallow water where the oxygen was adequately available, and the environment oxidizing. The oxygen would burn, or oxidize, all dead organisms (plants and animals settled at the bottom and mixed with clay/silt deposits). In the absence of oxygen, the organic remnants would color the sediment with a dark hue. Worthy to mention that the shallow water is not the only place where the oxygen is available. It could also be present at the sea bottom where a cold arctic oxygenated water current circulates. The cold water would usually retains the oxygen (Ehlers and Blatt, 1982). The carbon (C), H and N, in the dead organisms, would react with the oxygen to give the following:

$$C + O_2 \rightarrow CO_2$$
$$H_2 + O \rightarrow H_2O$$
$$N + O_2 \rightarrow NO_2$$

Nothing would be left, from the organic material to stain the sediments.

3.2. Dark Shale

The dark color, in the black shales, is caused by the carbon film left from the dead organic materials, mixed with the clay and silt, in a reducing environment where no adequate oxygen is present to react with C, H, N. The following equations would result.

$C + 2H_2 \rightarrow CH_4$ (methane gas)
$1\frac{1}{2}H_2 + N \rightarrow NH_3$ (ammonium gas)

In addition, an extra carbon film would remain mixed with the clays and silts to give the black shales.

In fresh lakes and lagoons, where dead trees and shrubs fall to the bottom, to be covered with clay/silt to prevent the oxygen from reaching them. With fermentation, in such a reducing environment, CH_4 and NH_3 gases would form pockets inside the formed coal seams. In addition, Fe^{+2} and S^{-1} ions would combine to form pyrite/marcasite FeS_2. Under compression/cementation, the clay/silt deposit would form shale laminae between the coal seams.

Pyritized fossils, found in black shales are commonly found around the world. Other sulfide ore deposits were beneficially mined in many locations. Potter, Maynard, and Pryer, in Raymond (1995), believed that a high content, of organic materials, would give the brown, gray, and black colors to rocks. Sheu and Presley, in Raymond (1995) stated that the dark gray and black shales, in the Gulf of Mexico, contain over 1.0 percent organic carbon, whereas the light gray shales contain less than 1.0 percent.

3.3. Red/Brown Shales

The oxidation/reduction conditions (Eh) play an important role in fixing the shale's color. This statement was confirmed by Krumbein and Garrels work (1952). Briefly, in an oxidizing environment, Fe^{+3} becomes higher than Fe^{+2}, and therefore, the red hematite Fe_2O_3 would form giving the red color to the shale. If the hematite adsorbed one water molecule, it would form the brown goethite ($Fe_2O_3{\cdot}H_2O$). When the environment became reducing, and Fe^{+2} dominating, the black magnetite Fe_3O_4, or the chocolaty-brown siderite $FeCO_3$, or the yellow pyrite/marcasite FeS_2, would form. All reactions and colors can take place during the deposition and the diagenetic process, i.e. ahead of the shale formation. Briefly, if Fe^{3}/Fe^{+2} was over 2:1, in rock with 3-6 percent iron, a red rock would prevail (ex: red shales and slates). The purple and black rocks (slates) would dominate when Fe^{+3}/Fe^{+2} was 1:2 or less. A.M. Thompson, in Raymond (1995), stated that: 1) the hematite gives red colored rocks; and pyrite and chlorite, gray-green colors. McBride concluded that, by decreasing Fe^{+3}/Fe^{+2}, the color changes from red to yellow to green. Furthermore, he mentioned that the red and brown shales would have their grains coated with Fe-oxide; and the green shales, with chlorite/illite (without hematite, organic matter, and sulfides). However, the olive and yellow shales have chlorite, illite, organic matter, and iron sulfides. The gray shales owe their color to the presence of organic matter and sulfides; whereas the blue-gray colored mudrocks, found in the California coastal ranges, to the presence of the iron-rich beidellite (an interlayered mixture of illite and montmorillonite, or a mixture of clay minerals and hydrated ferric oxide – Bates and Jackson, 1980), or to the iron-poor nontronite coatings (a montmorillonite where iron is replacing Al. Commonly occurs in weathered basaltic rocks. Pale yellow to green – Bates and Jackson, 1980, derived from the alteration of andesite. McBride has also found that the glauconite gives the green color (Raymond, 1995). In many red shales and mudrocks, green spots commonly occur surrounding decaying roots or other organic matter, suggesting that their environment was depleted in oxygen. This has forced Fe^{+3} in the red hematite to be reduced to Fe^{+2} which would be dissolved in groundwater.

4. USES

4.1. Beneficial Uses

The field observations reveal that shales commonly form impervious layers over oil accumulations, preventing the petroleum from escaping, as gases, to the atmosphere. When a light oil loses its gases, it becomes heavy, forcing its price to decrease. After losing their gases, many oil pools became very thick with asphaltic – waxy remnants. The asphalt is mined for street and highway paving. In folded rocks, the incompetent shales form disharmonious structures between the fractured competent rocks (sandstones and limestones), keeping thus the hydrocarbons inside the anticlines.

Shales also form impervious bases for perched and other aquifers (confined and unconfined). When metamorphosed to cleavable slates, the slaty sheets are commonly used in roofing. Moreover, those sheets are also used as tiles for floor paving, and as small framed boards for pupils. Many black shales were found hosting oil compounds; they are known as oil shales, which, in some occurrences, hold large quantities of hydrocarbons. The Oil Shale in the Green River Formation, Colorado, is an example; it is believed to contain between 2-3 trillion barrels of crude oil (about 10 times the reserve in Saudi Arabia). Some sixty years ago, several oil companies have tried to extract the crude oil from the Green River oil shale. A barrel has cost them about $55. At that time, the price of an imported barrel of crude hydrocarbon was $10 or less. This wide difference has force the oil companies to drop their extraction project. Today, the price of an imported barrel oscillates between $80 and $100; and that of an extracted one, may reach about $100. In other words, the two prices have become matchable; a good indication for oil companies to try again extracting the crude oil from the oil shales.

There is a similarity between the reserve in the oil shale of the Green River Formation, and that in the Athabasca sand to the north of Edmonton, Alberta, Canada. Both have giant oil reserves, originally formed abiogenically near the bottom of the lithosphere (Mahfoud and

Beck, 1995; Mahfoud, 2000). Both oil showings on the surface suggest that they are connected to their main pools (near the base of the lithosphere) through deep cracks (or faults). Both may have lost large amounts of light gases by being exposed to the atmosphere for a long time. However, both still maintain huge crude oil reserve below the surface.

Black shales were found, in places, to be mineralized with beneficial ore deposits. One example is the "Kupfershiefer" (copper deposit) which covers a large area in Central Europe. The copper sulfide was stratigraphically deposited with the black shale in a reducing environment. The sulfide was the reaction result between Cu^{+2}, Fe^{+2} and S^{-2} that were present in the depositional basin.

Another beneficial metallic deposit, in the black shale, is the uranium. The uranium ions (U^{+6}) has changed to (U^{+4}) by gaining 2 electrons from the reducing environment, has reacted with other ions and radicals, and has settled down with the black shale. The Black Sea still maintains its reducing environment at depth; and a uranium deposit was reported present at the bottom mixed with clay and shale layers.

4.2. The Hidden Danger in Shale

Ignoring the subsurface shale presence would be too risky to engineering projects, and to the human being health. As an example, the author would describe, in the following, the two major defects that he has found in the Toledo Bend dam. During his many visits to the dam site, with his engineering students, he has discovered that the dam was built without a deep geologic study to the area. Briefly, first, the earth-fill dam, about 3 km long, was built across a graben (which is the Sabine River Valley). This is a grave mistake. Second, the stratigraphy, as revealed on the western side of the graben (south of the bridge), clearly shows two formations: a dark gray shale section at the bottom, overlain by sandstone beds. The shale has become wet from the water which continuously seeps from beneath the dam, from a lake, about 80 km long, formed to the north. The shale resistance has weakened, and has allowed the sandstone beds to subsiding, jointing, and faulting. This, in turn, has allowed the water to continuously seeping, not too long after the dam was built. The engineers have taken the necessary precaution to save the dam from destruction. However, they were forced to ignore generating the electricity by lowering the lake's level to almost the dam's base in order to keep the structure safe for tourism.

Generally speaking, when a shale thickness adsorps water, it becomes muddy with a weak resistance, and an increase in its compactibility. This would lead the buildings to lean, and highways to fracture.

Being formed in a reducing environment, the black shales would likely have some uranium (and probably other radioactive isotopes) enclosed within the clays and carbon. With time, the uranium decay would release the radioactive radon gas, which may reach an aquifer, and may pollute its drinkable water. A high radon level is carcigenous if the water is used for drinking. State governments have recognized this danger; and are routinely checking the radon level in the consumption water.

CONCLUSION

From this brief description, one concludes the following: first, it would be to the benefit of the human being to thoroughly examine any shale, especially the black one which may have metallic deposits; and second, it should be imperative, for any engineering project, to drill a few shallow wells, in the site area, in order to study the stratigraphic section (and to look for discontinuities (fractures, faults,...) in those beds. Such an examination would prevent the problem occurrence in the future.

REFERENCES

Bates, R.L., and Jackson, J.A., 1980, Glossary of Geology: American Geological Institute, Falls Church, Virginia, 749pp.

Berry, L.G., Mason, B., and Dietrich, R.V., 1983, Mineralogy (2nd ed.): W.H. Freeman and Company, San Francisco, 561pp.

Boggs, S. Jr., 2001, Principle of Sedimentology: Prentice Hall, New Jersey, 720pp.

Ehlers, E.G., and Blatt, H., 1982, Petrology: W.H. Freeman and Company, San Francisco, 732pp.

Krumbein, W.C., and Garrels, 1952, Origin and classification of chemical sediments in terms of pH and oxidation – reduction potentials – Journal of Geology, Vol. 60, p. 1-33.

Mahfoud, R.F., and Beck, J. 1995, Why the Middle East fields may produce oil forever, Offshore Magazine, 55, No. 44, p. 56-106.

Mahfoud, R.F., 2000, Theory links lithospheric rotation to possible abiogenic oil re-charge, Offshore Magazine, 60, No. 5, p. 102-205.

McLean, A.C., and Gribble, C.D., 1988, Geology for Civil Engineers: Unwin Hyman, London, 314pp.

Nesse, W.D., 2000, Introduction to Mineralogy: Oxford University Press, Inc., New York, 442pp.

Perkins, D., 2002, Mineralogy: Prentice Hall, New Jersey, 483pp.

Raymond, L.A., 1995, Petrology: McGraw – Hill Book Company, New York, 742pp.

In: Horizons in Earth Science Research. Volume 11
Editors: Benjamin Veress and Jozsi Szigethy
ISBN: 978-1-63321-296-1

Chapter 3

PUBLICLY AVAILABLE OPEN ACCESS DATA AND MACHINE LEARNING MODEL-PREDICTIONS APPLIED WITH OPEN SOURCE GIS FOR THE ENTIRE ANTARCTIC OCEAN: A FIRST META-ANALYSIS AND SYNTHESIS FROM 53 CHARISMATIC SPECIES

Falk Huettmann[1] and Moritz S. Schmid[2]
[1]EWHALE lab, Institute of Arctic Biology, Biology and Wildlife Department, University of Alaska-Fairbanks (UAF), Alaska
[2]Takuvik Joint International Laboratory, Laval University (Canada) – CNRS (France), UMI3376, Département de biologie et Québec-Océan, Université Laval, Québec, Québec G1V0A6, Canada

ABSTRACT

The Antarctic Treaty is of global importance and it is very clear in its statement and spirit that science is to be shared with the global community. However, huge amounts of data still remain to be made freely available to the public. Available data is often in data formats that are not easily accessible or limited to just certain software platforms, and the associated metadata is often poorly done or non existent. There is tremendous progress brought by publicly available open access 'presence only' (location only) data sets though, e.g. for scientific and conservation management questions. While this is fully in line with ongoing data initiatives such as the International Polar Year (IPY 2007), the International Council for Science (ww.icsu.org) and the Cyber-infrastructure with U.S. NSF and its relatives, such compiled and ready-to-use datasets in a good format do not exist or are still very difficult to come by. Here we make such data sets publicly available for free for 53 species and for further use by the global community. Secondly, we provide a first application of such datasets using free tools such as machine learning software in R, as well as open source geographic information systems (such as QGIS, R and SAGA GIS). While these resulting data can be used and improved further, here we present a first and robust meta-analysis and synthesis from these experiences and models.

Keywords: Antarctic Ocean, Open Access Data, Open Source Geographic Information System (GIS), biodiversity, charismatic fauna, machine learning predictions

Introduction

Humans are bound by what the earth provides them (Czech 2006, Intergovernmental Panel on Climate Change (IPCC) (2007)). Antarctica is a remote and very precious region of such a relevance for the global climate (Martinson 20120) and human well-being (Huettmann 2012). It is governed by the international community and through an Antarctic Treaty System (Summerson 20012; http://en.wikipedia.org/wiki/Antarctic_Treaty_System). This region lacks an indigenous population as well as a long-term human history (e.g. beyond c. 200 years; Thomas et al. 2008, Ainley et al. 2010). However, the recent 50 years must have been among the most devastating for the Antarctic continent, for its species, habitats and the atmosphere (Huettmann 2012). Just the current foot print, e.g. done by scientists, tourists and administrative staff (Kerry and Riddle 2009, International Association Antarctica Tour Operations (IAATO) (2012)), warrants already serious attention due to the spoilage of wilderness, and it requires a change of business (Huettmann 2010a). The outlook for this continent is hardly a good one when considering alone the impacts brought by climate change, global growth of human population and consumption, and lack of an efficient sustainability governance scheme (Hilty et al. 2007, Friedman 2010). By now, Antarctica is not anymore a marginal place but it makes for an integral part of the global receipe of human well-being anywhere (Intergovernmental Panel on Climate Change IPCC 2007, Huettmann 2012; see Ainley et al. 2012 for toothfish harvest and global food supply) due to the free services this area provides. Its resources (marine, atmospheric, ecological services and slowly now also terrestrial) are at stake by global consumption, and also threatened due to related climate change (Thompson and Solomon 2002, Jenouvrier et al. 2009, McClintock et al. 2008, Turner et al. 2009).

Following the Antarctic Treaty and a recommended open access approach, here we present a data mining and assessment of the best-available data for 53 charismatic species of the Antarctic Ocean and across platforms. The compiled data in themselves are a scientific product and can be used in various ways. Their delivery forms the first objective of this contribution. Secondly, and as an applied example, we then present a first data mining analysis of spatial model predictions for these data and general patterns within the study area. These goals are meant to set a quantitative stage, and to enable a better science-based management of wilderness areas and their resources (sensu Ohse et al. 2009, Cushman and Huettmann 2010, Franklin 2010).

Methods

Raw Data

We compiled 'presence (location) only' data for 53 selected species living in the Antarctic Ocean. These species often catch public interest and tend to be charismatic, with widely fragmented data holders. While our data here are just a small subset of the marine biodiversity (see Brandt et al. 2006 for an example overall and the Antarctic webportal at http://www.scarmarbin.be/), no 'complete' and across-the-platform Antarctic compilations of these data exist, which are well-documented and digital as well as readily available GIS data sources for the public audience. Here we focus on 53 species because public users commonly request them. They are also powerful indicators for the Antarctic habitat, and they make for an impressive showcase for Antarctic research that shows what can be done and for a science-based conservation. To assure taxonomic validity, this list of 53 species is described and cross-matched with international and freely available online taxonomic databases (World Register of Marine Species WORMS www.marinespecies.org/).

Open Source GIS

Geographic Information Systems (GIS) are a widely used platform for data compilation, manipulation, analysis and presentation, e.g. Cushman and Huettmann (2010). Many software implementations exist to achieve. Here we used open source GIS to achieve all the tasks set in this study. We decided to use primarily QGIS (www.qgis.org/) for most operations, following closely ESRI ArcGIS 10 standards and campus-licensed software available to the global research community. In addition, we used the freely available Geospatial Model Environment (GME; www.spatialecology.com/ gme/) to carry out overlays, intersects and specific data manipulations. We used the R environment for many of these operations (R Development Core Team 2008) which also allowed to run SAGA GIS for instance.

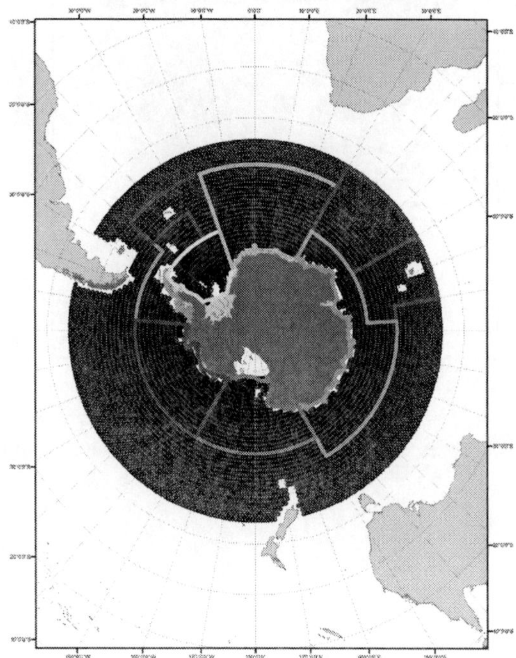

Figure 1. Overview of the Antarctic Ocean and some major domains. A 1*1 degree lattice (used for predictions) is shown.

Predictions Using Machine Learning

We followed a GIS-based machine learning approach presented in Huettmann and Diamond (2002), Yen et al. (2004), Ohse et al. (2009) and Huettmann et al. (2011).

As stated in the previous paragraph, we employed ArcGIS formats, the freely available Geospatial Modeling Environment (GME) and open source QGIS (sometimes SAGA GIS for R) to carry out all required GIS work. We then applied machine learning algorithms (randomForest; http://cran.r-project.org/web/packages/randomForest/index.html) in R to the training data (presence records combined with pseudo-absences).

The training data are shown in figure 2; the resulting model got applied to a lattice of 1*1 degrees (figure 1), and which was displayed in GIS (see appendices for results and details).

RESULTS

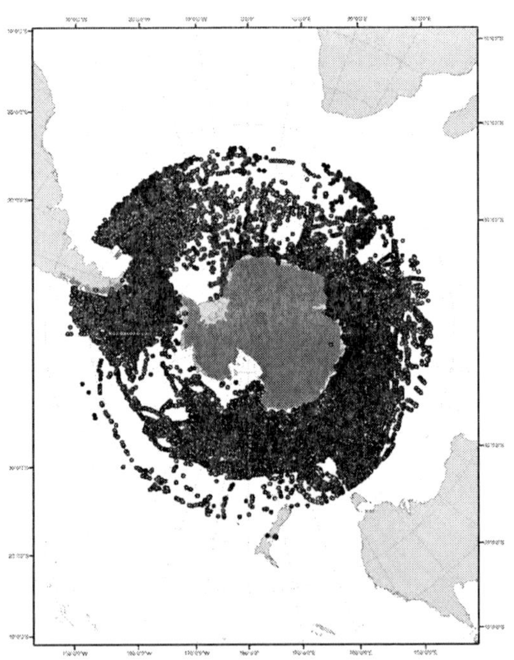

Figure 2. Data distribution and survey effort for the 53 species discussed (training data set; see table 1 for overview and sample sizes).

Raw Data

Here we presented for the first time the best available publicly available raw data for 53 species for the Antarctic Ocean (shapefiles; ASCII format is available upon request). Table 1 indicates the featured species and their sample sizes; figure 2 shows the overall distribution and the survey effort. More technical details can be seen in appendix 1.

In addition, we submitted ISO compliant metadata of this work to the USGS metadata clearinghouse, with the Polar Information Commons (PIK) and as well with our local institutional repository (dSPACE at UAF: https://scholarworks.alaska.edu/handle/11122/1010; also includes a full copy of the underlying data for free download). While these are primary scientific data, they are of great public interest, in high demand and carry a lot of weight for many conservation questions and subsequent applications such as Marine Protected Areas (MPAs; see Huettmann 2012 for an example).

Model Predictions

First model-predictions for these species were created using a widely published machine learning approach (Ohse et al. 2009, Drew et al. 2011 and references within; figure 3 for an example). Most predictions performed rather well and with high ROC curves; for specific details, please contact authors.

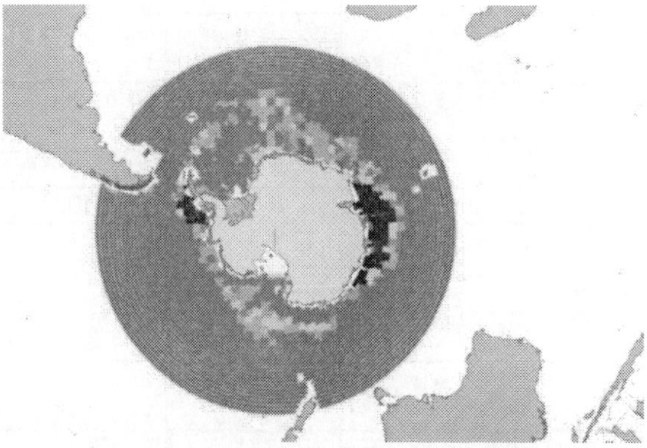

Figure 3. Example of a prediction of the ecological niche for Antarctic Crabeaster seal *Lobodon carcinophaga*, TSN 622023, APHIA ID 23416; based on 48,716 presence locations). Such a map represents the potential niche using best available empirical data with a consistent modelling method; it can be used to emphasize hot spots of habitats and occurrence indeces, as well as climate change and impact baselines.

However, these predictive models should be seen as a start, and like any new hypothesis they require more large-scale assessment, ideally with ground-truthing and new data for model training and testing. We hope that this effort triggers the discussion and release of more publicly available data to test and confront our models for further improvements and new model generations.

Nevertheless, these predictions are quantitative, exist as ASCII tables and in a common shapefile GIS format, and we find that overall patterns in distribution and diversity are rel. robust and can be used for a first and rapid assessment when other information is lacking (as it is frequently the case).

Overall, the results from this work are encouraging and can help to kick-off new model-prediction generations and scientific communications.

Table 1. List of 53 charismatic species of the Antarctic Ocean, with a valid taxonomic detail and sample size used in this study

Common Name	Scientific Name	Taxonomic Serial Number (TSN)	APHIA ID (WORMS)	Raw sample size (presence only locations)
Adelie Penguin	*Pygoscelis adeliae*	174445	225757	17859
Crabeater Seal	*Lobodon carcinophaga*	622023	231416	48716
Antarctic Petrel	*Thalassoica antarctica*	174608	212639	8821
Antarctic Fur Seal	*Arctocephalus gazella*	180630	231404	8670
Atlantic Petrel	*Pterodroma incerta*	174586	212641	14
Black-bellied Storm Petrel	*Fregetta tropica*	174655	212696	2503
Blackbrowed Albatross	*Thalassarche melanophrys*	723260	225756	14375
Blue Petrel	*Halobaena caerulea*	174597	212645	7789
Broad-billed Prion	*Pachyptila vittata*	174603	212647	3029
Cape Petrel	*Daption capense*	174603	212640	9118
South Polar Skua	*Catharacta maccormicki*	176801	159073	370
Brown Skua	*Catharacta skua lonnbergi*	176800	NA	8
Chinstrap Penguin	*Pygoscelis antarctica*	174446	225783	273
Greyheaded Albatross	*Thalassarche chrysostoma*	563381	225760	24898
Patagonian toothfish	*Dissostichus eleginoides*	642807	234700	991
Antarctic Toothfish	*Dissostichus mawsoni*	642808	234836	13
Emperor Penguin	*Aptenodytes forsteri*	174449	225773	3315
Southern Rockhopper Penguin	*Eudyptes chrysocome*	559996	212657	321
Royal Penguin	*Eudyptes schlegeli*	559998	225926	501
Antarctic Coastal Krill	*Euphausia crystallorophias*	95520	236216	196
Krill	*Euphausia superba*	95514	236217	74936
Fairy Prion	*Pachyptila turtur*	174602	212648	1229
Finwhale	*Balaenoptera physalus*	180527	137091	2566
Southern Fulmar	*Fulmarus glacialoides*	174540	212638	6204
Gentoo Penguin	*Pygoscelis papua*	174447	225777	396

Common Name	Scientific Name	Taxonomic Serial Number (TSN)	APHIA ID (WORMS)	Raw sample size (presence only locations)
Great Winged Petrel	*Pterodroma macroptera*	174589	212644	352
Greater Shearwater	*Puffinus gravis*	174549	137201	469
Humpback whale	*Megaptera novaeangliae*	180530	137092	682
Leopard Seal	*Hydrurga leptonyx*	180667	231417	151
King Penguin	*Aptenodytes patagonicus*	174450	212656	1211112
Kelp Gull	*Larus dominicanus*	176849	212624	172
Little Shearwater	*Puffinus assimilis*	174559	137200	8956
Macaroni Penguin	*Eudyptes chrysolophus*	174452	212658	25014
Minke Whale	*Balaenoptera bonaerensis*	612592	231405	455
Northern Giant Petrel	*Macronectes halli*	174595	212637	2847
Ross Seal	*Ommatophoca rossii*	180665	231412	463
Killer Whale	*Orcinus orca*	180469	137102	427
Antarctc Prion	*Pachyptila desolata*	174600	212649	3292
Imperial Shag	*Phalacrocorax atriceps*	174734	225759	82
Hooker's Sea Lion	*Phocarctos hookeri*	180617	231422	38
Light-mantled Albatross	*Phoebetria palpebrata*	174531	212631	13428
Pomarine Jaeger	*Stercorarius pomarinus*	176792	137173	4
Right Whale	*Eubalaena australis*	552771	220222	23
Southern elephant Seal	*Mirounga leonina*	180671	231413	61402
Sei Whale	*Balaenoptera borealis*	180526	137088	283
Snow Petrel	*Pagodroma nivea*	174614	225772	8136
Soft plumaged Petrel	*Pterodroma mollis*	174590	137199	1618
Southern Giant Petrel	*Macronectes giganteus*	174594	212636	4122
Sooty Albatross	*Phoebetria fusca*	174530	212632	249
Wandering Albatross	*Diomedea exulans*	174525	212583	17439
Weddel Seal	*Leptonychotes weddellii*	180669	195932	83
White-bellied Storm Petrel	*Fregetta grallaria*	174656	212695	346
White-chinned Petrel	*Procellaria aequinoctialis*	174610	212651	13435

DISCUSSION

Compiled value-added data presentations can make for a global heritage; making them freely available represents a best professional practice (Zuckerberg et al. 2011) and for the Antarctic Ocean (as mandated by the Antarctic Treaty). Such data are a true legacy and have inherent value, which is even enhanced when new information on virtually unexplored wilderness areas is presented and made freely available for the global community in a readily accessible digital data format. Such data bring great progress for ecosystem studies (e.g. Trivelpiece et al. 2010; see http://www.scarmarbin.be/ for great examples). While such value-added data compiled across platforms and sources might even be seen as competition with some existing online data services such as OBIS, GBIF, PIK, SCAR-MARBIN and initiatives like the Census of Antarctic Marine Life (CAML; www.caml.aq), Census of Marine Life (COML; www.coml.org) and the International Polar Year (www.ipy.org), they should not be seen as such. Instead they should be seen as a precious centralized baseline, a precious snapshot in time, provided directly by practioners and scientists and as a valuable research and assessment package. Arguably, webportals are very dynamic and will not remain online forever, with the same data content and quality, and when considering the global fluency of the internet and scientific funding (Costello et al. in press). Alone the American context and the shut-down of the US goverment in 2013 provide ample examples.

Here, 53 charismatic species data were presented as a basic compiled raw data package and with first GIS predictions and compliant metadata. While this is a very small portion of Antarctic biodiversity, these best-available data and the model predictions already present us with an example of what is possible for scientific projects and how data management and science of the future could look like in an openly networked world. Thus, it should be seen as a role model to work from but which was not achieved yet by all stakeholders involved in Antarctic research thus far.

For this work, we decided to follow standard taxonomic databases to define species names. The use of taxonomic serial numbers (TSNs and Aphia numbers) helps to stay consistent across projects and over time. ‘Splitting’ and ‘lumping’ of species is commonly observed in taxonomy, and the discovery of cryptic speciation shows the fluency of taxonomic systems and that presumably solid species information might not be at all. These topics matter if solid ecological niches are to be developed for a ‘species’, and if a proper species and ecosystem management is to be achieved.

Mining and predictions of unexplored data as used here are of great value and when trying to interpret species distributions (Huettmann et al. 2011). Being predictive and pro-active helps to follow the pre-cautionary principle, as promoted by the IUCN and other conservation management schemes, e.g. for an adaptive management as described in Silva (2012). Here we followed standard methodologies developed in the field that provide the first baseline information (*sensu* Ohse et al. 2007, Wei et al. 2011).

As shown in figures 1,2 and table 1, the coverage of samples of this dataset is opportunistic, limited, unequal and biased. This lies in the nature of data collection in remote areas and the state-of-art of uncoordinated data collection efforts in Antarctica. However, a general trend shows that western Antarctica (e.g. Ammundsen-Bellinghausen, Ross Sea) is undersampled, and that the current sampling scheme is lacking a well-coordinated and tested research design and with a sustainable monitoring scheme between institutions and countries.

It remains unclear how a valid scheme for resource management in this area can be developed for the greater good of the present wildlife (Constable et al. 2000) and its future set-up. Likely, much more data than we were able to compile exists (e.g. for pelagic seabird surveys from the U.S., the UK, France and Russia, or from many TOPP and disease surveillance projects). This however shows that the scientific community (marine researcher community ('ocean commons') and Antarctic researcher community) is still lacking proper data sharing (see Huettmann 2010b for a general assessment and next steps) and widely is in violation of IPY, national policies and the Antarctic Treaty itself (as boldly shown in Huettmann 2012).Albeit not perfect yet, here we argue for a change and present a solid case on how to proceed with such primary data that would be beneficial for 'anybody' and follows the intent and spirit of the Antarctic Treaty, managing a public resource for the benefit of mankind.

Acknowledgments

This project was only possible due to the foresight of David Carlson/IPY and great initiatives like GBIF, COML/OBIS, CAML and its contributors and supporters. FH is very grateful to all data providers and for data release to make this work possible. UAF provided a working platform. This is EWHALE lab publication # 126.

Appendix Placed at the Institutional DSPACE Repository of University of Alaska-Fairbanks UAF (https://scholarworks.alaska.edu/handle/11122/1010)

a) Compiled raw data (shapefiles and ISO metadata) of 53 charismatic species of the Antarctic Ocean
b) Compiled species predictions (ESRI grids and ISO metadata) for 53 charismatic species of the Antarctic Ocean

References

Ainley, D., Russell, J., Jenouvrier, S., Woehler, E., Lyver, P. O'B., Fraser, W. R. and Kooyman, G. L. (2010) Antarctic penguin response to habitat change as Earth's troposphere reaches 2°C above preindustrial levels. *Ecol. Monogr.* 80: 49-66.

Ainley, D., C.M. Brooks, J.T. Eastman, M. Massaro(2012) Unnatural Selection of Antarctic Toothfish in the Ross Sea Antarctica. Chapter 3 In: F. Huettmann (ed). *Protection of the Three Poles.* Springer, Tokyo Japan.Pp 53-76.

Brandt, A., De Broyer, C., De Mesel, I.G.,Ellingsen, K.E.,Gooday, A.J.,Hilbig, B.,Linse, K., Thomson, M.R.A., Tyler, P.A. (2006). The biodiversity of the deep Southern Ocean benthos. *Phil. Trans. R. Soc. Lond.* (B Biol. Sci.) 362(1477): 39-66.

Constable A.J., W.K. de la Mare, D.J. Agnew, I. Everson and D. Miller (2000). Managing fisheries to conserve the Antarctic marine ecosystem: practical implementation of the

Convention on the Conservation of Antarctic Marine Living Resources (CCAMLR). *ICES Journal of Marine Science* 57: 778–791.

Convey, P. (2001). *Antarctic Ecosystems.Encyclopedia of Biodiversity,* Volume 1. Academic Press. pp. 171 – 184.

Costello, M.J., W. Appeltans, N. Bailly, W.G. Berendsohn, Yde de Yong, M. Edwards, R. Froese, F. Huettmann, W. Los, J. Mees, H. Segers, F.A. Bisby (in press). Strategies for the sustainability of online open-access biodiversity databases. *Biological Conservation.*

Cushman, S. and F. Huettmann. (2010). Spatial Complexity, *Informatics and Wildlife Conservation.* Springer Tokyo, Japan.

Czech, B. 2006. Steady State Economy. *Encyclopedia of Earth.* Eds. Tom Tietenberg et al., National Council for Science and the Environment, Washington, DC.

Franklin J. (2010). Mapping Species Distributions: Spatial Inference and Prediction (*Ecology, Biodiversity and Conservation*). Cambridge University Press.

Friedman G. (2010). *The next 100 years: A forecast forthe 21st century.* Anchor Books.

Hilty J.A., C.C. Chester and M.S. Cross (2007). *Climate and Conservation: Landscape and Seascape Science, Planning, and Action.* Island Press.

Huettmann, F. (2007). Constraints, suggested solutions and an outlook towards a new digital culture for the oceans and beyond: experiences from five predictive GIS models that contribute to global management, conservation and study of marine wildlife and habitat, in: VandenBerghe, E. et al. (Ed.) *Proceedings of 'Ocean Biodiversity Informatics': an international conference on marine biodiversity data management Hamburg, Germany,* 29 November - 1 December, 2004. IOC Workshop Report, 202, VLIZ Special Publication 37: pp. 49-61. http://www.vliz.be/vmdcdata/imis2/imis.php?module=ref&refid=107201.

Huettmann, F. (2011a). From Europe to North America into the world and atmosphere: a short review of global footprints and their impacts and predictions. *The Environmentalist* 10.1007/ s10669-011-9338-5.

Huettmann F (2011b). Serving the Global Village through Public Data Sharing as aMandatory Paradigm for Seabird Biologists and Managers: Why, What, How, and aCall for an Efficient Action Plan. *The Open Ornithology Journal* 4:1-11.

Huettmann (ed) (2012). *Protection of the Three Poles.* Springer Japan.

Huettmann F., and Diamond A.W. (2001). Seabird colony locations and environmental determination of seabird distribution: A spatially explicit seabird breeding model in the Northwest Atlantic. *Ecological Modelling* 141: 261-298.

Huettmann F, Yu Artukhin, O. Gilg, and G. Humphries (2011). Predictions of 27 Arcticpelagic seabird distributions using public environmental variables, assessed withcolony data: a first digital IPY and GBIF open access synthesis platform. *Marine Biodiversity* 41: 141-179 DOI 10.1007/s12526-011-0083-2.

Intergovernmental Panel on Climate Change (IPCC) (2007). *Climate Change 2007: Impacts, Adaptation, and Vulnerability.* Cambridge University Press.

International Association Antarctica Tour Operations (IAATO) (2012). *Tourism Statistics.* URL: http://iaato.org/es/tourism-statistics.

Jenouvrier, S., H. Caswell, C. Barbraud, M. Holland, J.Strœveand H. Weimerskirch (2009). Demographic models and IPCC climate projections predict the decline of an emperor penguin population. *P. Natl. Acad. Sci. USA* 106: 1844-1847.

Kerry, K.R. and M. Riddle (2009) *Health of Antarctic Wildlife: A challenge for Science and Policy*. Springer.

Martinson D.G. (2012). Antarctic circumpolar current's role in the Antarctic ice system: An overview. *Palaeogeography, Palaeoclimatology, Palaeoecology,* 335–336: 71-74.

McClintock J, H.W. Ducklow and W. Fraser(2008) Ecological responses to climate change on the Antarctic Peninsula. *Am. Sci.* 96:302–310.

Ohse, B, Huettmann, F., S. Ickert-Bond and G. Juday (2009).Modeling the distribution of white spruce (*Piceaglauca*) for Alaska with high accuracy: an open access role-model for predicting tree species in last remaining wilderness areas. *Polar Biology* 32: 1717-1724.

R Development Core Team (2008) R: A language and environment for statistical computing. *R Foundation for Statistical Computing*, Vienna Austria, http://wwwR-projectorg.

Silva, N.J. (2012). *The Wildlife Techniques Manual: Research & Management.* 2 Volumes. The Johns Hopkins University Press; seventh edition.

Summerson R (2012) Protection of Wilderness and Aesthetic Values in Antarctica.In F. Huettmann (ed). *Protection of the Three Poles*. Springer Japan, pp. 77-112.

Thomas D.N., G.E. Fogg, P. Convey, C.H. Fritsen, J.M. Gili, R. Gradinger, J. Parry-Laybourn, K. Reid, D.W.H. Walton (2008). The biology of polar regions, 2nd edn. *Biology of habitats*. Oxford University Press, Oxford.

Thompson, D. W. J. and S. Solomon(2002). Interpretation of Recent Southern Hemisphere Climate Change. *Science* 296: 895–899.

Trivelpiece, W, J.T. Hinke, A.K. Miller, C. S. Reiss, S.G. Trivelpiece, and G.M. Watters (2010). Variability in krill biomass links harvesting and climate warming to penguin population changes in Antarctica. *Proc. National Academies of Sciences USA* 108:7625-7628.

Turner J., R. Bindschadler, P. Convey, G. di Prisco, E. Fahrbach, J. Gutt, D. Hodgson, P. Mayewski and C. Summerhayes (2009). *Antarctic Climate Change and The Environment. Scientific Committee on Antarctic Research*, Scott Polar Research Institute, Cambridge UK.

Wei C. et al. (15 co-authors) (2011). A global analysis of Marine Benthos Biomass using RandomForest. *Public Library of Science (PLOS)* 5:e15323.

Yen, P., F. Huettmann, and F. Cooke (2004).Modelling abundance and distribution of Marbled Murrelets (Brachyramphusmarmoratus) using GIS, marine data and advanced multivariate statistics. *Ecological Modelling*. 171: 395-413.

Zuckerberg, B, F. Huettmannand J. Friar (2011). Proper Data Management as a Scientific Foundation for Reliable Species Distribution Modeling. Chapter 3 In: C.A. Drew, Y. Wiersma and F. Huettmann (eds). *Predictive Modeling in Landscape Ecology.* Springer, New York.Pp 45-70.

In: Horizons in Earth Science Research. Volume 11
Editors: Benjamin Veress and Jozsi Szigethy
ISBN: 978-1-63321-296-1

Chapter 4

CAN ICE SHEETS SELF-DESTRUCT AND CAUSE RAPID CLIMATE CHANGE? A CASE STUDY: JAKOBSHAVN ISBRAE, GREENLAND

***Coen Hofstede*[1] *and Terence Hughes*[2]**
[1]Alfred Wegener Institute for Polar and Marine Research, Bremerhaven, Germany
[2]Earth Sciences and Climate Change, University of Maine, Fort Pierre, South Dakota, US

ABSTRACT

Recent global warming has greatly reduced the area of perennial sea ice in the Arctic Ocean, caused mountain glaciers to retreat and allowed major ice streams draining the Antarctic and Greenland ice sheets to approach and even exceed surge velocities. This has brought new urgency to the concept that large ice sheets, past, present, and future, may have the capacity to self-destruct causing rapid and permanent changes in global climate and sea level. This possibility has focused attention on these ice streams, beginning with Jakobshavn Isbrae, long the world's fastest ice stream and discharging seven percent of Greenland ice, a discharge rate that has now doubled following catastrophic disintegration of its buttressing ice shelf in Jakobshavn Isfjord in 1997. A geometrical solution to the longitudinal force balance was combined with a simple mass balance to obtain two solutions to the resulting force-and-mass balance from 1986 to 2006. The two solutions are obtained using two approaches to quantify the loss of buttressing. The best match to changes in surface elevations and velocities along the ice stream, especially the rapidly lowering surface during this timespan, is provided by the solution that has the strongest reduction of ice-bed coupling up the ice stream, quantified by a rapid increase in the "floating fraction" of ice. This pattern seems to be repeating for other Greenland and Antarctic ice streams. If rapid surface lowering and ice discharge spreads to all major ice streams and lasts for centuries, even accelerating over time, self-destruction of these two remaining ice sheets seems inevitable. Without the high albedo of ice sheets, global warming would increase until a new global equilibrium with the new boundary conditions was established. The other possibility is the observed rapid ice thinning and discharge may be transient and lead to a new stable equilibrium that keeps these ice sheets largely intact with no significant change in global climate and sea level. Our results from Jakobshavn Isbrae were unable to determine which scenario is most likely, but time will tell, probably over the next few decades as observational data accumulate and our modeling capability improves.

INTRODUCTION

Budd et al. (1971) developed the first computer model that simulated the dynamics of ice sheets and applied it to the Antarctic Ice Sheet. It was a steady-state model applied along surface flowlines normal to ice-elevation contour lines. They used surface accumulation rates, temperatures, and elevations, ice thicknesses, and bed topography to solve the equations of force, mass, and energy balance to obtain internal stresses, temperatures, and velocities, and basal melting rates for prescribed geothermal heat input. The model also computed time horizons at depth for specified surface accumulation rates that could be compared to internal radar-reflecting horizons and oxygen-isotope stratigraphy that were believed to be linked to former surface accumulation patterns along the flowlines. Modest changes in surface accumulation rates and the geothermal heat flux caused a largely frozen bed to become largely thawed with substantial basal melting rates. Over time, this buildup of basal water would decouple the ice sheet from its bed and cause the whole ice sheet to surge. Sugden (1977) corrected this defect by placing a basal freezing zone beyond the basal melting zone, so basal meltwater moving into the freezing zone would freeze onto basal ice and thereby couple the ice sheet to the bed for steady-state subglacial hydrology. He applied his model to the former Laurentide Ice Sheet at the last glacial maximum. These were essentially top-down models that used surface conditions to determine bed conditions.

Mahaffey (1976) made the next major breakthrough by extending flowline solutions to the map plane, so changing flow patterns through time could be simulated. Reeh (1982, 1984) used plasticity theory to reconstruct the Greenland Ice Sheet today and the former Innuitian Ice Sheet in Arctic Canada in the map plane. An early three-dimensional time-dependent model in the map plane was applied to the Antarctic (Huybrechts, 1990, 1992) and Greenland (Huybrechts 1994, 1996) ice sheets. These models produced flow that converged on "ice streams" that coincided with observed ice streams and discharged most of the ice, but ice-stream dynamics were not included in the models. Early models simulated slow sheet flow, which covers about 90 percent of ice sheets, but it became increasingly obvious that up to 90 percent of the ice was discharged by ice streams. Denton and Hughes (1981) made the first attempt to include ice streams in reconstructing ice sheets globally at the Last Glacial Maximum (LGM) 18,000 years ago. They mapped basal thermal zones (melted, melting, freezing, and frozen basal conditions) from glacial geology, and then calculated ice elevations from the strength of ice-bed coupling in these zones. This is essentially a bottom-up approach (Hughes, 1998; 2009a).

Like large rivers that discharge most precipitation over continents, large ice streams are fed by numerous tributaries and discharge most of the ice precipitated over ice sheets (Rignot et al., 2011). Ice streams are fast currents of ice that develop from slow interior sheet flow and deliver internal ice to terminal ice lobes grounded on land or to floating ice tongues that often merge to become floating ice shelves in water. Ice is lost by melting and calving from ice lobes and ice shelves. MacAyeal and Thomas (1982) and Morland (1987) were first to model ice shelves in the map plane. Antarctic ice shelves often lie in confining embayments and are locally grounded where they scrape over basal pinning points to produce surface ice rumples, or flow around pinning points to produce surface ice rises. These boundary conditions allow ice shelves to buttress ice streams that feed them (Thomas, 1973a, 1973b). Unlike sheet flow, where gravitational forcing is balanced against a basal drag force, in shelf flow gravitational forcing is balanced by a tensile force in ice arising from the imbalance in the longitudinal gravitational forces in ice and water, first recognized at the calving front (Weertman, 1957a; Robin, 1958), and later shown to exist from the calving front to the grounding line (Hughes, 2009b).

MacAyeal (1989) made the first attempt to model a transition from sheet flow to stream flow to shelf flow by including basal drag as an addition to stresses resisting gravitational shelf flow. He applied his model to Whillans Ice Stream (formerly ice stream B) that supplies ice to the Ross Ice Shelf in Antarctica. Hughes (1992) introduced a "floating fraction" of ice that decreases from unity at the ice-shelf grounding line to zero at the head of the ice stream. The floating fraction partitioned forces resting gravitational flow in ice streams between basal drag where the ice stream is locally grounded and a tensile force where the ice stream is locally floating.

By the turn of the millennium, Earth-orbiting satellites were mapping substantial reductions of Arctic sea ice, occasionally opening the Northeast and Northwest Passages to the Orient, glacial geologists were mapping retreat of mountain glaciers worldwide, and glaciologists were reporting rapid accelerations of some Greenland and Antarctic ice streams and outlet glaciers, even exceeding "surge" velocities that occur periodically in some mountain glaciers. This was attributed to global warming, and it focused attention of glaciologists on the dynamics of ice streams. A great advance in gathering field data and in developing ice-sheet models followed.

In the field, initial attention was focused on Pine Island Glacier (Rignot, 1998a) and Thwaites Glacier (Rignot, 2001) in Antarctica, and on Jakobshavn Isbrae (Thomas et al., 2003) and on Kangerdlugssuaq Glacier and Helheim Glacier (Stearns and Hamilton, 2007) in Greenland. Hughes (1986) predicted the positive-feedback mechanisms leading to the surge of Jakobshavn Isbrae would move northward along both Greenland coasts as climate warmed, and eventually cause surges of northern Greenland ice streams. He called this "The Jakobshavn Effect" and it has apparently moved to several northern ice streams (Rignot et al., 2001). These positive feedbacks included disintegration of buttressing ice shelves and surface meltwater reaching the bed through crevasses and moulins. Acceleration caused by meltwater reaching the bed has now been confirmed for Jakobshavn Isbrae (Zwally et al., 2002).

Accelerated discharge of ice streams has followed disintegration of buttressing ice shelves, both in Antarctica when the central Larsen Ice Shelf disintegrated (DeAngelis and Skvarka, 2003) and in Greenland when the floating ice tongue of Jakobshavn Isbrae disintegrated (Joughin et al., 2008). Catastrophic ice-shelf disintegration has been linked to warm ocean water passing under Antarctic ice shelves (Jacobs et al., 1996; Shepherd et al., 2004) and entering Greenland fjords (Holland et al., 2008; Rignot and Steffen, 2008). Bottom melting of ice shelves destabilized their grounding lines in both Antarctica (Rignot, 1998a, 1998b, 2001) and Greenland (Rignot, 1998c; Rignot and Steffen, 2008).

Modeling advances have paralleled field studies. The stability of ice shelves is linked to stress fields that produce crevasses (Rist et al., 1996, 1999, 2002), to catastrophic disintegration of ice shelves along crevasses transverse to ice flow (MacAyeal et al., 2003), to ice dynamics at the grounding line (Hindmarsh and LeMeur, 2001; Schoof, 2007; Schoof and Hindmarsh, 2010) and at the calving front (Kenneally and Hughes, 2006; Hindmarsh, 2012), and to ocean swells that pass under ice shelves (MacAyeal et al., 2009). Modeling basal conditions has advanced from slow sheet flow sliding on a wet bed (Weertman, 1957b; Kamb, 1970) to sliding of fast ice streams that supply ice shelves (Dupont and Alley, 2005a, 2005b). Models now treat the initiation of stream flow (e.g., Hindmarsh, 2009; Hughes, 2009c) and how basal meltwater uncouples ice from the bed beneath ice streams (e.g., Alley, 1989a, 1989b; MacAyeal, 1989; Anandakrishnan and Alley, 1997; Kamb, 2001; Hughes, 2009b; Bueler and Brown, 2009; Bassis, 2010; Schoof and Hindmarsh, 2010, Hughes et al., 2011). Modeling ice-bed decoupling by basal water over time requires modeling subglacial hydrology as basal water flows from sources to sinks (Johnson, 2002; Johnson and Fastook, 2002). In ice streams, the full equilibrium/momentum equations must eventually be solved because all six stresses resist gravitational flow (Sargent, 2009).

A HOLISTIC ICE-SHEET MODEL

A holistic ice-sheet model must provide smooth transitions from sheet flow to stream flow to shelf flow in the longitudinal force balance in the direction of gravitational flow of ice. If this force balance is done for flowbands having the width of the ice stream, assumed to be constant, the six resisting stresses reduce to four, a longitudinal tension stress σ_T that pulls upslope ice, a longitudinal compression stress σ_C that pushes downslope ice, a basal shear stress τ_O due to basal drag, and side shear stress τ_S due to side drag. Transverse stresses caused by converging and diverging flow that changes the flowband width can then be ignored. This allows a simple geometrical force balance in the longitudinal direction of ice flow, along which all of these stresses vary with changing floating fraction ϕ of ice in the flowband. For sheet flow, $\phi = 0$ when the bed is dry (frozen) and $\phi \to 0$ when the bed is wet (thawed). For stream flow, $0 < \phi < 1$ with ϕ generally increasing downstream. For shelf flow, $\phi = 1$ for a freely-floating ice shelf and $\phi \to 1$ when a confined and locally pinned ice shelf provides buttressing for the ice stream.

Figure 1 is a cartoon showing places where $\phi = 0$ for grounded ice, wet under an ice stream, and $\phi = 1$ for places where floating ice exists under the ice stream. These places generally correspond to hills and hollows in bedrock topography, or to soft sediments or till that are unsaturated and supersaturated with water, respectively. Bedrock hills and an unsaturated soft bed resist gravitational motion. Taking Cartesian coordinates with x horizontal and positive against ice flow, y horizontal and transverse to ice flow, and z vertical and positive above sea level, at distance x from the ice-shelf grounding line, a flowband of width w_I has floating segments that add up to width $w_F < w_I$ in the ice stream. Floating fraction ϕ can then be defined for the horizontal longitudinal force balance as:

$$\phi = \frac{w_F}{w_I} = \frac{h_F}{h_I} = \frac{(\rho_W / \rho_I) h_W}{h_I} = \frac{\rho_W g h_W}{\rho_I g h_I} = \frac{P_W^*}{P_I} \tag{1}$$

where $h_F = h_I (w_F/w_I) = h_I \phi$ is the part of ice thickness h_I supported by basal water, ρ_W is water density, ρ_I is ice density, $P_W^* = \rho_W g h_W$ is an effective basal water pressure that resists ice flow, $P_I = \rho_I g h_I$ is the ice overburden pressure, and g is gravity acceleration. In a vertical force balance, apply Newton's second and third laws of motion to the base of columns having basal area $A_O = w_I \Delta x$. Gravity forces $\rho_I g h_I A_O$ and $\rho_W g h_W A_O$ are balanced by pressure forces $P_I A_O$ and $P_W A_O$, respectively, giving $P_W = \rho_W g h_W$ as the actual basal water pressure and $P_I = \rho_I g h_I$ as the basal ice pressure. For ice shelves, $P_W = P_I$ everywhere. For ice streams $P_W \approx P_I$ because basal water flowing from sources to sinks causes variations in P_W that do not coincide everywhere with P_I.

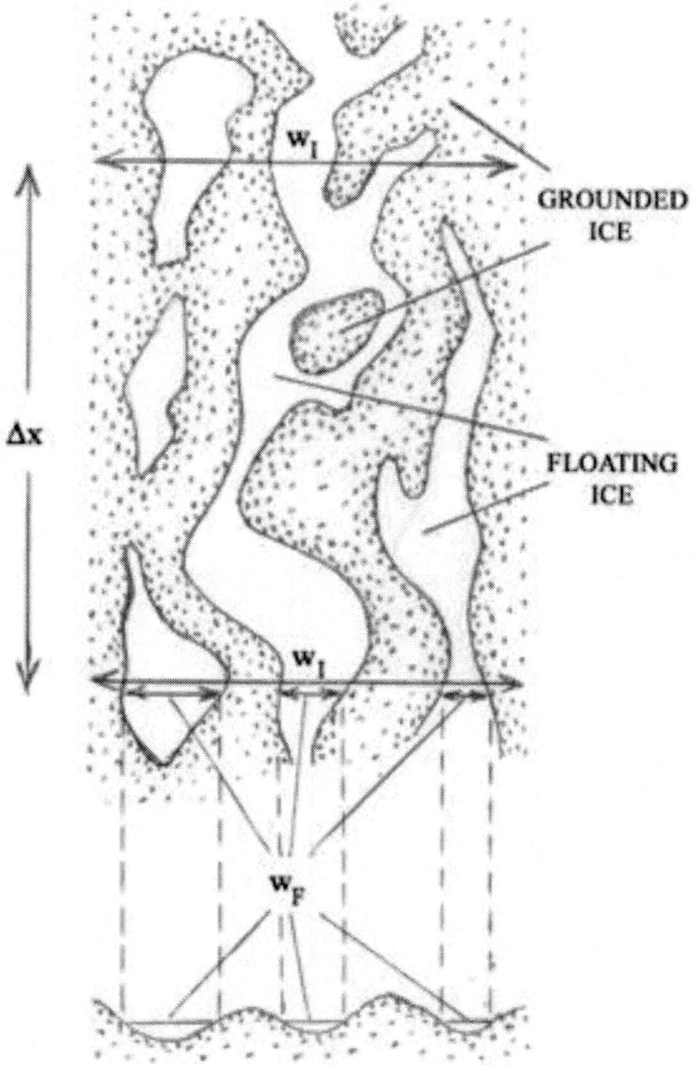

Figure 1. A cartoon of the bed under an ice stream. Ice flow is along incremental length Δx in plan view (top) and at x in transverse cross-section (bottom). Ice is either floating above bedrock or supersaturated sediments and till (undotted areas) or grounded on bedrock or unsaturated sediments and till (dotted areas) for respective floating flowband widths w_F and grounded flowband widths $w_I - w_F$. Floating fraction ϕ of ice over area $w_I \Delta x$ becomes $\phi = w_F / w_I$ at x when $\Delta x \to 0$.

Figure 2 shows an exaggerated vertical longitudinal cross-section of a flowband from the ice divide to an ice stream and ending at the calving front of a confined and pinned ice shelf. Flow is from right to left. The top panel shows in shading the part of the flowband that rests on the bed. Solid, broken, and dashed lines show respective heights h_I, h_F, and h_W above basal ice. The ice shelf lies in a confining embayment grounded along side lengths L_S, at an ice rise of circumference C_R, and at ice rumples of area A_R, so it buttresses the ice stream. Stresses resisting gravitational flow are shown at distance x from the ice-shelf grounding line and averaged over the distance from 0 to x. The middle panel shows a large triangular area equal to the gravitational driving force ½ $P_I h_I$.

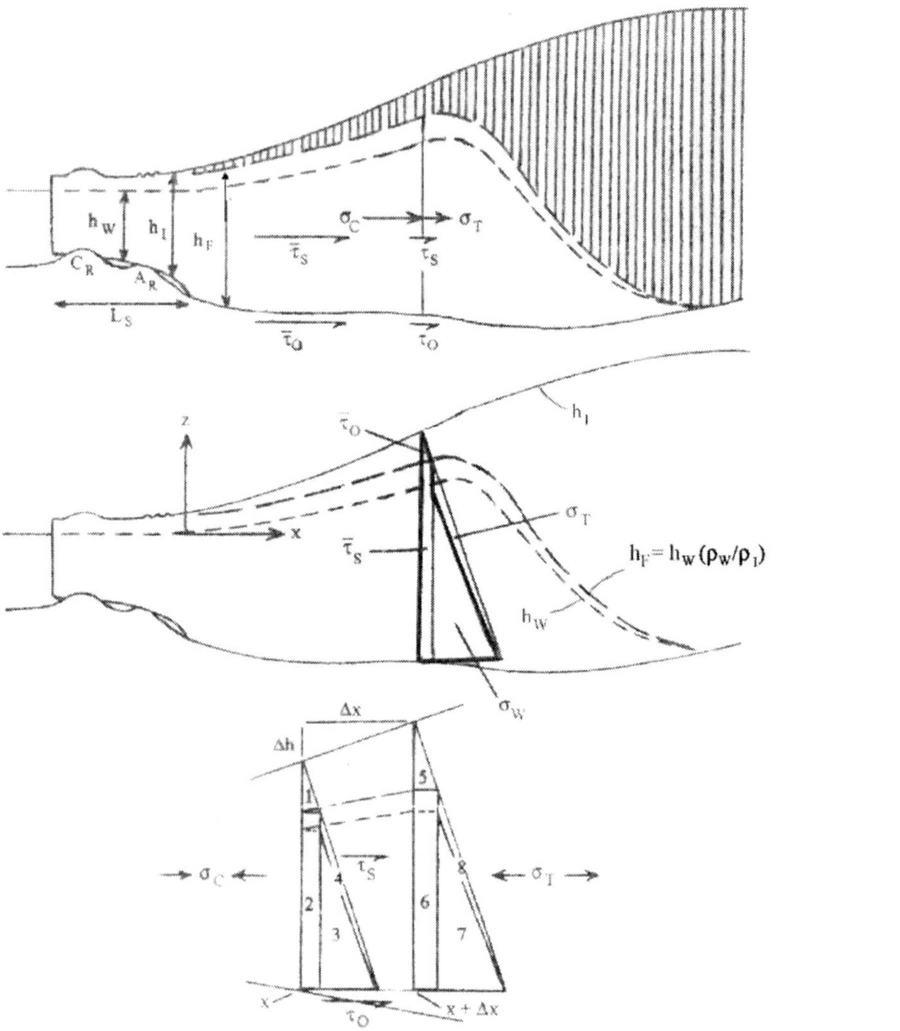

Figure 2. The geometrical force balance on an ice stream ending as a confined ice shelf. Top: Stresses that resist gravitational flow along x. The bed supports ice in the shaded area. Ice in the unshaded area is supported by "effective" basal water pressure P_W*. Middle: Gravitational forces at x represented as triangles and a rectangle are linked to specific resisting stresses. The area inside the thick border is linked to compressive stress σ_C. Heights h_I, h_W, and h_F are measured from the bed for $x > 0$. Bottom: Resisting stresses and gravitational forces along Δx. Resisting and gravitational forces are balanced along x and Δx, see Hughes (2009a). From Hughes (2009a).

Within that triangle are areas linked to resisting forces, with the area inside the bold border linked to compressive force $\sigma_C h_I$ and the remaining small triangular area linked to tensile force $\sigma_T h_I$.

This force balance gives:

$$\tfrac{1}{2} P_I = \overline{P}_I = \sigma_C + \sigma_T \tag{2}$$

Note that $\sigma_C >> \sigma_T$ because area 1+2+3 enclosed by the bold border greatly exceeds triangle area 4, so a minor downslope change in resistance to ice flow causes a small change in σ_C but a large change in σ_T because P_I is initially unchanged. This illustrates the ability of σ_T to pull more ice out of ice sheets for a small decrease in downslope resistance to ice flow (Hughes, 1992).

The bottom panel equates areas 1, 2, and 3 with compressive force $\sigma_C h_I$, triangular area 4 to tensile force $\sigma_T h_I$, triangular area 3 to water-buttressing force $\sigma_W h_I$, area 3+4 to flotation force $\sigma_F h_I$, the difference between triangular areas 5 and 4 to basal drag force $\tau_O \Delta x$, and the difference between rectangular areas 6 and 2 to side drag force $2\tau_S \Delta x$ for two sides. Balancing these longitudinal forces as $\Delta x \rightarrow 0$ gives (Hughes, 2011):

$$\sigma_T = \tfrac{1}{2}\rho_I g h_I \left(1 - \rho_I / \rho_W\right)\phi^2 = \overline{P}_I \left(1 - \rho_I / \rho_W\right)\phi^2 \tag{3}$$

$$\sigma_C = \tfrac{1}{2}\rho_I g h_I \left[1 - \left(1 - \rho_I / \rho_W\right)\phi^2\right] = \overline{P}_I \left[1 - \left(1 - \rho_I / \rho_W\right)\phi^2\right] \tag{4}$$

$$\sigma_W = \tfrac{1}{2}\rho_I g h_I \left(\rho_I / \rho_W\right)\phi^2 = \overline{P}_I \left(\rho_I / \rho_W\right)\phi^2 \tag{5}$$

$$\sigma_F = \sigma_T + \sigma_W = \overline{P}_I \phi^2 \tag{6}$$

$$\tau_O = \tfrac{1}{2}\rho_I g h_I \left(1 - \phi\right)\left[\alpha_I - \partial\left(h_I \phi\right) / \partial x\right] \approx P_I \left(1 - \phi\right)^2 \alpha \tag{7}$$

$$\tau_S = 2\rho_I g h_I \left(w_I / h_I\right)\left[\phi\alpha_I - \left(1 - 2\phi\right)\partial\left(h_I \phi\right) / \partial x\right] \approx P_I \left(w_I / h_I\right)\phi\left(1 - \phi\right)\alpha \tag{8}$$

$$\partial\left(\sigma_F h_I\right) / \partial x = \partial\left(\sigma_T h_I\right) / \partial x + \partial\left(\sigma_W h_I\right) / \partial x = P_I \left(\rho_I / \rho_W\right)\phi\alpha_W \tag{9}$$

Here $\Delta h / \Delta x \rightarrow \alpha$ is the ice surface slope, $\Delta h_I / \Delta x \rightarrow \alpha_I$ is the ice thickness gradient, and $\Delta h_W / \Delta x \rightarrow \alpha_W$ is the gradient of basal water height giving effective basal water pressure P_W^* resisting gravitational ice flow as $\Delta x \rightarrow 0$. Water buttressing stress σ_W in ice is due to P_W^* in the force balance $\sigma_W h_I = P_W{}^* h_W$. Flotation stress σ_F in ice is due to $\sigma_W + \sigma_T$ in the force balance $\sigma_F h_I = \sigma_W h_I + \sigma_T h_I$. These are real stresses. They are inaccessible using continuum mechanics in conventional ice-sheet models, but they visibly emerge from the geometrical force balance in the holistic ice-sheet model.

The longitudinal force balance equates gravitational driving force gradient $\Delta(\bar{P}_I h_I)/\Delta x = P_I \alpha$ as $\Delta x \rightarrow 0$ from the difference between area 5+6+7+8 and area 1+2+3+4 in incremental length Δx with the areas inside these triangles linked to resisting stresses τ_O and τ_S and flotation force gradient $\partial(\sigma_F h_I)/\partial x$ to obtain (Hughes, 2011):

$$P_I \alpha = \tau_O + 2\tau_S (h_I / h_W) + \partial(\sigma_F h_I)/\partial x \tag{10}$$

Equation (10) is satisfied using substitutions from Equations (7), (8) and (9).

Now approximate bed topography with an up-down staircase in which Δx is the constant step length and $\pm\Delta h_B$ is the variable gain or loss in step height. A normal stress σ_N in the direction of ice flow pushes against $-\Delta h_B$ and pulls away from $+\Delta h_B$ with force $F_N = \pm\sigma_N \Delta h_B$ compared to gravitational driving force $F_G = \bar{P}_I h_I$, so that $\sigma_N \Delta h_B/\Delta x$ and $P_I \Delta h/\Delta x$ are force gradients with σ_N close to plastic yield stress σ_O = 100 kPa for ice (Hughes, 1998, Figure 8.3). Then F_N is much less than F_G until the bed slope exceeds ±30 degrees (Hughes, 2012, Appendix E), so F_N can be ignored for lesser bed slopes. Then $\Delta h = \Delta h_I$ can be used for each Δx step.

Substituting Equations (7), (8), and (9) into Equation (10), ignoring the terms containing $\partial\phi/\partial x$, dividing by P_I, solving for surface slope α, and returning to the incremental form:

$$\frac{\Delta h}{\Delta x} = \frac{\Delta(\sigma_F h_I)\Delta x}{P_I} + \frac{\tau_O}{P_I} + \frac{2\tau_S (h_I / h_W)}{P_I}$$
$$= \phi^2 \left(\frac{\Delta h_I}{\Delta x}\right)_F + (1-\phi)^2 \left(\frac{\Delta h_I}{\Delta x}\right)_G + 2\phi(1-\phi)\frac{\Delta h}{\Delta x} \tag{11}$$

Here $\Delta h_I = \Delta h$ so $(\Delta h/\Delta x)_F$ is for the floating fraction of the ice column linked to σ_F and $(\Delta h/\Delta x)_G$ is for the grounded fraction of the ice column linked to τ_O.

A simple mass balance is shown in Figure 3 for constant ice accumulation rate a and ice thinning rate r, at $h_I = h_L$ where ice velocity u = 0 at the ice divide (x = L), $h_I = h_S$ where $u = u_S$ and stream flow begins ($x = S$), and $h_I = h_O$ where $u = u_O$ at the ice-shelf grounding line (x = 0) so that:

$$(a - r)(L - x) = h_I\, u \tag{12}$$

Using a conventional sliding law for ice (Weertman, 1957b), where B includes bed roughness and physical properties of temperate ice and m = 2 for grounded ice:

$$u = u_S = \left(\frac{\tau_O}{B} \right)^m \tag{13}$$

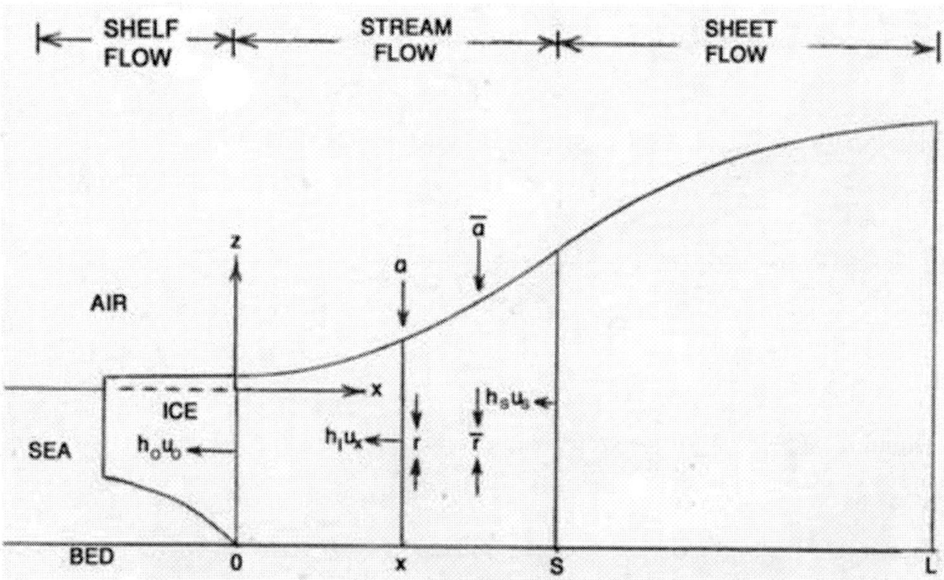

Figure 3. A longitudinal profile of an ice-sheet flowband of constant width showing components of the mass balance for sheet, stream, and shelf flow from right to left. Ice thickness h_I and mean ice velocity u_x are shown at x positions 0, x, S, and L for mean accumulation rate $\bar{a}$ and ice thinning rate $\bar{r}$ averaged along x, and rates a and r at x.

Here is how $(\Delta h/\Delta x)_G$ is obtained. Equate ice elevation h with ice thickness h_I for a horizontal bed at sea level. Combine equations (12) and (13), with $\tau_O = \rho_I g\, h_I\, dh_I/dx$:

$$(a-r)(L-x) = h_I u = h_I (\tau_O / B)^m = h_I \left[(\rho_I g h_I / B) dh_I / dx \right]^m \tag{14}$$

Integrating from the ice divide at $x = L$ to the ice-shelf grounding line at $x = 0$ where $h_I = h_O$, with a and r constant along x:

$$\int_O^x \left[(a-r)(L-x) \right]^{\frac{1}{m}} dx = \int_{h_O}^{h_I} (\rho_I g / B) h_I^{\frac{m+1}{m}} \, dh_I \tag{15}$$

Solving for h_I:

$$h_I = \left\{ h_O^{\frac{2m+1}{m}} + \frac{B}{\rho_I g} \left(\frac{2m+1}{m+1} \right) (a-r)^{\frac{1}{m}} \left[L^{\frac{m+1}{m}} - (L-x)^{\frac{m+1}{m}} \right] \right\}^{\frac{m}{2m+1}} \tag{16}$$

In Equation (16), only a and r determine change in h_I along x. Now let h_I vary with bed topography, using h_I in Equations (14) and (16). Solve for surface slope $\alpha = dh/dx$:

$$\alpha = \frac{dh}{dx} = \frac{B}{\rho_I g h_I}\left[\frac{(a-r)(L-x)}{h_I}\right]^{\frac{1}{m}} \tag{17}$$

Taking $\tau_O = \rho_I\, g\, h_I\, \alpha$ and solving for τ_O in Equations (16) and (17):

$$\tau_O = \frac{B\left[(a-r)(L-x)\right]^{\frac{1}{m}}}{\left\{h_O^{\frac{2m+1}{m}} + \frac{B}{\rho_I g}\left(\frac{2m+1}{m+1}\right)(a-r)^{\frac{1}{m}}\left[L^{\frac{m+1}{m}} - (L-x)^{\frac{m+1}{m}}\right]\right\}^{\frac{1}{2m+1}}} \tag{18}$$

Setting $\alpha = (\Delta h/\Delta x)_G$ for ice grounded in incremental length Δx, Equations (16) and (17) give:

$$\left(\frac{\Delta h_I}{\Delta x}\right)_G = \frac{\tau_O}{\rho_I g h_I} = \frac{(B/\rho_I g)\left[(a-r)(L-x)\ \right]^{\frac{1}{m}}}{h_I^{\frac{m+1}{m}}}$$

$$= \frac{(B/\rho_I g)\left[(a-r)(L-x)\right]^{\frac{1}{m}}}{\left\{h_O^{\frac{2m+1}{m+1}} + \left(\frac{2m+1}{m-1}\right)\frac{B(a-r)^{\frac{1}{m}}}{\rho_I g}\left[L^{\frac{m+1}{m}} - (L-x)^{\frac{m+1}{m}}\right]\right\}^{\frac{m+1}{2m+1}}} \tag{19}$$

Using the flow law of ice (Glen, 1955), where A is an ice-hardness parameter dependant on temperature and $n = 3$ for ice, $\dot{\varepsilon} = \Delta u/\Delta x$ is the extending strain rate for stress σ_T in floating ice, and R is a dimensionless scalar that takes account of other strain rates in addition to $\dot{\varepsilon}$:

$$\dot{\varepsilon} = \dot{\varepsilon}_{xx} = \frac{\Delta u}{\Delta x} = R\left(\frac{\sigma'_{xx}}{A}\right)^n = R\left(\frac{\sigma_T}{2A}\right)^n \tag{20}$$

The mass balance in Figure 3 also applies for stream flow, with ice thickness $h_I = h_S$ where ice velocity $u = u_S$ at the beginning of stream flow ($x = S$) so that $(\Delta h/\Delta x)_F$ is:

$$\left(\frac{\Delta h}{\Delta x}\right)_F = \frac{(a-r)-h_I\dot{\varepsilon}}{u_x} = \frac{h_I(a-r)-h_I^2\dot{\varepsilon}}{h_O u_O + (a-r)\ x} \tag{21}$$

Collecting terms in Equation (11):

$$\left(1-2\phi+2\phi^2\right)\frac{\Delta h}{\Delta x} = \phi^2\left(\frac{\Delta h}{\Delta x}\right)_F + (1-\phi)^2\left(\frac{\Delta h}{\Delta x}\right)_G \tag{22}$$

Writing as a quadratic equation:

$$\left[2\left(\frac{\Delta h}{\Delta x}\right)-\left(\frac{\Delta h}{\Delta x}\right)_F-\left(\frac{\Delta h}{\Delta x}\right)_G\right]\phi^2-\left[2\left(\frac{\Delta h}{\Delta x}\right)-2\left(\frac{\Delta h}{\Delta x}\right)_G\right]\phi+\left[\left(\frac{\Delta h}{\Delta x}\right)-\left(\frac{\Delta h}{\Delta x}\right)_G\right]=0 \tag{23}$$

Setting $C_1 = (\Delta h/\Delta x)$, $C_2 = (\Delta h/\Delta x)_F$, and $C_3 = (\Delta h/\Delta x)_G$ and solving for ϕ:

$$\phi = \frac{(2C_1-2C_3)\pm\left[(2C_1-2C_3)^2-4(C_1-C_3)(2C_1-C_2-C_3)\right]^{1/2}}{2(2C_1-C_2-C_3)} \tag{24}$$

In a flowline solution, $w_I = 0$ so $\tau_S = 0$. Yet side drag remains and contributes to the ice elevation needed to overcome resistance to ice flow, so it must be taken into account in some way. Our way is to enlarge τ_O to effective basal shear stress $\tau_O{}^*$ linked to areas 5+6 minus areas 1+2 as incremental length $\Delta x \rightarrow 0$ in Figure 2. Then $\tau_O{}^*$ is:

$$\tau_O{}^* = \rho_I g h_I\left(1-\phi^2\right)\Delta h/\Delta x - \rho_I g h_I^2 \phi \Delta\phi/\Delta x \tag{25}$$

and the longitudinal force balance becomes, ignoring the $\Delta\phi/\Delta x$ term (Hughes, 2011):

$$\frac{\Delta h}{\Delta x} = \frac{\Delta(\sigma_F h_I)/\Delta x}{P_I} + \frac{\tau_O{}^*}{P_I} = \phi^2\left(\frac{\Delta h}{\Delta x}\right)_F + \left(1-\phi^2\right)\left(\frac{\Delta h}{\Delta x}\right)_G \tag{26}$$

Collecting terms:

$$\left[\left(\frac{\Delta h}{\Delta x}\right)_G-\left(\frac{\Delta h}{\Delta x}\right)_F\right]\phi^2-\left[\left(\frac{\Delta h}{\Delta x}\right)_G-\left(\frac{\Delta h}{\Delta x}\right)\right]=0 \tag{27}$$

Solving for ϕ:

$$\phi = \pm\left[\frac{C_3 - C_1}{C_3 - C_2}\right]^{1/2} \tag{28}$$

In Equations (24) and (28), the correct solution puts ϕ in the range $0 \leq \phi \leq 1$.

Application to Jakobshavn Isbrae

Jakobshavn Isbrae becomes afloat in Jakobshavn Isfjord, which empties into Disko Bugt on the west coast of Greenland at 69° 10' N. It has been observed longer than any other ice stream on the planet, and has traditionally been the fastest, moving about 7 km/a at its calving front (Weidick et al., 1990; Weidick, 2007). The calving front and grounding line of its floating ice tongue began a retreat of 30 km at the end of the Little Ice Age in Greenland in 1850, when the Industrial Revolution began and contributed to global warming (Sohn et al., 1998).

The first glaciological study was a photogrammetric determination of ice elevations and velocities over an area of 10,000 km^2 of ice converging on Jakobshavn Isflord (Brecher, 1986, Fastook et al., 1995). They found that grounded ice converges strongly as it enters Jakobshavn Isflord, so the flowband width decreases sharply. If the flowband width is taken to be constant in an ice-stream model, the flux of converging ice must be distributed over the flowband surface as an ice accumulation rate. Actual accumulation and ablation rates over this region were made by Echelmeyer et al. (1992).

Photogrammeric measurements were subsequently made in much greater detail over the floating ice tongue by Prescott et al. (2003). We will call the floating tongue an ice shelf because it is supplied by two ice streams, one being the main trunk of Jakobshavn Isbrae which continues eastward for some 100 km into the Greenland Ice Sheet following what is informally called Gogineni Gorge, see Figure 4, and the other being a much smaller local ice stream which enters Jakobshavn Isfjord from the northeast. By the time ice reaches the calving front, the smaller ice stream supplies only about 10 percent of the calving front and the main trunk retains almost a constant width from its grounding line to the calving front (Prescott et al., 2003). Ice velocities do not change much over this distance, perhaps due to resistance from side shear along the fjord walls, to local grounding near the calving front that produces ice rumples on the surface (length L_S and area A_R in Figure 1), substantial ice spilling over the south fjord wall and terminating as a grounded ice lobe, and high rates of surface and especially basal melting. The ice shelf rapidly disintegrated in 1997, probably for a variety of reasons. It was so heavily crevassed its dynamics may have been understood better using the theory of granular materials passing through a funnel, rather than using continuum mechanics (Johnson et al., 2004). Grains form "bridges" that arc upflow and retard ice flow, but when bridges collapse under the pressure of moving ice, they lose all cohesion, disperse catastrophically, and allow a sudden increase in discharge velocity.

Zwally et al. (2002) found that surface meltwater reached the bed by way of crevasses and moulins near the head of the smaller local ice stream, and this coincided with an increase in ice velocity during the most intense hours of sunlight. This probably occurs generally in the lower ablation zone of the Greenland Ice Sheet, and has been called The Zwally Effect. Iken et al. (1993) drilled three holes into Gogineni Gorge in the main trunk of Jakobshavn Isbrae and found a layer of temperate ice about 300 m thick above the bed. This ice can still be frozen to the bed if latent heat is not supplied to melt the ice. For example, if all ice crystals are taken to be cubes of the same size, and dislocations due to ice deformation pile up at boundaries of these crystal grains, supplying strain energy that can be converted into latent heat, melting begins first at points where eight grains meet, then progresses along lines where four grains meet, then spreads over areas where two grains meet, and finally melts the grains. This progression of melting from points to lines to areas to volumes allows a wide range of water to be stored in temperate ice. Temperate ice is soft, especially if it contains substantial meltwater, and this alone could account for the fast velocity of Jakobshavn Isbrae. In addition, if gravity supplies this internal meltwater to the bed, the water can drown bedrock bumps that pin ice to a hard bed or supersaturate and mobilize a soft bed consisting of sediments or till.

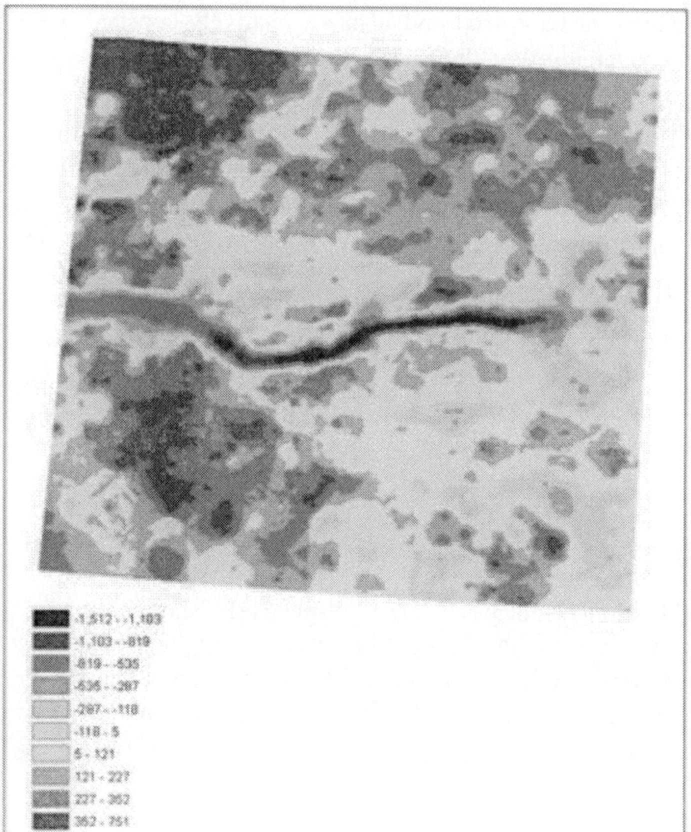

Figure 4. The bedrock map of Jakobshavn Isbræ produced from airborne radar profiles by the Center for Remote Sensing of Ice Sheets (CReSIS) at the University of Kansas. We used a digitized version for the central flightline along Gogineni Gorge for all data sets. The map is 100 km on each side.

Both processes can account for the fast velocity of Jakobshavn Isbrae. Thomas (2004) made the first attempt to model the recent doubling of velocity and resulting surface lowering of Jakobshavn Isbrae following catastrophic disintegration of its buttressing ice shelf. His approach was similar to ours, except that he considered floating fraction $\phi \approx 1$ all the way up Gogineni Gorge. He modeled the whole ice drainage system of Jakobshavn Isbrae, including the ice catchment area that is about 7 percent of the Greenland Ice Sheet. This required taking diverging and converging flow into account, diverging from the ice divide and converging on Jakobshavn Isbrae. Because ice-bed decoupling was almost total up Gogineni Gorge, Thomas (2004) obtained a faster upflow response due to the loss of downflow buttressing when the ice shelf disintegrated. Thomas et al. (2011) conluded that this process, part of The Jakobshavn Effect (Hughes, 1986), could occur generally for all Greenland ice streams if climate continues to warm.

In our modeling attempt, we chose to model Jakobshavn Isbrae using a version of the flowline model and Equation (28) instead of the flowband model leading to Equation (24).

Our reasons are purely practical. Prasad Gogineni (personal communication, 2011) provided us with a radar flightline down the center of Gogineni Gorge, see Figure 4, which gave us continuous surface and bed topography down the centerline of Jakobshavn Isbrae. Side shear vanishes along the centerline of ice steams, so we could ignore side sear stress τ_S. Also, Gogineni Gorge is very deep and it has a relatively constant width along its entire length. Laterally converging flow does not affect flow of this deep ice. Converging ice flow above the gorge can be accommodated by increasing the surface accumulation rate a in Equations (19) and (21).

Equations (19) and (21) includes $(\Delta h/\Delta x)_F$ for floating fraction ϕ of ice in our model, linked to longitudinal strain rate $\dot{\varepsilon}$ as seen in Equation (21), and also the flow law of ice given by Equation (20) which links $\dot{\varepsilon} = \dot{\varepsilon}_{xx}$ to σ_T given by Equation (3) and R given by (Hughes et al., 2011):

$$R = \left[1 + \left(\frac{\dot{\varepsilon}_{yy}}{\dot{\varepsilon}_{xx}} \right) + \left(\frac{\dot{\varepsilon}_{yy}}{\dot{\varepsilon}_{xx}} \right)^2 + \left(\frac{\dot{\varepsilon}_{xy}}{\dot{\varepsilon}_{xx}} \right)^2 + \left(\frac{\dot{\varepsilon}_{xz}}{\dot{\varepsilon}_{xx}} \right)^2 \right]^{\frac{n-1}{2}} \tag{29}$$

Where $\dot{\varepsilon}_{xx}$, $\dot{\varepsilon}_{yy}$, $\dot{\varepsilon}_{xy}$, and $\dot{\varepsilon}_{xz}$ are strain rates associated with longitudinal extension, lateral compression, side drag, and basal drag, respectively. Lateral compression occurs in converging thin upper ice but not in thick lower ice occupying Gogineni gorge. There is no lateral shear down the centerline, and there is little basal shear if the bed is wet and ϕ is high. So $\dot{\varepsilon}_{xx}$ is the dominant strain rate and $R \approx 1$ for $n = 3$. The longitudinal strain rate is therefore, using Equation (3) for σ_T, following Hughes et al. (2011):

$$\begin{aligned} \dot{\varepsilon} = \dot{\varepsilon}_{xx} &= (\sigma_T / 2A)^n = \left[(\rho_I g h_I / 4A)(1 - \rho_I / \rho_W)\phi^2 \right]^n \\ &= \left[(\rho_I g h_I / 4A)(1 - \rho_I / \rho_W) - (\sigma_B / 2A) \right]^n \end{aligned} \tag{30}$$

Here σ_B is a back-stress due to buttressing by the ice shelf in Jakobshavn Isfjord and given by:

$$\sigma_B = f_B\left[\tfrac{1}{2}\rho_I g h_O\left(1-\rho_I/\rho_W\right)\right] \tag{31}$$

where f_B is a buttressing fraction with $f_B = 0$ for no buttressing and $f_B = 1$ for full buttressing.

Equations (30) and (31) allow two possibilities for $\dot{\varepsilon}$ variations along x. If ϕ is also allowed to vary along x in Equation (30) the solution is:

$$\dot{\varepsilon} = \left[\left(\rho_I g h_I / 4A\right)\left(1-\rho_I/\rho_W\right)\phi^2\right]^n \tag{32}$$

with ϕ at $x = 0$ being a measure of ice-shelf buttressing such that $\phi = 1$ if the ice shelf has disintegrated. If ϕ is replaced by ice-shelf buttressing at $x = 0$, then Equation (31) requires that:

$$\dot{\varepsilon} = \left[\tfrac{1}{2}\rho_I g h_I\left(1-\rho_I/\rho_W\right)\right]^n\left[1-f_B\left(h_O/h_I\right)\right]^n/(2A)^n \tag{33}$$

Equation (33) emphasizes that ice-shelf buttressing is transmitted upstream. With either Equation (32) or Equation (33) substituted for $\dot{\varepsilon}$ in Equation (21), we see that $(\Delta h/\Delta x)_F$ varies with either ϕ^6 or $[1 - f_B(h_O/h_I)]^3$ for $n = 3$. Both possibilities will be considered. In the case of Equation (33), $0 \leq f_B \leq 1$ is chosen to match the observed h_O at the ice-shelf grounding line. When the ice shelf has disintegrated, $f_B = 0$ is expected. Equation (33) is not so different from one used by Thomas (2004) in modeling the ongoing surge of Jakobshavn Isbrae following disintegration of its buttressing ice shelf in Jakobshavn Isfjord.

Before our approach can be applied to Jakobshavn Isbrae, it is necessary to address a concern that arises from approximating bed topography using an up-down staircase. As shown by Hughes (2012, Appendix E), this is acceptable for bed slopes not exceeding 30 degrees in the force balance. In the mass-balance, however, the staircase produces abrupt velocity changes between Δx steps, an abrupt decrease at "down" steps and increase at "up" steps, because ice thickness h_I increases and decreases abruptly between steps in the mass balance. Therefore gradients $(\Delta h/\Delta x)_G$ and $(\Delta h/\Delta x)_F$ used as surface slopes in Equations (19) and (21) for the force balance should be retained as ice-thickness gradients $(\Delta h_I/\Delta x)_G$ and $(\Delta h_I/\Delta x)_F$ in the mass balance. This is done by setting:

$$\frac{\Delta h}{\Delta x} = \frac{\Delta h_I}{\Delta x} - \frac{\Delta h_B}{\Delta x} \tag{34}$$

where $\Delta h_B/\Delta x$ is the bed slope. Bed slopes are known along Jakobshavn Isbrae so this information should be used.

Since $\Delta h_B/\Delta x$ due to changing bed topography represented by h_B did not affect the derivation of Equation (19), see Denton and Hughes (1981, Chapter 5), Hughes (1998, Chapter 5), and Hughes (2012, Chapter 17), Equation (19) remains:

$$\left(\frac{\Delta h}{\Delta x}\right)_G = \frac{\left(B / \rho_I g\right)\left[(a-r)(L-x)\right]^{\frac{1}{m}}}{\left\{h_O^{\frac{2m+1}{m+1}} + \left(\frac{2m+1}{m+1}\right)\frac{B(a-r)^{\frac{1}{m}}}{\rho_I g}\left[L^{\frac{m+1}{m}} - (L-x)^{\frac{m+1}{m}}\right]\right\}^{\frac{m+1}{2m+1}}} = K_1 \tag{35}$$

where $K_1 = C_3$ in Equation (23).

Derivation of Equation (35) includes $\Delta h_B/\Delta x$. For the mass balance with $h_O\, u_O$ at the ice-shelf grounding line, $x = 0$, and $h_I\, u$ at distances upstream from the grounding line in Figure 3, ice thinning rate r in time t is:

$$\frac{\delta h_I}{\delta t} = r = a - \frac{\Delta(h_I u)}{\Delta x} = a - h_I \frac{\Delta u}{\Delta x} - u\frac{\Delta h_I}{\Delta x} \tag{36}$$

Solving for ice thickness gradient $\Delta h_I/\Delta x$; using Equation (36):

$$\left(\frac{\Delta h_I}{\Delta x}\right)_I = \frac{a-r}{u} - \frac{h_I(\Delta u / \Delta x)}{u} = \frac{h_I(a-r) - h_I^2\dot{\varepsilon}}{h_O u_O + (a-r)x} = \left(\frac{\Delta h}{\Delta x} + \frac{\Delta h_B}{\Delta x}\right)_I \tag{37}$$

where ice flux $h_I u = h_O u_O + (a-r)x$ and $\dot{\varepsilon} = \Delta u/\Delta x$ is the longitudinal strain rate, both at x in Figure 3. The ice surface slope for floating ice is then:

$$\left(\frac{\Delta h}{\Delta x}\right)_F = \left(\frac{\Delta h_I}{\Delta x}\right)_F - \frac{\Delta h_B}{\Delta x} = \frac{h_I(a-r) - h_I^2\dot{\varepsilon}}{h_O u_O + (a-r)x} - \frac{\Delta h_B}{\Delta x} \tag{38}$$

For an ice shelf, $\Delta h_B/\Delta x$ is the bottom surface of floating ice and must therefore satisfy the buoyancy condition:

$$\left(\frac{\Delta h_B}{\Delta x}\right)_F = \left(\frac{\Delta h_I}{\Delta x} - \frac{\Delta h}{\Delta x}\right) = \frac{\Delta}{\Delta x}\left(\frac{h}{1-\rho_I / \rho_W}\right) - \frac{\Delta h}{\Delta x} = \left[\frac{\rho_I / \rho_W}{1-\rho_I / \rho_W}\right]\frac{\Delta h}{\Delta x} \tag{39}$$

where positive $\Delta h_I/\Delta x$ and $\Delta h/\Delta x$ require negative $\Delta h_B/\Delta x$. Note that $\dot{\varepsilon}$ in Equation (38) has two solutions, Equation (32) and Equation (33).

We have applied three models, based on the geometrical force balance, to Jakobshavn Isbrae: *GFB*1, *GFB*2, and *GFB*3, where *GFB*1 uses Equation (33) for $\dot{\varepsilon}$, *GFB*2 uses

Equation (32) for $\dot{\varepsilon}$, and *GFB*3 shows the consequences of ignoring bed slope by setting $\Delta h_B/\Delta x = 0$. With Equation (33), similar to Thomas (2004), Equation (26) including $\Delta h_B/\Delta x$ becomes:

$$\frac{\Delta h}{\Delta x}=\left(\frac{\Delta h}{\Delta x}\right)_F \phi^2+\left(\frac{\Delta h}{\Delta x}\right)_G \left(1-\phi^2\right)=\left[\left(\frac{\Delta h_I}{\Delta x}\right)_F-\frac{\Delta h_B}{\Delta x}\right]\phi^2+\left(\frac{\Delta h}{\Delta x}\right)_G \left(1-\phi^2\right) \tag{40}$$

Using Equation (33) for $\dot{\varepsilon}$ gives the expression for *GFB*1:

$$\begin{aligned}\frac{\Delta h}{\Delta x}&=\left[\frac{\Delta h_I}{\Delta x}-\frac{\Delta h_B}{\Delta x}\right]\phi^2+K_1\left(1-\phi^2\right)=\left[\frac{h_I(a-r)-h_I^2\dot{\varepsilon}}{h_O u_O+(a-r)}-\frac{\Delta h_B}{\Delta x}\right]\phi^2+K_1\left(1-\phi^2\right)\\&=\left[\frac{h_I(a-r)-h_I^2\left[\tfrac{1}{2}\rho_I g h_I\left(1-\rho_I/\rho_W\right)\right]^n\left[1-f_B\left(h_O/h_I\right)\right]^n/(2A)^n}{h_O u_O+(a-r)x}-\frac{\Delta h_B}{\Delta x}\right]\phi^2+K_1\left(1-\phi^2\right)\\&=\left(K_2-\Delta h_B/\Delta x\right)\phi^2+K_I\left(1-\phi^2\right)\end{aligned} \tag{41}$$

Uisng Equation (32) for $\dot{\varepsilon}$ gives the expression for *GFB*2:

$$\begin{aligned}\frac{\Delta h}{\Delta x}&=\left[\frac{\Delta h_I}{\Delta x}-\frac{\Delta h_B}{\Delta x}\right]\phi^2+K_1\left(1-\phi^2\right)=\left[\frac{h_I(a-r)-h_I^2\dot{\varepsilon}}{h_O u_O+(a-r)x}-\frac{\Delta h_B}{\Delta x}\right]\phi^2+K_1\left(1-\phi^2\right)\\&=\left[\frac{h_I(a-r)}{h_O u_O+(a-r)x}\right]\phi^2+\left[\frac{-h_I^2\left[\left(\rho_I g h_I/4A\right)\left(1-\rho_I/\rho_W\right)\phi^2\right]^n}{h_O u_O+(a-r)x}\right]\phi^2-\left(\frac{\Delta h_B}{\Delta x}\right)\phi^2+K_1\left(1-\phi^2\right)\\&=\left[\frac{h_I(a-r)}{h_O u_O+(a-r)x}-\frac{\Delta h_B}{\Delta x}\right]\phi^2+\left[\frac{-h_I^{n+2}\left[\left(\rho_I g/4A\right)\left(1-\rho_I/\rho_W\right)\right]^n}{h_O u_O+(a-r)x}\right]\phi^{2n+2}+K_1\left(1-\phi^2\right)\\&=\left(K_3-\Delta h_B/\Delta x\right)\phi^2+K_4\phi^{2n+2}+K_1\left(1-\phi^2\right)\end{aligned} \tag{42}$$

The consequences of ignoring bed slope are found by setting $\Delta h_B/\Delta x = 0$ in Equations (41) and (42). This is *GFB*3.

The recent and sustained surge of Jakovshavn Isbrae began when its buttressing ice shelf in Jakobshavn Isfjord disintegrated suddenly in 1997. This has been attributed to warm ocean water entering the fjord and passing under the ice shelf (Holland et al., 2008), when the melting rate was already 109 m/a, the highest measured for any ice shelf on Earth (Prescott et al., 2003). There was also substantial surface melting, with most surface meltwater reaching the bed through crevasses (Hughes 1986; Echelmeyer et al., 1992).

Therefore it is prudent to include the ice shelf in our model. This was done by moving the origin of x, y, z coordinates to the calving front of the ice shelf and using Equation (39) for evaluating the buoyancy condition using known top and bottom slopes measured on the ice shelf.

RESULTS FOR JAKOBSHAVN ISBRAE

For our analysis of Jakobshavn Isbræ, we used ice-surface data sets from 1985, 1993, 2003 and 2006. This gave us four sequential time shots and three sequential time steps with ϕ distributions along the central flowline. The data sets represent approximately the last 70 km along the central flowline of the Jakobshavn Isbræ ice drainage system, which has a total length of about 570 km from the ice divide to the calving front (Bindschadler, 1984). Surface data sets have been collected by different means, so they have different resolutions. Surface data for 1985 were obtained using gridded data from 1985 and 1986 measured photogrammerically, with an accuracy of about 2 m (Brecher, 1986). Surface data for 1993 and 2003 were collected by NASA glaciologists using an airborne laser scanning system with an accuracy of 0.1 m. Surface data for 2006 along the center flowline were collected by the Center for Remote Sensing of Ice Sheets (CReSIS) at the University of Kansas using airborne radar with an accuracy of about 1 m. CReSIS also provided bed topography obtained by radar sounding along the center flowline and shown in Figure 4. This topography is used in all data sets and is accurate within a few tens of meters.

Until approximately 1997 Jakobshavn Isbræ still ended as an ice shelf approximately 7 km long. The length of the ice shelf in 1985 and 1993 was determined from surface data using a flotation criterion in which the depth of basal ice below sea level is $b = -h/(1 - \rho_W / \rho_I)$ for surface elevation h. The position of the grounding line was determined when bedrock depth h_B was reached by basal ice.

Ice deformation is largely determined by ice hardness parameter A in Equation (20) that varies with ice temperature. There is quite some variation of ice temperatures in Jakobshavn Isbræ depending on time and place. Borehole studies showed a temperature variation with depth from -2 °C to a minimum of -22 °C (Iken et al., 1993; Lüthi et al., 2002).

Thomas et al. (2011) preferred ice temperatures from -4 °C to -6 °C for Jakobshavn. We fitted A to calculated strain rates for the ice stream determined in 1985 (Fastook et al., 1995), giving $A = 1.4\times10^8$ Pa s$^{1/3}$ corresponding to a temperature of -15 °C (Paterson, 1994) and kept A constant through the different time shots.

As ice motion is very fast and the bed is likely to be thawed, we assumed ice is sliding. Whether this is due to high deformation in a temperate ice layer (Lüthi et al., 2009), high creep deformation in a layer with ice crystals having vertically orientated c-axes (Horgan et al., 2008), or ice sliding over bedrock (Weertman, 1957b) is not important. Hence in Equation (13) we use a value of B that simulates fast-moving ice over bedrock, $B = 1.123 \times 10^7$ Pa m$^{1/2}$ s$^{-1/2}$ (Reusch and Hughes, 2003).

Outlet ice velocity u_O at h_O varied from one data set to another. From Joughin et al. (2004) we know the outlet velocities from 1985, 1993 and 2003 varied from 6700 m a^{-1}, 5600 m a^{-1}, to 12600 m a^{-1}, values we used for those data sets. For 2006 we assumed an outlet velocity of 14000 m a^{-1} based on an annual linear increase of 5% (Joughin et al., 2008).

Actual estimates of accumulation rate vary between 0.28 m a^{-1} to 0.59 m a^{-1} (Bindschadler, 1984). For calculalting ϕ variations, uncertainties in accumulation rates hardly matter. We used an accumulation rate of 0.59 m a^{-1}.

Until 1985, Jakobshavn Isbræ apparently was close to steady state. The calving front did not move much and the terminal velocity was rather constant (Bindschadler, 1984; Pelto et al., 1989; Echelmeyer et al., 1992). Seasonality of the outlet velocity was hardly noticeable.

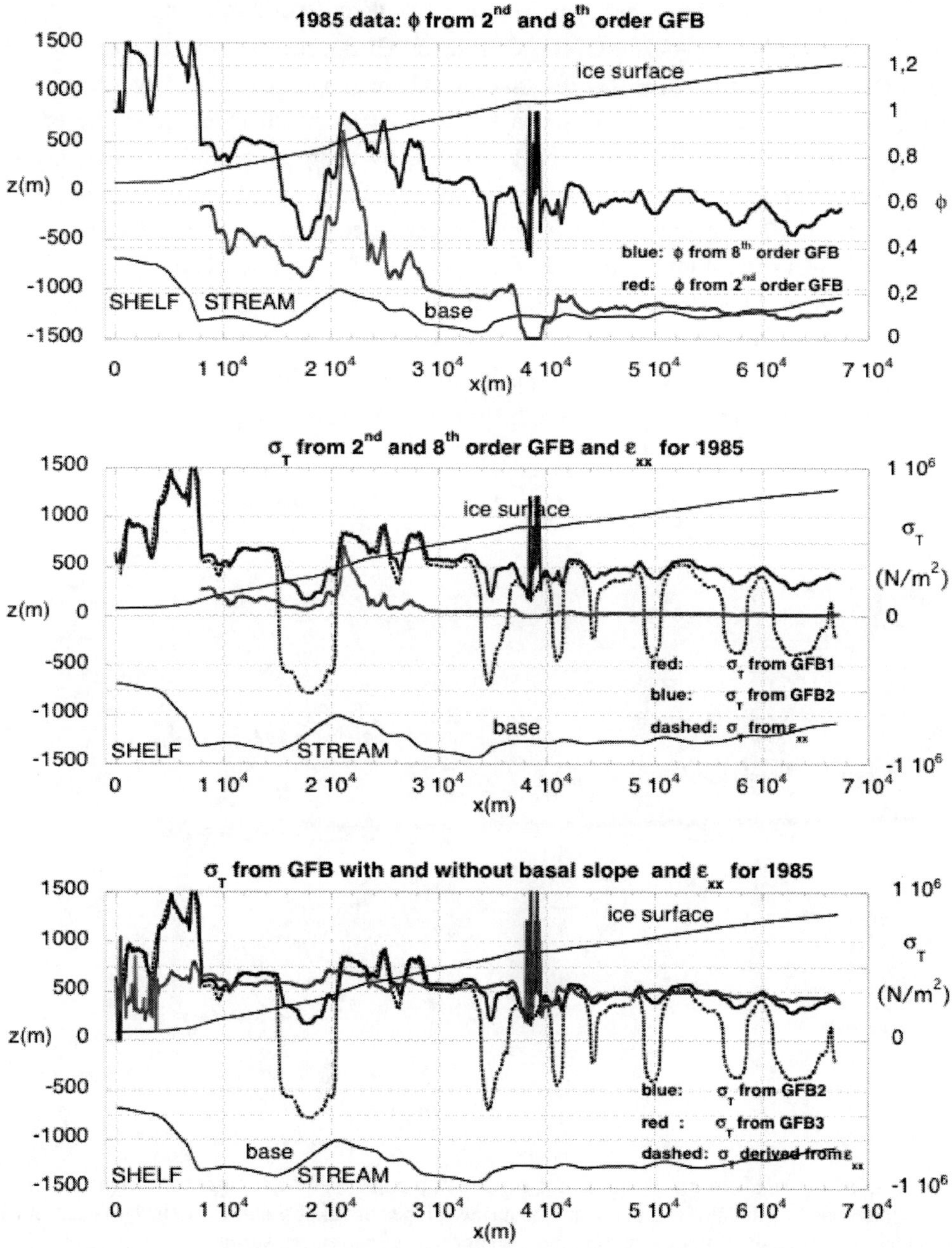

Figure 5. Results for the three different versions of the GFB using data from 1985 for Jakobshavn Isbræ. Profiles of basal and surface ice, shown in black, are quantified on the left vertical axis, while ϕ or σ_T is quantified on the right vertical axis. The top graph investigates the effect of including ϕ in σ_T by comparing GFB1 in red, when σ_T is independent of ϕ, with GFB2 in blue, when σ_T depends on ϕ. The middle graph compares σ_T from GFB1 in red with σ_T from GFB2 in blue. The black dashed graph is σ_T calculated from $\dot{\varepsilon}_{xx}$ and is taken as the correct σ_T. The bottom graph investigates the effect of including the basal slope in the floating part of the GFB by comparing GFB2 with GFB3, which is GFB2 when $\Delta h_B/\Delta x = 0$.

To determine what force balance is best, we calculated ϕ and σ_T using Equations (41), (42), and (42) with $\Delta h_B/\Delta x = 0$ for GFB1, GFB2, and GFB3, respectively, all using the 1985 data set. Results for testing GFB1, GFB2, and GFB3 are shown in Figure 5, where ϕ or σ_T is plotted along the centerline of Jakobshavn Isbræ. The top graph investigates the effect of using Equation (3) for σ_T when $\phi = 1$ or $1 \geq \phi \geq 0$ by comparing GFB1 in red, given by Equation (41), with GFB2 in blue given by Equation (42). The vertical left axis displays the values of ϕ, the horizontal and right axes represent distance from the origin, the 1985 calving front, and ice elevation above sea level. Both graphs have similar shape but GFB2 gives higher values for ϕ. This affects σ_T as shown in the second graph. The blue graph represents σ_T calculated from GFB2 and the red graph from GFB1. These values are plotted on the left vertical axis. As a reference signal, σ_T was calculated from longitudinal strain rate $\dot{\varepsilon}_{xx}$, shown as the black dashed graph, assuming σ_T is the true tensile stress.

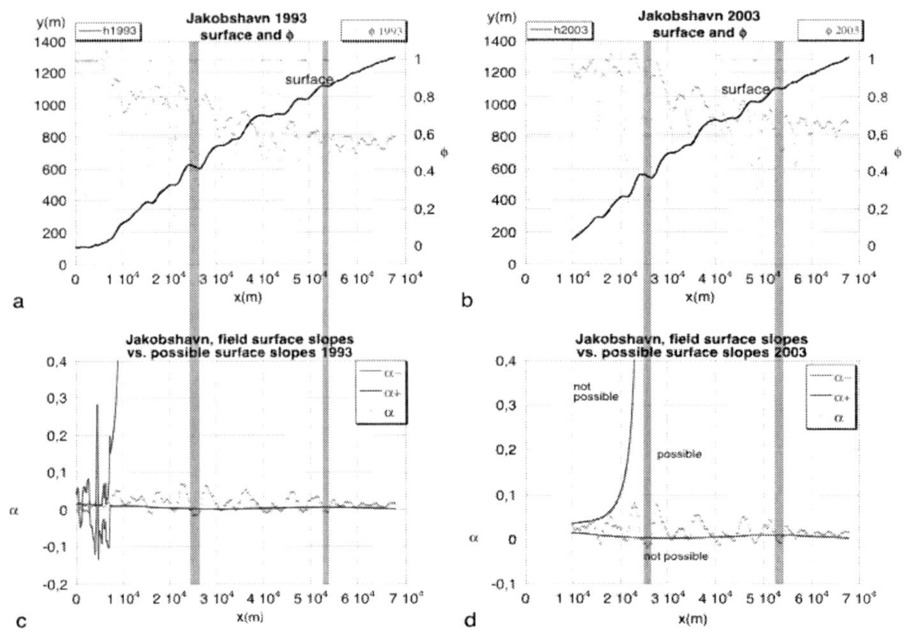

Figure 6. A comparison between ϕ and the ice surface (top graphs) and a comparison between the surface slope limits of the model and the actual surface slopes of Jakobshavn Isbræ in 1993 and 2003 (bottom graphs). When the model can no longer reproduce the measured surface slope (mostly when the slope is negative) the calculated value of ϕ is meaningless. (a) Shown here are the calculated ϕ values (red crosses) and ice surface surface of 1993 plotted along the flow line. The surface is quantified against the left vertical axis and ϕ is quantified against the right vertical axis. (b) The same data representaion as in (a) but now for the 2003 data. (c) Shown here are the minimum (red continuous line) and maximum (green continuous line) surface slope limits of the model and the actual surface slopes from the field data (blue crosses) from 1993. (d) The same data representation as in (c) but now for the 2003 data. In the shaded areas the surface slope of the field data cannot be reproduced by the model. In the bottom graphs the blue crosses lie outside the red and green slope limits of the model. As a result, ϕ shown in these graphs is meaningless here. The spiky lines at the start of the 1993 data are caused by the irregular ice-shelf profile due to many deep crevasses. In 2003 the ice shelf is gone and the terminus retreated about 10 km.

By using Equation (10) with $\tau_S = 0$, we restricted ourselves to a two-dimensional flowline solution, whereas Jakobshavn Isbræ is a complex three-dimensional system. To get meaningful results, two-dimensional strain rates were calculated from the ice thickness data. Ice thickness gradient $\partial h_I/\partial x$ and the outlet ice flux were calculated assuming a steady state. Calculating strain rates $\dot{\varepsilon}_{xx}$ in this way gives negative values for the tensile stress whenever the ice thickness gradient diverges in the flow direction. As mentioned, the ice hardness parameter A was tuned such that the values of σ_T , the blue line, from GFB2 matched the true values of σ_T, the black dashed line. Especially in the ice shelf, the GFB2 estimate is very good. This causes unrealistic values of ϕ for the ice shelf in the top graph where $\phi > 1$. Obviously σ_T needs to be higher than linear shelf flow can produce. This is somewhat surprising, because the ice shelf of Jakobshavn Isbræ consisted of nearly parallel flow in 1985 (Prescott et al, 2003). So we would expect a flowline model to be suitable. Prescott et al. (2003) calculated a high surface-and-basal melting rate of 109 m/a (mostly basal) that we did not take into account. The high melting rate producces a large ice thickness gradient which over-estimates σ_T. This unrelistic high value of σ_T requires that $\phi > 1$. The bottom graph investigates the effect of including the basal slope in the GFB by comparing Equation (42) with Equation (42) when $\Delta h_B/\Delta x = 0$. On average, σ_T is comparable in magnitude but not in shape. Including the basal slope in the GFB makes σ_T more similar to the true tensile stress obtained from $\dot{\varepsilon}_{xx}$. Whenever the true tensile stress becomes negative, as for diverging flow, σ_T from GFB2 follows this trend but remains positive. This is because GFB2 links the magnitude of a force to an area in Figure 2 that cannot be negative.

A second shortcoming of the GFB is it accommodates only a limited range of surface slopes. Figure 2 shows that the idealized sheet, stream, and shelf slopes along the flowline have only positive surface slopes, whereas the field data often have negative slopes too. The model cannot account for local drastic slope changes or negative slopes because $\Delta h/\Delta x$ is too strong as a variable. At these places the predicted ϕ values are meaningless. This problem is particularly clear for the highly resolved surface data from 1993 and 2003. Figure 6 shows the boundaries of possible surface slopes from the model versus the measured surface slopes from 1993 and 2003. The red graph is the lower limit of the model's surface slopes and the green graph the upper limit. The red crosses represent the actual surface slopes along the flow line and they fall occasionally outside the area of possible surface slopes, as marked by the shaded rectangles. Here the calculated ϕ values fall outside the model's range and are meaningless. Using a running mean for surface slopes can smooth the surface enough to eliminate false local variations in ϕ.

In general GFB2 gave best results for σ_T so Equation (42) is the equation we solved for the consecutive surface profiles of Jakobshavn Isbræ. Floating fraction ϕ is presented in Figure 7 and its tensile stress σ_T is shown in Figure 8. The top graph shows the total change, from 1985 to 2006. The three graphs under this represent the sequential time steps leading to the total time interval. In all time steps, the blue graphs represent the older data set and the red graphs represent the newer data set. These values are plotted against the right vertical axis. Ice surface and bed data are plotted in black, where the dashed section is the older data.

In Figure 7 the general trend of all graphs is that $\phi \approx 1.0$ at the calving front except for the 1985 data set where ϕ is a bit higher, suggesting almost completely buoyant ice to the grounding line, with gradually more grounding going upstream.

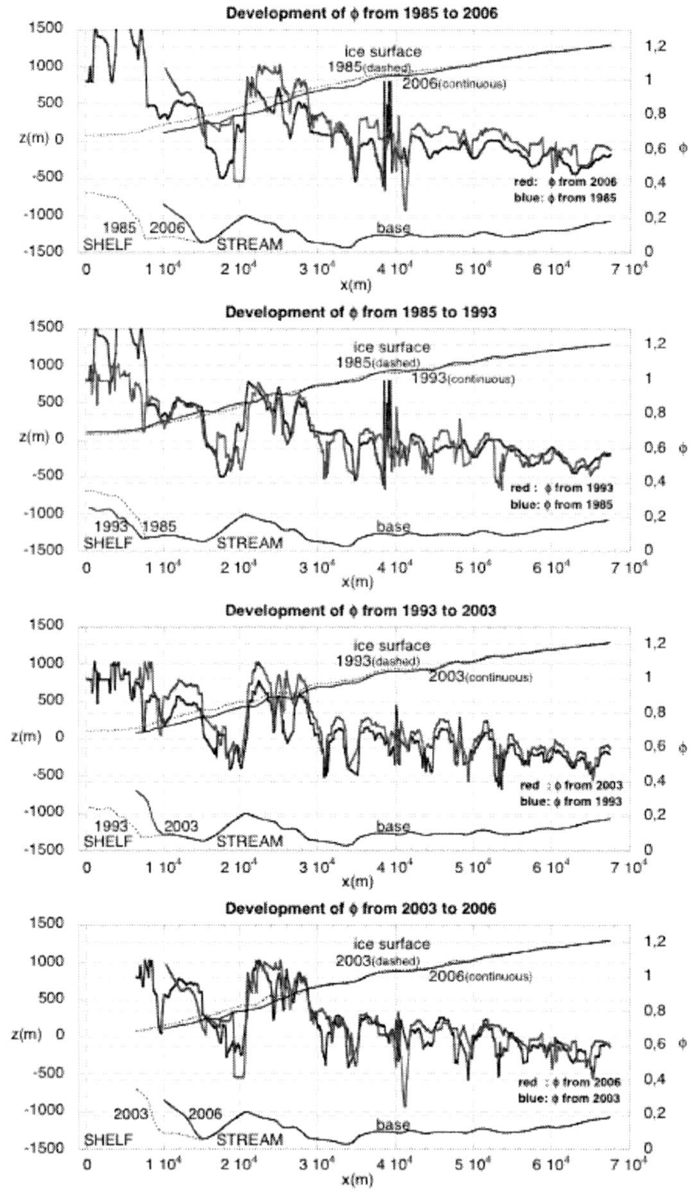

Figure 7. The development of ϕ using GFB2 for Jakobshavn Isbræ. The first (top) graph shows the overall development from 1985 to 2006. The blue graph shows the old values of ϕ and the red graph shows the new values plotted against the right vertical axis. The surface and bedrock data are plotted in black against the left vertical axis where a possible alternative bed profile is dashed. The next three graphs show the consecutive time steps, the second graph is time step 1985 to 1993, the third graph is from 1993 to 2003, and the fourth (bottom) graph is from 2003 to 2006.

Even 70 km upstream from the grounding line, $\phi \approx 0.55$. The ice stream is largely buoyant but becomes more grounded in the upstream direction. This suggests the presence of basal water throughout the modeled flowline and confirms the possibility of a temperate ice layer suggested by Lüthi et al. (2002, 2009) and Iken et al. (1993). Since Gogineni Gorge under Jacobshavn Isbræ is roughly 800 to 1500 m below sea level, a wet base can be expected. The top graph, covering the period from 1985 to 2006, shows a significant increase in ϕ over the entire flowline, decreasing even further upsteam, suggesting an increased buoyancy over this period. The largest increase in buoyancy took place near the ice-shelf grounding line.

If we then look at the consecutive time steps, we see that between 1985 and 1993 there is no clear change in ϕ except for the ice shelf. The ice shelf thickened in this period and ϕ decreased. Upstream from the grounding line, ϕ is relatively unchanged. Local differences are to be expected, as resolution of the 1993 data is much higher and thus the surface profile is different (more rugged). In general the shape and magnitude of ϕ is the same. By then the shelf had disintegrated.

From 1993 to 2003 we see an overall increase in ϕ along the entire flowline with the largest change near the calving front. The shape of the graphs is very similar but so is the resolution. The bottom graph covers the period from 2003 to 2006. The resolution of the 2006 surface data is about a meter, making a comparison difficult. There is a small increase in ϕ starting 13 km upstream from the grounding line, suggesting an increase in buoyancy. The higher resolution of the 1993 and 2003 surface data sets makes the ϕ values noisier. The surface data are bumpier and especially low surface slopes fall outside the range the model can produce.

Development of tensile stress σ_T, shown in Figure 8, is very similar to the ϕ development. Increasing σ_T downstream seen in the first (top) graph is in accordance to the observed velocity increase over the period 1985 to 2006. From 1985 to 1993 not much change was observed. The first 20 km of the ice thickened somewhat, in particular for the ice shelf, reducing ice thickness gradient $\partial h_I/\partial x$. In the same period the outlet velocity reduced somewhat from 6700 m a^{-1} to 5700 m a^{-1} (Joughin et al., 2004). Both effects contribute to a smaller σ_T as is registered. This slowing down has been attributed to possible jamming in the ice shelf (Johnson et al., 2004) and could explain why thickening is observed only at the front. To pinpoint this hypothesis, velocity data for 1985 and 1993 upstream from the thickening are compared in the second graph. If they remained unchanged (and the values for σ_T suggest this), jamming may well have occurred in this period.

Over the next period, from 1993 to 2003, Figure 8 shows a general increase in the value of σ_T all along the flowline. This time interval is the period in which most changes in ice dynamics for Jakobshavn Isbræ have been documented.

After the outlet velocity slowed down somewhat in the early nineties, the ice stream started speeding up around 1998 (Joughin et al., 2008). This velocity increase came in steps, at least initially. Around 2003 the small ice shelf disintegrated (Podlech and Weidick, 2004), possibly followed by increased velocity (Thomas et al., 2003). By this time the outlet velocity increased from 5600 m a^{-1} in 1992 to 12600 m a^{-1} (Joughin et al., 2004) and had significant seasonal variation. The ice-shelf breakup in its turn was probably caused by an unusual influx of warmer ocean water (Holland et al., 2008). The observed increase in σ_T took place just upslope from the grounding line and gradually decreased in the upstream direction. In general, the ice stream became more uncoupled or more shelf-like in its behavior. One reason for this is ice thinning due to an increased outlet velocity. If the ice thins, the ice overburden pressure P_I decreases and thus ϕ and σ_T increase. This is known as the Jakobshavn Effect (Hughes, 1986). As shown in the fourth (bottom) graph of Figure 8, for σ_T, from 2003 to 2006 there was a slight increase in σ_T mainly after the first 13 km. A somewhat blocky character of the 2006 data set is due to the decreased resolution and the linear extrapolation that is used between surface data points. This time step only covers three years so not as much change compared to the previous periods can be expected.

The data would let us believe that most velocity changes took place about 13 km upstream from the front, but according to Joughin et al. (2008) the velocity of Jakobshavn Isbræ has increased linearly about 5 percent per year since 2003. At the same time the seasonal variation of the outlet velocity increased by about 1000 m a^{-1}. This is not observed in the modeled data.

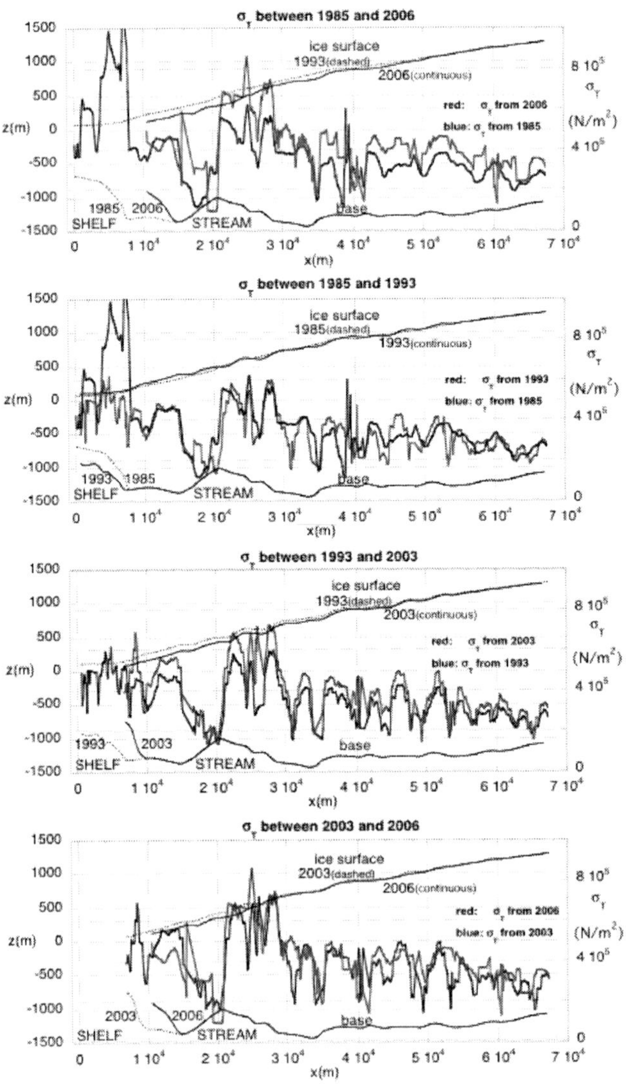

Figure 8. The development of σ_T using GFB2 for Jakobshavn Isbrae. The first (top) graph shows the overall development from 1985 to 2006. The blue graph shows the old values of σ_T and the red graph shows the new values plotted against the right vertical axis. The surface and bedrock data are plotted in black against the left vertical axis where a possible alternative bed profile is dashed. The next three graphs show the consecutive time steps, the second graph is time step 1985 to 1993, the third graph is from 1993 to 2002, and the fourth (bottom) graph is from 2003 to 2006.

CONCLUSION

Our holistic model was applied to transitions from sheet flow to stream flow to shelf flow along the central flowline of Jakobshavn Isbrae in Greenland. It quantifies a horizontal force balance along flow using the geometrical force balance (GFB) coupled to a simplified mass balance. The model needs little more than an outlet ice flux at the calving front of the terminal ice shelf in Jakobshavn Isfjord and profiles of both the top and bottom surfaces along the flowline. These were provided by Prasad Gogineni, Director of the Center for Remote Sensing of Ice Sheets (CReSIS) at The University of Kansas. Surface accumulation and ablation rates for sheet and stream flow, and top and bottom rates for shelf flow, have been measured but they have no major effect on the model. This justifies using a simplified mass balance equation. Jakobshavn Isbrae lies in the ablation zone of the Greenland Ice Sheet.

However ice flow is strongly convergent on Jakobshavn Isfjord in this zone, and the flowline model can take this ice flux into account by converting it into an ice precipitation flux along the flowline. This puts the flowline in an effective accumulation zone. The converging ice flux affects only the upper part of Jakobshavn Isbrae because the lower part occupies a deep extension of Jakobshavn Isfjord we have called "Gogineni Gorge" to honor Prasad Gogineni, who mapped the gorge and the whole region by radar sounding.

The GFB includes a longitudinal tensile stress that "pulls" ice out of the ice sheet and exists because of a large "floating fraction" of ice we quantify as ϕ and which gives stream flow its fast velocity. Three versions of the model were tested, GFB1 in which the main cause of increased flow was sudden disintegration of the buttressing ice shelf, GFB2 in which increased flow was caused mainly by increasing ϕ upstream due to surface meltwater reaching the bed by way of crevasses and moulins after the ice shelf disintegrated, and GFB3 which applied the model to a hypothetical condition that ignored the effect of changing bed topography upstream. The best match with observed surface velocities was provided by GFB3. However, both processes, reduced buttressing and an increasing floating fraction, are related and responsible for the doubling of discharge velocity from 1997 to 2006.

The surface of Jakobshavn Isbrae is quite bumpy, for reasons not fully understood. Some bumps have negative slopes, which will give negative values of ϕ, and some bumps have slopes too steep to keep ϕ in the acceptable range $0 \leq \phi \leq 1$. Therefore a running mean of surface elevation data should be employed to keep ϕ in this range.

The main shortcomings of the model are (1) it applies to a limited range of surface and bed slopes, (2) flowlines cannot treat diverging and converging ice flow, (3) the effects of side drag in ice steams has to be absorbed into basal drag along flowlines, (4) the model has no provision for changing flowlines through time, (5) the energy balance is not included so temperatures in the ice and thermal conditions at the bed are not obtained, (6) the model is not readily adaptable to the map plane, and (7) the flowband version of the model would have to convert the force triangles and rectangle in Figure 1 into volumes for flowbands of variable width. Our results show the surge of Jakobshavn Isbrae that began in 1997, when its buttressing ice shelf disintegrated, has not abated and in fact is accelerating. This is confirmed by ongoing field observations (Thomas et al., 2003, 2011). The main issue we wanted to address was the short-term stability of the Greenland Ice Sheet, as measured over the next 300 years (Hughes, 2004; Pfeffer, 2008). From our modeling results, there is no inherent reason why the ongoing surge of Jakobshavn Isbrae cannot continue well into the future and be joined by surges of other important Greenland ice streams, as predicted by positive feedback mechanisms that make The Jakobshavn Effect propagate northward along both the east and west coasts of Greenland (Hughes, 1986). However, there are probably negative feedback mechanisms that are not captured by our model and which could stabilize the ice sheet far short of a major collapse. If partial collapse of the Greenland Ice Sheet lowered the ice surface by 30 percent in 300 years, sea level would rise about 2 m and icebergs would flood the North Atlantic and produce a lid of cold fresh water as they melted (Hughes, 2004).

The Polar Front would move toward the Equator and push the North Atlantic Current toward Africa. The lid of melting icebergs would be semi-trapped in a gyre of cold fresh water north of the Polar Front that would suppress the ocean-to-atmosphere heat transport in the North Atlantic, so cold westerly winds would cause substantial cooling in northwest Europe not known since the Little Ice Age, from 1250 to 1850 in Greenland. The lid would suppress formation of North Atlantic Deep Water (NADW), which today forms in the Labrador Sea and north of Iceland. Here the "ocean conveyor belt" proposed by Broecker and Denton (1989) originates and brings NADW to the world ocean, providing our worldwide mild Holocene interglacial climate. If the lid of cold fresh water prevents NADW from forming, Little Ice Age conditions will prevail in western Europe. If the Greenland Ice Sheet continues to collapse, sea level could rise up to 5 m more and the new Little Ice Age could become a new 90,000-year worldwide glaciation cycle of the Quaternary Ice Age in which we have lived for the past million years.

ACKNOWLEDGMENTS

Our work was supported by NSF and NASA through a grant from the Center for Remote Sensing of Ice Sheets (CReSIS) at The University of Kansas. We thank CReSIS Director, Prasad Gogineni, for this support and for providing the radar record of surface and bed topography down the centerline of Jakobshavn Isbrae from the interior ice divide to the grounding line of the terminal ice shelf in Jakobshavn Isfjord.

We also thank Joel Plummer for contributing to collecting these data. We thank Beverly Hughes for helping to prepare our manuscript, Olaf Eisen at the Alfred Wegener Institute for editorial input, Aitbala Sargent and James Fastook at The University of Maine for assistance with computer programming, and anonymous referees for sharpening our presentation.

REFERENCES

Alley, R. B. (1989a), Water-pressure coupling of sliding and bed deformation: I. Water system, *Journal of Glaciology*, *35*(119), 108-118.

Alley, R. B. (1989b), Water-pressure coupling of sliding and bed deformation: II. Velocity-depth profiles, *Journal of Glaciology*, *35*(119), 119-129.

Anandakrishnan, S. and R. B. Alley (1997), Tidal Forcing of Basal Seismicity of ice stream C, West Antarctica Observed Far Inland, *Journal of Geophysical Research*, *102*(B7), 15,183–115,196.

Bassis, J.R. (2010). Hamilton's Principle applied to ice sheet dynamics: New approximations for the large-scale flow of ice sheets, *Journal of Glaciology*, 56(97), 497-513.

Bindschadler, R.A. (1984). Jakobshavn Glacier drainage basin: a balance asessment. *Journal of Geophysical Research*, 89, 2066-2072.

Bindschadler, R., and C. Hyengu (2007). Increased water storage in ice-stream onsets: a critical mechanism? *Journal of Glaciology*, 53(181), 163-171.

Brecher, H. H. (1986), Surface velocity determination on large polar glaciers by aerial photogrammetry, *Annals of Glaciology*, *8*, 22-26.

Broecker, W.S., and G.H. Denton (1989), The role of ocean-atmosphere reorganiztions in glacial cycles, *Geochimica et Cosmochimica Acta*, 53, 2465-2501.

Budd, W. F., D. Jenssen, and U. Radok (1971), Derived physical characteristics of the Antarctic Ice Sheet, Australian National Antarctic Research Expeditions (ANARE) Interim Reports, Series A (IV), *Glaciology*, 178 pages, University of Melbourne, Meteorology Department, Melbouorne, Australia.

Bueler, E., and J. Brown (2009), Shallow shelf approimation as a "sliding law" in a thermodynamically coupled ice sheet model, *Journal of Geophysical Research*, 114, F03008, doi: 10.1029/2008JF001179.

De Angelis, H. and P. Skvarca (2003), Glacier Surge After Ice Shelf Collapse, *Science*, *299*(Report), 1560–1562.

Denton, G. H., and T. J. Hughes (1981), *The Last Great Ice Sheets*, 484 pp., Wiley-Interscience, New York.

Dupont, T. and R. B. Alley (2005a), Assessment of the Importance of Ice-Shelf Buttressing to Ice-Sheet Flow, *Geophysical Research Letters*, *32*(4), LO4503.

Dupont, T. and R. B. Alley (2005b), The importance of small ice shelves in sea-level rise, *Geophysical Research Letters*, *33*, LO9503.

Echelmeyer, K., W. D. Harrison, T. S. Clarke and C. Benson 1992. Surficial glaciology of Jakobshavns Isbræ, West Greenland: Part II. Ablation, accumulation and temperature. *Journal of Glaciology*, *38*(128), 169-181.

Fastook, J. L., H. H. Brecher, and T. J. Hughes (1995), Derived bedrock elevations, strain rates, and stresses from measured surface elevations and velocities: Jakobshavn Glacier, Greenland. *Journal of Glaciology*, *41*(137), 161-173.

Glen, J. W. (1955), The creep of polycrystalline ice, *Proceedings of the Royal Society of London*, *Series A*(228), 519-538.

Gogineni, P. (2011), personal communication.

Hindmarsh, R. C. A. (2009), Consistent generation of ice-streams via thermo-viscous instabilities modulated by membrane stresses, *Geophysical Research Letters*, *36*(6), L06502. doi: 06510.01029/02008GL-036877.

Hindmarsh, R. C. A. (2012), An observationlly validated theory of viscous flow dynamics at the ice-shelf calving front. *Journal of Glaciology*, *58*(208), 1-13.

Hindmarsh, R. C. A., and E. LeMeur (2001), Dynamic processes involved in the retreat of marine ice sheets, *Journal of Glaciology*, *47*(157), 271–282.

Holland, D. M., R. H. Thomas, B. deYoung, M. H. Ribergaard, and B. Lyberth (2008), Acceleration of Jakobdshavn Isbrae triggered by warm subsurface ocean waters. *Nature Geoscience*, *1*(10), 659-664.

Hughes, T. (1986), The Jakobshavns Effect, *Geophysical Research Letters*, *13*(1), 46-48.

Hughes, T. (1992), On the pulling power of ice streams, *Journal of Glaciology*, *38*(128), 125-151.

Hughes, T. (1998), *Ice Sheets*, 343 pages, Oxford University Press, New York.

Hughes, T. (2009a), Modeling ice sheets from the bottom up, *Quaternary Science Reviews*, *28*(19-20), 1831-1849.

Hughes, T. (2009b), Variations in ice-bed coupling beneath and beyond ice streams: The force balance, *Journal of Geophysical Research*, *114*(B01410), doi:10.1029/2008JB005714.

Hughes, T. (2009c), Thermal Convection and the Origin of Ice Streams, *Journal of Glaciology*, *55*(191), 524-536.

Hughes, T. (2011), A simple holistic hypothesis for the self-destruction of ice sheets, *Quaternary Science Reviews*, *30*(15-16), 1829-1845.

Hughes, T. (2012), *Holistic Ice Sheet Modeling:* A First-Order Approach, 261 pages, Hauppauge, New York, Nova Science Publishers, Inc.

Hughes, T., A. Sargent, and J. L. Fastook (2011), Ice-Bed Coupling Beneath and Beyond Ice Streams: Byrd Glacier, Antarctica, *Journal of Geophysical Research*, *116*, 1-17.

Huybrechts, P. (1990), A 3-D model for the Antarctic Ice Sheet: a sensitivity study on the glacial-interglacial contrast, *Climate Dynamics*, *5*(2), 79-92.

Huybrechts, P. (1992), The Antarctic Ice Sheet and environmental change: a three-dimensional modelling study, *Ber. Polarforschung*, *99*, 241 pages.

Huybrechts, P. (1994), The present evolution of the Greenland Ice Sheet: an assessment by modelling, *Global and Planetary Change*, *9*, 39–51.

Huybrechts, P. (1996), Basal temperature conditions of the Greenland ice sheet during the glacial cycles, *Annals of Glaciology*, *23*, 226-236.

Iken, A., and M. Truffer (1997). The relatioship between subglacial water pressure and velocity of Findelengletscher, Switzerland, during its advance and retreat. *Journal of Glaciology*, 43(144), 328-338.

Iken, A., K. Echelmeyer, W. Harrison, and M. Funk (1993), Mechanics of fast flow in Jakobshavns Isbrae, West Greenland: Part I. Measurements of temperature and water level in deep boreholes, *Journal of Glaciology*, *39*(131), 15-25.

Jacobs, S. S., H. H. Hellmer, and A. Jenkins (1996), Antarctic Ice Sheet melting in the Southeast Pacific, *Geophysical Research Letters*, *23*(9), 957-960.

Johnson, J. (2002), A Basal Water Model for Ice Sheets, Ph.D. thesis, 187 pp, University of Maine, Orono.

Johnson, J., and J. L. Fastook (2002), Northern Hemisphere glaciation and its sensitivity to basal melt water, *Quaternary International*, *95–96*, 65–74.

Johnson, J., P. P. Prescott, and T. Hughes (2004), Ice dynamics preceding catastrophic disintegration of the floating part of Jakobshavn Isbrae, Greenland, *Journal of Glaciology*, *50*(171), 492–504.

Joughin, I., W. Abdalati, and M. Fahnestock (2004). Large fluctuations in speed on Greenland's Jakobshavn Isbrae glacier, *Nature*, 432, 608-610.

Joughin, I. R., I. M. Howat, M. A. Fahnestock, B. Smith, W. B. Krabill, R. B. Alley, H. Stern and M. Truffer (2008), Continued evolution of Jakobshavn Isbrae following its rapid speedup. *Journal of Geophysical Research*, *113*(F04006), doi:10.1029/2008JF001023.

Kamb, B. (1970), Sliding motion of glaciers: theory and observation, *Reviews of Geophysics and Space Physics*, *8*, 673-728.

Kamb, B. (2001), Basal Zone of the West Antarctic Ice Streams and its Role in Lubrication of Their Rapid Motion, in *The West Antarctic Ice Sheet: Behavior and Environment*, edited by R. B. Alley and R. A. Bindschadler, pp. 157–200, American Geophysical Union (Antarctic Research Series), Washington, DC.

Kenneally, J. P., and T. Hughes (2006), Calving giant icebergs: old principles, new applications, *Antarctic Science*, *18*(3), 409–419.

Lüthi, M., M. Funk, A. Iken, S. Gogineni, and M. Truffer (2002). Mechanisms of fast flow in Jakobshavn Isbrae, West Greenland, Part III: measurement of temperature and water level in deep boreholes, *Journal of Glaciology*, 46, 369-385.

Lüthi, M., M. Fahnestock, and M. Truffer (2008). Calving icebergs indicate a thick ice layer of temperate ice at the base of Jakobshavn Isbrae, Greenland, *Journal of Glaciology*, 55, 563-566.

MacAyeal, D. R. (1989), Large-scale ice flow over a viscous basal sediment: Theory and application to Ice Stream B, Antarctica, *Journal of Geophysical Research*, *94*(B4), 4071-4087.

MacAyeal, D. R., and R. H. Thomas (1982), Numerical modeling of ice-shelf motion, *Annals of Glaciology*, *3*, 189–194.

MacAyeal, D. R., T. A. Scambos, C. I. Hulbe, and M. A. Fahnestock (2003), Catastrophic Ice-Shelf Break-Up by an Ice-Shelf-Fragment-Capsize Mechanism, *Journal of Glaciology*, *49*(164), 22–36.

MacAyeal, D. R., E. A. Okal, R. C. Aster, and J. N. Bassis (2009), Seismic observations of glaciogenic oceanwaves (micro-sunamis) on icebergs and ice shelves *Journal of Glaciology*, *55*(190), 193-206.

Mahaffy, M. W. (1976), A three-dimensional numerical model of ice sheets: tests on the Barnes Ice Cap, Northwest Territories, *Journal of Geophysical Research*, *81*, 1059–1066.

Morland, L. W. (1987), Unconfined ice-shelf flow, in *Dynamics of the West Antarctic Ice Sheet*, edited by C. J. Van der Veen and J. Oerlemans, pp. 99–116, Reidel Publishing Company, Dordrecht.

Pelto, M.S., T.J. Hughes, and H. H. Brecher (1989), Equilibrium state of Jakobshavn Isbrae, West Greenland, *Annals of Glaciology*, 12, 127-131.

Podlech, S., and A. Weidick (2004), A catastrophic break-up of the front of Jakobshavn Isbrae, West Greenland, 2002/2003. *Journal of Glaciology*, 50, 153-154.

Prescott, P. R., J. P. Kenneally, and T. J. Hughes (2003), Relating crevassing to nonlinear strain in the floating part of Jakobshavns Isbrae, West Greenland, *Annals of Glaciology*, *36*, 149–156.

Reeh, N. (1982), A plasticity theory approach to the steady-state shape of a three-dimensional ice sheet., *Journal of Glaciology*, *28*(100), 431-455.

Reeh, N. (1984), Reconstruction of the glacial ice covers of Greenland and the Canadian Arctic islands by three-dimensional, perfectly plastic ice-sheet modeling, *Annals of Glaciology*, *5*, 115-121.

Reusch, D., and T.J. Hughes (2003). Surface "waves" on Byrd Glacier, Antarctica. *Antarctic Science*, 15(4), 547-555.

Rignot, E. J. (1998a), Fast recession of a West Antarctic glacier, *Science*, *281*, 549–551.

Rignot, E. J. (1998b), Radar interferometry detection of hinge-line migration on Rutford Ice Stream and Carlson Inlet, Antarctica, *Annals of Glaciology*, *27*, 25–32.

Rignot, E. J. (1998c), Hinge-line migration of Petermann Gletscher, north Greenland, detected using satellite-radar interferometry, *Journal of Glaciology*, *44*(148), 469-476.

Rignot, E. J. (2001a), InSAR studies of Pine Island and Thwaites Glaciers: A review of changes and possible causes, paper presented at West Antarctic Ice Sheet Initiative (WAIS), Eight Annual Workshop, Algonkian Meeting Center, Stirling, VA.

Rignot, E. J. (2001b), Evidence for rapid retreat and mass loss of Thwaites Glacier, West Antarctica, *Journal of Glaciology*, *47*(157), 213–222.

Rignot, E., and K. Steffen (2008), Channelized bottom melting and stability of floating ice shelves, *Geophysical Research Letters*, *35*(L02503), doi:10.1029/2007GL031765.

Rignot, E. J., S. Gogineni, I. Joughin, and W. Krabill (2001), Contribution to the glaciology of northern Greenland from satellite radar interferometry, *Journal of Geophysical Research*, *106*(D24), 34,007–034,019.

Rignot, E., J. Mouginot, and B. Scheuchl (2011), NASA research yields field map of Antarctic ice flow, in *NASA News*, edited, www.jpl.nasa.gov/news/news.cfm? release=2011-256 andcid=release.

Rist, M. A., P. R. Sammonds, S. A. F. Murrell, P. G. Meredith, H. Oerter and C. S. M. Doake (1996), Experimental fracture and mechanical properties of Antarctic ice: preliminary results. *Annals of Glaciology*, *23*, 85–102.

Rist, M. A., P.R. Sammonds, S.A.F. Murrell, P.G. Meredith, C.S.M. Doake, H. Oerter and K. Matsuki (1999), Experimental and theoretical fracture mechanics applied to Antarctic ice fracture and surface crevassing. *Journal of Geophysical Research*, *104*(B2), 2973–2987.

Rist, M. A., P. R. Sammonds, H. Oerter and C. S. M. Doake (2002), Fracture of Antarctic shelf ice. *Journal of Geophysical Research*, *107*(B1), ECV 1–13.

Robin, G. d. (1958), Glaciology III: Seismic shooting and related investigations, *Scientific Results of the Norwegian, British, Swedish Antarctic Expedition, 1949-1952*, *5*, 111-125.

Sargent, A. (2009), Modeling Ice Streams, Ph.D. thesis, 106 pp, University of Maine, Orono.

Schoof, C. (2007), Ice sheet grounding line dynamics: steady states, stability and hysteresis, *Journal of Geophysical Research Earth Surface*, *112*(FO3S28), doi: 10.1029/2006JF000664.

Schoof, C., and R. C. A. Hindmarsh (2010), Thin-film flows with wall slip: an asymptotic analysis of higher order glacier flow models, *Quarterly Journal of Mechanics and Applied Mathematics*, *63*(1), 73-114, doi:1093/qjmam/hbp1025.

Shepherd, A., D. J. Winkham, and, E. J. Rignot (2004), Warm ocean is eroding West Antarctic Ice Sheet, *Geophysical Research Letters*, *31*, L23402.

Sohn, H.-G., K. C. Jezek, and C. J. van der Veen (1998), Jakobshavn Glacier, West Greenland: 30 years of spaceborne observations, *Geophysical Research Letters*, *25*(14), 2699–2702.

Stearns, L. A., and G. S. Hamilton (2007), Rapid volume loss from two East Greenland outlet glaciers quantified using repeat stereo satellite imagery, *Geophysical Research Letters*, *34*(L05503), doi:10.1029/2006GL028982.

Sugden, D. E. (1977), Reconstruction of the morphology, dynamics, and thermal characteristics of the Laurentide Ice Sheet at its maximum, *Artic and Alpine Research*, *9*, 21-47.

Thomas, R. H. (1973a), The creep of ice shelves: theory, *Journal of Glaciology*, *12*, 45-53.

Thomas, R. H. (1973b), The creep of ice shelves: interpretation of observed behaviour, *Journal of Glaciology*, *12*, 55-70.

Thomas, R. H. (2004), Force-perturbation analysis of recent thinning and acceleration of Jakobshavns Isbrae, Greenland, *Journal of Glaciology*, *50*(168), 57-66.

Thomas, R. H., W. Abdalati, E. B. Frederick, W. B. Krabill, S. Manizade and K. Steffen (2003), Investigation of surface melting and dynamic thinning of Jakboshavns Isbrae, Greenland. *Journal of Glaciology*, *49*, 231–239.

Thomas, R.H., E. Frederick, J. Li, W. Krabill, S. Manizade, J. Paden, J. Sonntag, R. Swift and J. Yungel (2011), Accelerating ice loss from the fastest Greenland and Antarctic glaciers. *Geophysical Research Letters*, *3*, doi:10.1029/2011GL047304.

Weertman, J. (1957a), Deformation of floating ice shelves, *Journal of Glaciology*, *3*(62), 38-42.

Weertman, J. (1957b), On the sliding of glaciers, *Journal of Glaciology*, *3*(21), 33-38.

Weideck, A., H. Oerter, N. Reeh, H. H. Thomsen, and L. Thorning (1990), The recession of the Inland Ice margin during the Holocene climatic optimum in the Jakobshavns Isfjord area of West Greenland, *Palaeogeography, Palaeoclimatology, Palaeoecology (Global and Planetary Change Section)*, *82*, 389–399.

Weideck, A. and O. Bennike (2007), Quaternary glaciation history and glaciology of Jakobshavn Isbrae and the Disko Bugt region, West Greenland: a review, *Geological* Survey of Denmark and Greenland Bulletin 14, Ministry of Climate and Energy.

Zwally, H. J., W. Abdalati, T. Herring, K. Larson, J. Savba and K. Steffen (2002), Surface Melt-Induced Acceleration of Greenland Ice-Sheet Flow, *Science*, *297*(5579), 218–222.

In: Horizons in Earth Science Research. Volume 11
Editors: Benjamin Veress and Jozsi Szigethy
ISBN: 978-1-63321-296-1

Chapter 5

Conservation Planning and Management of Limited Water Resources in Arid and Semi-Arid Areas As Economic Resources

Abdeen Mustafa Omer*
Energy Research Institute, University Park
Nottingham, Nottinghamshire, UK

Abstract

Thirty-four million people live in Sudan, and environmental pollution is a major concern throughout the country; therefore industry, communities, local authorities and central government, to deal with pollution issues, should adopt an integrated approach. Most polluters pay little or no attention to the control and proper management of polluting effluents. This may be reflect lack of enforceable legislation and/or the fear of spending money on the treatment of their effluent prior to discharge. Present resources must be strictly monitored and managed effectively if further deterioration is to be avoided. Community participation in water management programmes should fully involve women, who play a central role in the provision and safeguarding of water for domestic use in rural areas. The pricing system recognises the economic value of water, as stated in the principles of the policy.

Keywords: Sudan, water resources development, Community water supply, effective water-supply management, environment

Introduction

Sudan is the largest country of the African continent, with an area of approximately one million square miles (2.5 x 10^6 km^2). Sudan is a federal republic located in the eastern Africa.

* Corresponding author: Abdeen Mustafa Omer, Energy Research Institute, University Park, Nottingham NG7 2RD, Nottinghamshire, UK. E-mail: abdeenomer2@yahoo.co.uk.

It extending between latitudes 3°'N and 23°'N, and longitudes 21° 45'E and 39°'E. Sudan is a relatively sparsely populated country. The total population according to the 1999 census was 34 x 10^6 inhabitants. The growth rate is 2.8%/y, and population density is 14 persons per square kilometres. The country is divided into 26 states and a federal district, in which the capital, Khartoum is located. Sudan is known as a country of plentiful water, rich in land, with highest total and renewable supply of fresh water in the region (eastern Africa).

Sudan is considered one of the least developed countries, with a per capita income of less than US $ 400, and a real growth rate of 0.2 % of real gross domestic product (GDP) during the last ten years. However, during 1980s the real growth rate of the GDP was negative mainly due to drought and desertification. The backbone of Sudan's economy is its agricultural sector. The agricultural sector determines to a great extent the economic performance of the Sudanese economy. In fact the country can be rescued by proper organisation and utilisation of its agricultural potential. Recent development due to rehabilitation and improvement in the agricultural sector has raised the share to 41%.

Agriculture continues to play a pivotal role in economy. It is directly influences the level of activities of all other sectors. It provides 90% of the raw materials for local industries, accounts 80% of export earning, and provides income and employment for more than 80% of the population. The agricultural sector is composed of three distinct modes: Irrigated agriculture, mechanised rain-fed agriculture, and traditional rain-fed agriculture and livestock raising. Sudan is an agricultural country with fertile land, plenty of water resources, livestock, forestry resources, and agricultural residues.

To ensure a better quality of life for all people, now and in the future, through the implementation of sustainable development initiatives that promote:

- Food and water security
- Environment integrity
- Economic efficient that helps to eliminate inequalities
- Social equity for all, regardless of race, gender, disability or creed
- Democracy
- Effective education for environmentally and socially responsible citizenship
- Environmental justice, and
- Mutual understanding between people.

In a country with a high population density, there are extreme pressures on water and waste systems, which can stunt the country's economic growth. However, Sudan has recognised the potential to alleviate some of these problems by promoting renewable water and utilising its vast and diverse climate, landscape, and resources, and by coupling its solutions for waste disposal with its solutions for water production. Thus, Sudan may stand at the forefront of the global renewable water community, and presents an example of how non-conventional water strategies may be implemented.

In Sudan, more than ten million people do not have adequate access to water supply, twenty million inhabitants are without access to sanitation, and a very little domestic sewage is being treated. The investment needed to fund the extension and improvement of these services is great. Most governments in developing countries are ready to admit that they lack the financial resources for proper water and sanitation schemes.

Moreover, historically, bilateral and multilateral funding accounts for less than 10% of total investment needed. Thus, the need for private financing is imperative.

Water utilities in developing countries need to work in earnest to improve the efficiency of operations. These improvements would not only lead to better services but also to enhanced net cash flows that can be re-invested to improve the quality of service. Staff productivity is another area where significant gains can be achieved. Investment and consumption subsidies have been predicated on the need to (a) help the poor, which have not an access to basic services and (b) improve the environment. Failure of subsidies to reach intended objectives is in part, from lack of transparency in their allocation. Subsidies are often indiscriminately assigned to support investment programmes that benefit more middle and high-income families that already receive acceptable service. Consumption subsidies often benefit upper-income domestic consumers much more than low-income ones. Many developing countries (Sudan is no exception) are encouraging the participation of the private-sector as a means to improve productivity in the provision of water and of wastewaters services. Private-sector involvement is also needed to increase financial flows to expand the coverage and quality of services. A key element to successful private participation is the allocation of risks. How project risks are allocated and mitigated determines the financial and operational performance and success of the project, under the basic principle that the risk should be allocated to the party, which is best able to bear it. Many successful private-sector interventions have been undertaken. Private operators are not responsible for the financing of works, nonetheless they can bring significant gains in productivity, which would allow the utility to allocate more resources to improve and extend services. Redressing productivity, subsidy and cross-subsidy issues before the private-sector is invited to participate, has proven to be less contentious. I have previously sought to encourage more private-sector involvement (Omer, 1995).

Sudan is geo-politically well located, bridging the Arab world to Africa. Its large size and extension from south to north provides for several agro-ecological zones with a variety of climatic conditions, rainfall, soils and vegetation. These agro-ecological zones and the large mass of arable land support a variety of food and industrial crops. Vast natural pastures and forests support a large herd of livestock including cattle, sheep, goats and camels. It has plenty of solar radiation, and vast productive land estimated at 84×10^6 hectares (ha) while the actual area under crops has not exceeded 12%. Water resources available to Sudan from the Nile system, together with groundwater resources, provide a potential for a three-fold increase in the irrigated sub-sector (Omer, 1995). There are also opportunities for increased hydroelectrical generation. The strategy of Sudan at the national level aims at the multi-purpose use of water resources to ensure water security for attaining food security, drinking-water security, fibre-security, hydro-energy security, industrial security, navigation, waste disposal and the security at the regional levels within an environmentally sustainable development context and in harmony with the promotion of basin-wide integrated development of the shared water resources (Noureddine, 1997).

The government has continued to pay for the development and operation of water systems, but attempts are being sought to make the user communities pay water charges. In order to ensure the sustainability of water supplies, an adequate institutional and legal framework is needed. Funds must be generated (a) for production, (b) for environmental protection to ensure water quality, and (c) to ensure that water removal remains below the annual groundwater recharge.

At present, there are private-sector providers who do not have an enabling environment to offer the services adequately. There is a need for the government to have a mechanism to assist in the regulation and harmonisation of the private-sector providers. Privatisation is part of a solution to improve services in delivery of water and of sanitation sector. At present, there is a transitional situation characterised by: (i) A resistance to payment of water charges; (ii) Insufficient suitable law and inadequate law enforcement; (iii) Insufficient capacities to provide services; and (iv) Inadequate interaction between the private citizen, business opportunities, and government.

WATER RESOURCES

Sudan is rich in water (from the Nile system, rainfall and groundwater) and lands resources (Table 1). Surface water resources are estimated at 84 billion m^3. The annual rainfall varies from almost nil in the arid hot north to more than 1600 mm in the tropical zone of the south. The total quantity of groundwater is estimated to be 260 billion m^3, but only 1% of this amount is being utilised. Water-resources assessment in Sudan is not an easy task because of uncertainty of parameters, numerous factors that affect the variables, lack of information and inaccurate measurements. However, according to seasonal water availability, Sudan can be divided into three zones: (a) areas with water availability throughout the year are the rainy regions (equatorial tropical zones); (b) areas with seasonal water availability; and (c) areas with water deficit throughout the year. The last occupy more than half the area of Sudan.

Table 1. Land use, land-resource zones and water resources

(a) Land use (x 10^6 hectares)

Geographical area	250.6
Land area	237.6
Cultivable area	8.4
Pastures	29.9
Forests and woodland	108.3
Uncultivatable land	81.0
Area under crop	10.0

(b) Land-resource zones

Zone	Area as % to total area of Sudan	Persons per km^2	Mean average rainfall range (mm)
Desert	44	2	0-200
QOS sands (sand dunes)	10	11	200-800
Central clay plains	14	19	200-800
Southern clay plains	12	8	800-900
Ironstone plateau	12	7	800-1400
Hill area and others	8	16	Variable

(c) Water resources

Water resource	Available number	Static water level (m)	Number
Haffirs (excavated land)	824	0-0	824
Slow sand filters	128	0-0	128
Open shallow wells	3000	0-10	3000
Boreholes (deep wells)	2259	0-25 26-50 51-75 76-100	1248 478 287 246

(d) Geological formations

Basins	Amount of water recharged (10^6 m^3)	Water level below land (m)	Aquifer thickness (m)	Velocity (m/year)	Abstraction (10^6 m^3/year)
Sahara Nile	136	30-100	300-500	1-2.5	7.3
Sahara Nubian	20.6	10-50	300-500	0.8-1.5	1.5
Central Darfur	47.6	25-100	250-550	0.3-6.0	5.5
Nuhui	15.4	75-120	200-400	1.0-2.75	1.6
Sag El Na'am	13.5	50-1000	300-500	1.0-25.0	2.5
River Atbara	150	100-150	250-300	0.3-5.0	2.3
Sudd	341	10-25	200-400	0.1-1.8	1.8
Western Kordofan	15	50-70	300-500	0.1-0.3	1.7
Baggara	155	10-75	300-500	0.1-2.4	11.9
Blue Nile	70.9	10-50	250-500	0.1-2.5	10.2
The Alluvial	N.A	Shallow	N.A	N.A	N.A
Gedaref	41.7	50-75	200-500	0.1-2.0	1.2
Shagara	1.1	25-30	200-300	0.1-2.5	0.7

(e) Annual rainfall pattern (10^9 m^3 $year^{-1}$)

Average range (mm)	Annual rainfall (mm)	Percent (%)
< 100	41.7	3.8
100-300	76.5	6.9
300-600	199.5	18.2
600-1000	515.5	47.2
> 1000	261.0	23.9
Total	1094.2	100.0

Water resources in the Sudan include surface water and ground water spread over large parts of the country. Both resources are utilised for agricultural production by the use of canal system from dams, pumps, embankments (flood), and wells. *Surface water produced internally* includes the average annual flow of rivers generated from endogenous precipitation and base flow generated by aquifers.

The Nile System

The Nile system travels Sudan from south to north. The Nile basin system in Sudan comprises the Blue Nile, Rahad, and Dender system. The White Nile and its branches Bahr El Gabal, Zaraf, Bahr El Ghazal and Sobat. In addition, the Atbara system.

Bahr El Gabal	$26 \times 10^9 m^3$
Bahr El Ghazal	$15 \times 10^9 m^3$
Sobat at Malakal	$13 \times 10^9 m^3$
Total at Malakal	$54 \times 10^9 m^3$
Losses in swamps	$27 \times 10^9 m^3$
White Nile at Malakal	$26 \times 10^9 m^3$
Blue Nile at junction	$53.4 \times 10^9 m^3$
Atbara system	$11.6 \times 10^9 m^3$
Losses along river	$77 \times 10^9 m^3$
Net available at Aswan	$84 \times 10^9 m^3$

In 1959, the Nile water agreement was concluded between Egypt and Sudan. The agreement is based on average Nile yield of 84×10^9 m^3 (measured at Aswan). After construction of High Aswan Dam (saving 22×10^9 m^3): 14.5×10^9 m^3 for Sudan and 7.5×10^9 m^3 for Egypt. As a result Sudan share has become 18.5×10^9 m^3 and 55.5×10^9 m^3 for Egypt, leaving 10^{10} m^3 for evaporation from Aswan High Dam reservoir.

Ground Water

The ground water quality is suitable for animal and human consumption as well as for agriculture and other uses. The potential renewable ground water suitable is estimated at 6×10^9 m^3. The main aquifer is Nubian sandstone covering 28% of surface of the country. Groundwater is mainly used to satisfy animal and human needs in rural areas while small areas are irrigated from groundwater. About 5×10^9 m^3 may be used for agriculture during the coming ten years. Current use for both agriculture and drinking water is about 1.2×10^9 m^3. The groundwater potential is believed much more than presently estimated.

Uses of Water in Sudan

The utilisation of water in Sudan is widely estimated in agriculture, human use, domestic, animal uses, industrial, hydropower generation and navigation. The agricultural sector is the major source of water consumption in Sudan. Sudan is presently utilising 16.5×10^9 m^3 annually from its share in irrigated agriculture sub-sector, currently covering an area of 1.7×10^6 hectares. The potential is three-fold. The irrigated crops include: cotton, wheat, sorghum, groundnuts, sugarcane, vegetables and fruits. Mechanised and traditional rain-fed farming sectors cover an area estimated annually at about $8\text{-}10 \times 10^6$ hectares. The total area cropped in the rain-fed areas varies from year to year depending on rainfall.

Crops grown in the rainfall sub-sector include: sorghum, millet, sesame, sunflower, and groundnuts. The potential in the rainfall sector is more than fivefold. The southern tropical zone of Sudan is distinct with red lateritic soil and heavy rainfall ranging between 1000-1500 mm over 7-8 months. Agriculture is largely subsistence with a wide range of food crops including: maize, sorghum, millet, root crops, banana, pulses, tea, and coffee.

Because of the marked fluctuation between the flood discharge and the low season period in the Nile system, storage reservoirs in Sennar, Rosaries, Griba and Jebel Awlia were constructed to ensure the availability for water during the recession period. These dams are used for irrigation, navigation and hydropower generation. Drinking water supplies have been provided for people and animal in most of the urban and rural areas, but still more than 40% of the rural areas, and more than 25% in urban areas are in need of safe drinking water supplies. Water use for industry and sanitation is still very limited. The demand for water in industries at the present estimated at about 0.24×10^9 m^3 per year, and most of which from the surface water.

The most important research and development policies which have been adopted in different fields of water resources are: (i) the water resource; (ii) irrigation development; (iii) the re-use of drainage water and groundwater; (iv) preventive maintenance of existing facilities, including canals; (v) aquatic weed control and river channel development, and (vi) protection plans for water resources in general. The base of physical and human-resources can provide for sustainable agriculture growth and food security for itself and for others in the region. Failure to plan for sustainability in the past derives from several causes and constraints, which are manageable. These include misguided policies, poor infrastructure, low level of technology use, recurring droughts and political instability. Perhaps the biggest challenge is that of finding resources for capital improvements in the light of changing water-quality regulations and ageing systems (James, 1994). Environment institutions and regulators are already working to solve the next set of issues in their constant pursuit of better quality of live. Increasing watershed and source protection that combat microbial and organic contaminations is possible with new detection techniques that enable identification, and control of pesticide runoff, and reduce chlorine by-products. Upgrades of ageing infrastructure are receiving unprecedented attention. Surface-water systems are continuing to rely on conventional treatment (coagulation, flocculation, sedimentation, filtration and disinfection) for particulate removal, but other treatment processes such as ozone and granular activated carbon will see increased use, while new disinfection strategies will be needed to minimise both microbial risks and unwanted by-products. Sudan is, therefore, moving into a new era in the protection of drinking-water supplies. It is now time for water utilities to combine creative management, dependable treatment methods and new technologies to ensure that drinking water is as safe as possible.

Water is one of the most fundamental of natural resources that a country must harness in its efforts for rapid economic development. The role of water in the development process cannot be over-emphasised. The demand for water in Sudan has increased tremendously over the years and will continue to increase in view of the accelerating pace of population growth, urbanisation and industrialisation. There are many approaches and programmes through which human resource can be properly developed for the water sector. These programmes are not mutually exclusive and include:

- Non-formal education through media and non-governmental groups.
- Formal education through universities and technical colleges.
- On-the-job training and regular refresher courses.

Sudan has to build a bigger and better base of science and technology. Also, Sudan is strengthening the national science and technology systems. Sudan should collaborate and achieve collective strength within the framework of the perennial Islamic values and ideals. Education and training play key role in human resource development in water and other sectors. Education has to include the moral, spiritual and ethical values of Islam. These values promote peace, contentment and inner happiness, which are not dependent on material wealth and prosperity. Religion is the fountain, which irrigates the tree of virtue, goodness and truth. Human resource development in Sudan should emphasise education rooted in religion to inculcate and promote fellow feelings, love and respect for all creatures, provide succour and support to those who are poor and destitute, and discourage hoarding and exploitation. Education based on Islamic teachings goads to simple living making minimal demands on natural resources. In the case of water attention must pay due to how it is used and how it can be reused.

Community Water Quality and Sanitation Management

Community water supply and sanitation management is a new form of cooperation between support agencies in the water and sanitation sector and communities. It involves a common search to identify problems with the local water supply and sanitation systems, and the possibilities for, and constraints on, management by communities, and to find possible solutions that may be tested. Some fundamental principles of community water and sanitation management are:

i Increased management capacities as the basis for improved water and sanitation systems', wherein each community develops its own specific management systems; and

ii Communities' ownership of the process of water charges; facilitators and local researchers participate in the community's projects, not the other way around.

Through that approach, the support agency is no longer the provider of technical goods or solutions, but the facilitator of process to enhance the capacity of the community to manage its own water and sanitation systems.

Constraints include:

i A lack of funds or long delays in allocating funds for requirements such as operation and maintenance of irrigation and drainage projects

ii Deterioration of data-collection

iii A lack of appropriate and consistent policies for water development for both large and small-scale projects

iv Serious delays in completing water projects after major investments, have been made, e.g., dams, other hydraulic structures, and main canals are finished
v An absence or inadequacy of monitoring, evaluation, and feedback at local, national and international levels
vi A lack of proper policies on cost recovery and for water pricing or, if policies exist, lacks implementation
vii A shortage of professional and technical personal, and training facilities
viii A lack of beneficiary participation in planning, implementation, and operation of projects
ix Inadequacy of knowledge, and absence of appropriate research to develop new technologies and approaches, and an absence of incentives to adopt them
x General institutional weaknesses and lack of coordination between irrigation, agriculture, energy, human health services, environment, and planning
xi Inappropriate project development by donor agencies, e.g. irrigation development with drainage, supporting projects which should not have been supported; and
xii A lack of donor coordination resulting in differing approaches and methodologies, and thus conflicting advice.

As developing nations strive to provide safe and reliable drinking-water supplies to their growing and increasingly urbanised population, it is becoming more evident that new approaches to this problem are needed. To meet this challenge, new methods of reclaiming and re-using water have been developed in cost-effective and environmentally sound ways (ODA, 1987, Seckler, 1992, Salih, et al., 1992). Measuring or assessing total river flow occurring in a country on a yearly basis usually computes surface water resources.

Despite the constraints, over the last decade the rate of implementation of rural and peri-urban water supply and sanitation programmes has increased greatly, and many people are now being served more adequately. The following are Sudan's experiences in water supply and sanitation projects:

At the community level:

- Participatory approaches in planning, implementation and monitoring.
- Establishment and training of water committees.
- Clear ownership of improved water supply and sanitation systems.
- Selection of technology and level of service by consumers.
- Sensitive timing of education in hygiene and sanitation.
- Establishment and training of reliable financial and maintenance management.

At the district and national level:

- Integrated and multi-sectoral approach to development.
- Training and material development for district and extension staff.
- Continuing support from well-trained, integrated and multi-sectoral extension teams.
- Establishment of technical support system.
- Multi-sectoral advisory group to include representation from membership training and research institutions.

- Development and dissemination of relevant information to district and extension staff.

Water is a substance of paramount ecological, economical, and social importance. Interrelationships inherent in water use should encourage integrated management. To better manage water resources the following are needed:

1 Ensure more reliable availability of water and efficient use of water in the agricultural sector.
2 Mitigate flood damage by ground reservoirs and terraces.
3 Control water pollution.
4 Prevent development of soil salinity and water logging.
5 Reduce the spread of water-borne diseases.

The development of new, modern, and complete water resources information systems is one of the basic needs for the implementation of the water resources management system in Sudan. Decisions relating to drought, flood or over-exploitation, can only be correct if based on reliable information.

A comprehensive database on water availability, users, water quality monitoring, current technologies (like geographical information systems), is certainly the way to produce an efficient framework for decision-making. Lack of information is one of the most critical points regarding global development and implementation of new management systems (FAO, 1999). For example the types of data related to flood management in Sudan and globally include:

- Topographic data (elevations, land use, and soils).
- Imagery (satellite images).
- Administrative data (political boundaries).
- Infrastructure data (roads, levels, wells, utilities, bridges and culverts, and hydraulic structure).
- Environmental data (threatened and endangered species, critical aquatic and wildlife habitat).
- Hydrometeorology data (flood plain delineation, stream flows, precipitation, temperature, wind, solar radiation, soil moisture, discharge rating curves, and flood frequency).
- Economic data (stage-damage relationships), and
- Emergency management data (emergency plans, and census data).

The emerging water crisis, in terms of both water quantity and quality, requires new approaches and actions. Priority areas needing concerted action in various sectors in Sudan are:

(a) Water use efficiency
(b) Flood control
(c) Water resources management
(d) Water quality management and provision of safe drinking water, and

(e) Coordination and integration of various aspects of water management, and water management with other related resources and societal concern.

The following are recommended:

- The community must be the focus of benefits accruing from restructures. The most legal framework to protect community interest on the basis of equity and distribution, handover of the assets to the community should be examined. Communities should encourage the transfer of the management of water schemes to a professional entity.
- The private-sector should be used to mobilise, and strengthen the technical and financial resources, from within and without the country to implement the services, with particular emphasis on utilisation of local resources.
- The government should provide the necessary financial resources to guide the process of community management of water supplies. The government should be a facilitator through setting up standards, specifications and rules to help harmonise the private-sector and establish a legal, independent, body by an act of parliament to monitor and control the providers. The government should assist poor communities that cannot afford service cost, and alleviate social-economic negative aspects of privatisation.
- All sector actors should create awareness in the community to the roles of the private business and government in the provision of water and sanitation services.
- Support agencies should assist with financial and technical assistance, the establishment of training facilities, coordination, development and dissemination of water projects, and then evaluate of projects with a view to future improvements.

The institutional framework provides the basis by which all actions are taken, and an assessment of its functional character helps determine the collaborative potential. The resulting criteria for measuring a given community's institutional capacity can be found in Table (2).

GROUNDWATER AND ENVIRONMENT

In northern Sudan, the desert environment is fragile and highly affected by man's activities. One of the geotechnical hazards that may occur under desert conditions and of special interest in Sudan, is sand drifting and dune movement. Moving sand covers approximately one-third of the country. Because sand poses natural erosional-depositional hazards on the existing structures, such as roads and urbanised areas, it become necessary to study the behaviour of the sand forms in the different parts of the country. Deserts are known to be simply barren areas; they are scientifically defined in terms of aridity, soil type, topography and vegetation. Most of the Middle Eastern countries lie within the semi-arid, arid, and hyper-arid desert zones, with an aridity index (ratio between annual precipitation and mean annual potential evapotranspiration) ranging between 0.03 and 0.02 (Anon, 1979). The heavy mineral content ranges between 11.8 and 15.4%. The desert environment is being a fragile ecosystem, needs to be treated with care. Intercommunications between different national and international agencies and education of the layman should help to keep the system balanced and reduce environmental hazards. In addition, any suggested remedial measures should be planned with nature in mind and be engineered with natural materials.

Table 2. Capacity assessment for community management: institutional factors (Salih, 1992)

High capacity (plans etc. in places)	• Basin-wide management plan has been drafted. • Natural mitigation strategy in place. • Basin-wide coordination and communications strategy instituted. • Trained emergency management staff coordinating at the regional level. • Effective regulatory policies that address floodplain occupancy. • Decentralised decision-making with a high degree of local autonomy.
Medium capacity (evidence of activity on- going)	• Evidence of an updated national response plan. • Bilateral response agreements. • Evidence of regional preparedness and response training. • Some trained emergency management staff at the local and/or national level. • Evidence of some regulatory policies designed to address floodplain occupancy. • Attempts to decentralise decision-making, moderate local discretion.
Low capacity (no formalisation in place nor apparently evolving)	• No existing flood response plan. • No evidence of mitigation-related activities. • Poor local-and national- level coordination and communications. • Little or no evidence of flood preparedness and response training. • No regulatory policies addressing floodplain occupancy. • Centralised decision-making, no evidence of local autonomy.

The Policy Regime in Water Quality Management

Apart from effluent regulations, and sometimes, national water quality guidelines, a common observation are that few developing countries (Sudan is no exception). Although water supply is seen as a national issue, pollution is mainly felt at, and dealt with, at the local level. With few exceptions, national governments in developing countries have little information on the relative importance of various types of pollution (agriculture, municipal, industrial, animal husbandry, and aquaculture), and therefore, have no notion of which is of greatest economic or public health significance (Salih, 1992). Usually freshwater quality management is completely divorced from coastal management even through they are intimately linked. Consequently, it is difficult to develop a holistic management plan for assuring water quality or to efficiently focus domestic and donor funds on priority issues. A Sudanese national water policy should include the following water quality components:

- A policy framework that provides broad strategic and political directions for future water quality management.
- A strategic action plan for water quality management based on priorities that reflect an understanding of economic and social costs of impaired water.

This plan should include the following components:

- A mechanism for identifying national priorities for water quality that will guide domestic and donor investment.
- A plan for developing a focused and cost-effective data programme for water quality and related uses, as a basis for economic and social planning.
- A consideration of options for financial sustainability including donor support, public-private sector partnerships, and regional self-support initiatives.
- A regulatory framework that includes a combination of appropriate water quality objectives (appropriate to that country and not necessarily based on Western standards) and effluent controls. This must include both surface and groundwater.
- A mechanism for public input into goals and priorities.
- A process for appointing specific agencies with responsibility for implementation so that accountability is firmly established and inter-agency competition is eliminated.
- Specific mechanisms for monitoring the quantity of drinking water, at the community level if necessary.
- National data standards that must realistically reflect national needs and capabilities. The objective is to ensure reliable data from those organisations that provide information for national water management and for community- level drinking water quality.

The design criteria in any water quality programme in Sudan should include the management issues for which water quality data are required. Generally, there are four categories of data objectives:

- Descriptive data that are typically used for government policy and planning, meeting international obligations, and for public information (Appendix 1).
- Data specific to public health.
- Regulatory mechanism and concerns, and
- Aquatic ecosystem health.

The last category is not normally included in many developing countries for reasons of cost and complexity (UNESCO, 1999).

In most developing countries, countries with transitional economies, and some developed countries, the technology of monitoring has changed little since 1970s, yet some of the largest advances in monitoring in recent years involve technical innovation that serve to reduce costs and increase efficiency (Salih, 1992).

Admittedly, not all are inexpensive; however when deployed appropriately, they may eliminate traditional monitoring, or reduce costs by increasing the efficiency of more traditional approaches to chemical monitoring.

Types of innovation include: biological assessment, use of enzymatic indicators, miniaturisation, automation, and simplification of laboratory analytical methods. Water quantity is highly variable in Sudan reflecting regionally different levels of development and is indirectly responsive to different needs for water quality programmes (Tables 3). *Groundwater recharge* is the total volume of water entering aquifers within a country's borders from endogenous precipitation and surface water flow.

In Sudan, financial and sustainability issues include cost avoidance and cost reduction, local and accountability frameworks that encourage good business practices by senior programme managers, the use of new cost-effective technologies for monitoring, and a variety of donor/public/private-sector linkages that focus on commercial benefits that permit the transfer of certain parts of water quality programmes to the private-sector.

Sustainable Development

In Sudan there are a limited water resources and increased demands to cope with the rapid development in petroleum production. It is paramount to inaugurate strategies that control augmentation and conservation of water. Such measures essentially include uses of water, minimising losses, quality protection, exploration, artificial recharge and water harvesting techniques.

A schematic technological advancement of low cost water supply systems such as dug wells, roof top catchments, haffirs (in western and eastern of Sudan) and small dams combined with development of guidelines for settlement policy will hopefully lead to an improvement of water supply systems, water quality and reduction of the distance to the supply points.

Table 3. The present water management of Sudan

Use of resources	Sources	Administrative	Pricing Principle	Price Details
Urban	Surface and groundwater	National Water Corporation (NWC)	Full cost recovery	Progressive rate with increasing uses. Rates lower in the north
Major rural villages	Mostly groundwater	Rural Water Corporation (RWC)	Stand pipe free, recovery of recurrent costs, charges for yard and house connections	Progressive rates but less comparative to urban cities
Rural villages	Groundwater	District Councils	As above	Not available
Livestock	Surface and groundwater	Rural Water Corporation (RWC)	All investments and recurrent costs	Regressive, no charges on relatively small use
Mines	Surface and groundwater	National Water Corporation (NWC)	Full cost recovery	Progressive rates
Wildlife	Mostly surface	Rural Water Corporation (RWC)	Full cost of boreholes	Regressive

The following problems encountered so far need to be solved in the near future:

- The establishment of efficient operation, maintenance and repair of water yards.
- Community participation in operation and maintenance of water yards.
- The extent to which capital costs can or should be recovered from water users.

Goals and Challenges

In the past decades, sustainability has increasingly become a key concept and ultimate global criterion for socio-economic development in the modern world. Without a doubt, sustainable development and management of natural resources fundamentally controls the survival and welfare of human society. Water is an indispensable component and resource for life and essentially all human activities rely on water in a direct or in direct way. Yet supplying water of sufficient quantity and safe quality has seldom been an easy task.

Although sustainability is still a loosely defined and evolving concept, researchers and policy-makers have made tremendous efforts to develop a working paradigm and measurement system for applying this concept in the exploitation, utilisation and management of various natural resources. For water resources, recent development has been synthesised and presented in two important documents published by ASCE, (1998), and UNESCO, (1999). Those attempt to give a specific definition and a set of criteria for sustainable water use. When considering the long-term future as well as the present, sustainability is concept and goal that can only be specified and implemented over a range of spatial scales, of which urban water supply is a local problem with great reliance on the characteristics and availability of regional water resources.

Sudan needs assistance in developing and implementing (a) river-basin management, (b) diffuse source pollution, (c) environmental restoration, and (d) urban storm drainage. At present the international, bilateral donor agencies, and relevant United Nations bodies provide such assistance, e.g., UNICEF, UNDP, and WHO. The international associations constitute an additional, but as yet untapped, source of assistance.

Advanced research and technology contribute to resolving water shortage and sanitation problems, and non-conventional, reliable water supplies cannot be provided unless environmental impacts are considered. Looking to the future, Sudan has set the following priorities for water-resource research and development until the year 2020 (Amir, 1998):

(i) Maximise overall water-use efficiency. This could be achieved by (a) improving irrigation and assuring its flexibility to cope with modern farming (b) drawing up a proper mechanism for water pricing; (ii) Modify the cropping pattern; for example (a) planning different cropping patterns according to water quality, (b) gradually replacing sugarcane by beet, (c) introducing and developing of salt tolerance crops, and (d) reducing the area of clover (*Berseem*); (iii) Re-using all the possible agricultural drainage water using proper technical means to deal with its quality, especially after implementing the irrigation development; (iv) Plan properly the re-use of sewage effluent after drawing up guidelines for its use; (v) Research agreements of losses of water and suggest conservation projects; (vi) Use and management of reservoirs and groundwater sources in the Nile valley in an integrated fashion, giving special consideration to drought conditions; (vii) Develop on a sustainable basis, sustainable groundwater resources in the deserts; (viii) Harvest rainfall in desert areas and make full use of torrential streams and flash floods; (ix) Use new economical technology of seawater desalination; (x) Raise public awareness about water resource scarcity and government management plans; and (xi) Consider laws to match with required development and existing scarcity; (xii) Establish efficient operation, maintenance and repair procedures; (xiii) Assure community participation in operation and maintenance; (xiv) Assess the extent to which initial government investment can or should be recovered from water uses; (xv) Aim to assure domestic potable water supply reaches at least 25 litres per day per person; (xvi) Aim to assure water should be available for ten livestock units at 450 litres per day (L/d); (xvii) Assure that potable water available within two kilometres of individual residences.

The challenges of enhancing water supplies in the twenty-first century are as follows: (a) Drinking-water sources should be treated chemically; (b) Suitable toilet facilities should be provided along the main roads to minimise pollution; (c) Proper arrangements should be made for litter dumping and waste disposal; (d) Local people should be fully educated about environment matters and hygiene; (e) Previous damage should not be allowed to continue while planning for a balanced development in the future; (f) The concept of the ecosystem (involving education and interpretation of the natural environment) must be promoted; (g) The polluters must be introduced.

POLLUTION FROM RIVER NILE

From a visual investigation of the River Nile (Table 4), the major sources of pollution are industrial effluents, crude sewage from blocked, broken or overloaded sewers, sewage effluents, surface runoff, and solid wastes which have been dumped into the river. Therefore remedial and improvement measures must be taken before the environment becomes further polluted and the natural resources are completely over-exploited (Omer, 2000).

THE CHALLENGE OF OVERCOMING THE COUNTRY'S DIVERSITY

Table (5) shows main water resource issues in Sudan by region. Proper water management is essential to sustain development. Competing for water are municipal supply, industry, and agriculture.

Table 4. Types and sources of wastes in River Nile water

Materials sources	(%)
Paper, and wood	50.0
Ferrous residues	12.5
Glass	11.0
Organic wastes	10.0
Plastics	5.0
Non-ferrous residues	1.5
Other	10.0

Table 5. Main water resource issues in Sudan by region

Region	Water resource issues
South	1. Abundant water resources 2. Localised scarcity of water and untapped water supplies 3. High hydropower potential 4. Water conflicts arising from immigration of Bagara Arabs (nomadic) from north to south 5. Water-borne diseases 6. International water conflicts (Upstream and downstream countries)
Central	1. Water quality problems from untreated sewage and other pollution 2. Water-borne diseases 3. Potential use of rivers for navigation and recreational purposes 4. Intensive erosion and sedimentation from agriculture 5. High hydropower potential 6. Excessive use in large urban and industrialised areas 7. Frequent urban floods
North	1. Good water quality 2. Scarcity of water resources 3. Intensive erosion and sedimentation from agriculture 4. Frequent urban floods
Northeast	1. Scarcity of water resources 2. Water quality problems from untreated sewage 3. In mining areas, water quality problems from effluent
West	1. Scarcity of water resources 2. Water conflicts between nomadic and non-nomadic tribes 3. Water-borne diseases 4. Soil erosion and degradation caused by agriculture

The National Water Act of 1994 (Law No. 1155) defines the objectives, principles, and instruments of the National Water Resources Policy and the National Water Resources Management system. The law establishes the institutional arrangement under which the country's water policies are to be implemented. The National Water Resources Policy was proposed to achieve:

- Sustainability: to ensure that the present and future generations have an adequate availability of water with suitable quality.
- Integrated management: to ensure the integration among uses in order to guarantee continuing development.
- Security: to prevent and protect against critical events, either arising from natural causes or inappropriate uses.

To achieve such objectives, water management must be implemented according to the following principles:

- Water is a public good.
- Water is a finite resource that has economic value.

- The use of water required to meet people's basic needs shall have priority, especially in critical periods.
- Water management shall comprise and include multiple uses.
- The river basin is the appropriate unit for water management, and
- Water management shall be decentralised, with participation by government, stakeholders and society.

Water Scarcity Impacts and Potential Conflicts

The failure of water resources to meet the basic requirements of society has a host of social, economic, environmental, and political impacts. Water scarcity is man-made phenomenon brought about by population pressures. The imbalance in the population- water resources equation strains society and has an adverse impact on domestic hygiene, public health, and cost of domestic water, and could impart political problems as a serious as bringing down government (Amir, 1998). On the social side, water scarcity adversely impacts job opportunities, farm incomes, credibility and reliability of agricultural exports, and ability of the poor to pay for domestic water. Economically, the adverse impact is displayed in the loss of production of goods, especially agricultural goods, the loss of working hours because of the hardships society faces as a result of water scarcity. The impacts of water scarcity on regional stability are addressed with reference to water in the East Africa (Egypt and Sudan downstream and (Ethiopia, Uganda, Kenya, etc. Upstream), taking into account the serious impacts of conflicts and potential water war (Amir, 1998).

Conditions of scarcity propel an increase in competition among the different sectors of water use with results, invariably, at the expense of irrigated agriculture (Amir, 1998). Pure market forces create a pressure gradient under which water flows from the poor to the rich. Tough decisions await politicians, and the consequences are expected to displease some parties, and please others. The scene of domestic politics becomes as fluid as water itself, with politicians shifting positions continuously in response to domestic pressures. The political fallout from water scarcity on the domestic scene is parallel to the impact the scarcity has on domestic households in terms of basic needs for drinking and food preparation, on domestic hygiene, and on public health.

Other important factors have a delayed response to water scarcity, and these pertain to the integrity of the environment, and deterrence in development investment and economic credibility of the country. The cost of mitigating these problems and of the provision of services to increased urbanisation could very well be beyond the ability of government to bear. The political consequences resulting from this will not be in favour of domestic stability, and social explosions can be anticipated (Omer, 2000).

Water relations can be transformed into a positive sum game by which all parties can win. One common gain to all is environmental protection of the common watercourse or water body. Lack of cooperation and agreement generally leads to environmental neglect and water quality degradation, which is loss o all. International encouragement to attain cooperation can, therefore, be brought to bear on the regional parties, and efforts of international lending agencies can be called upon to pool with the regional and international efforts to achieve this objective (e.g., Great lakes Canada and US).

Today's advanced societies heavily depend on energy. The sources of energy and electricity generation today in Sudan are waterfalls, natural and artificial, and fossil fuel. Energy from waterfalls is short of meeting the current or future energy requirements, and the fossil fuel resources, being depleted with time, will eventually run out. For human civilisation to continue at its natural pace, new forms of affordable and clean energy must come on line. Failure of human civilisation to introduce new forms of energy will render that civilisation doomed, and the quality of life will deteriorate. If this unlikely scenario actually takes place, the requirements will decrease because the mechanism of making it available for use (pumping) diminishes.

The more likely scenario is more optimistic one, and it is that a new form of energy generation will be introduced in which case water desalination becomes affordable and its pumping from the coastal desalination plants become possible at reasonable cost. The way out of the looming water crisis rests, therefore, in the invention of new forms of energy generation that will make possible the reliance on desalination and in the recycling of wastewater for reuse in agricultural production and for environmental reasons. Integrated management of the three resources of water, energy, and the environment, will result in better results with a positive sum for society.

Prospects for Water Resource Planning in Nile Basin

Common Language and Culture

A common language and similar culture simplify communication and reduce the potential for misunderstandings. In Nile basin where several languages are spoken, an international language English is used with some success by multi-jurisdictional basin management authorities.

Factors Promoting Data and Information Exchange

Data and information exchange is more probable when needs are compatible and when there is potential for mutual benefit from cooperation (Table 6). Where countries are working on developments that are beneficial to both countries as well as other riparians, there is little incentive to hide project impacts.

This means that because data and information exchange are unlikely to lead to pressure from surrounding countries that might restrict developments, countries have less reason to restrict access to their data and information resources. It is important for there to be no perceived clash of interests in development plans and needs. An example of this might be in developing their part of the basin primarily for hydroelectric development, while the lower riparians are more interested in developing the irrigation potential of their portion of the basin. By constructing large storage dams in the upper part of the basin, the river Nile seasonal flow might be evened out, reducing flooding downstream while increasing irrigation water supplies and even making downstream run-of-the-river hydroelectric projects more profitable.

Table 6. Summary of the situation relating to data and information exchange in Nile basin

River basin	Nile basin
Basin states or territories	Burundi, Democratic Republic of Congo, Egypt, Eritrea, Ethiopia, Kenya, Rwanda, Sudan, Tanzania, and Uganda
Cooperative frameworks in place	Nine of the countries of basin are pursuing the development of a cooperative framework
Major languages spoken in	More than 6 official languages and numerous unofficial languages
Major water issue facing the basin	Rapid population growth, environmental degradation, under development
External funding of cooperative basin initiatives	Extensive external funding of cooperative initiative
Range of GDP *per capita* of the basin	\$550-\$3000
Extent of data/information exchange	Information exchange through the cooperative framework being developed is beginning to occur

Sufficient levels of economic development

Sufficient levels of economic development across a basin are needed to permit joint funding of cooperative processes, particularly data collection and dissemination. Although countries with differing levels and forms of economic development may, at times, have more complementary needs than countries with similarly structured economies, the overall level of economic development is still significant. A weathier country in a river basin may be able to assist with the funding of data collection activities in the neighbouring country with much needed data and helping to build confidence between the two countries.

Increasing Water Resources Stress

A bilateral agreement between Egypt and the Sudan in 1959 by which the two countries share the Nile flow: 55.5 billion cubic meters to Egypt, 18.5 billion to Sudan, and 10 billion were allocated to evaporation. Hopes are high for achieving a more extensive participation by the other riparian parties in what could be a multilateral treaty on the Nile encompassing the other riparian states in addition to Egypt and Sudan.

The above agreement is incomplete, lack of the entry of other legitimate riparian states, lack of the water quality components, and the focus on quantity and misses important management issues.

It is to be noted that regional relations, including those among the riparian parties, are not disconnected from the political, economic, and trade network of international relations. Water is not the only determinant factor in shaping the nature of bilateral, regional, or international relations.

As *per capita* water resources availability decreases (Table 7), tensions between riparian nations may rise and make cooperation difficult. Stress may, therefore, reduce cooperation and data sharing rather than strife.

The historical background of the basin may have a lasting effect on current negotiations. Past conflicts can have a deleterious effect on the prospects for establishing cooperative practices, such as data sharing. Where there is a history of conflict between two nations, both nations may view the present situation primarily as competitive and focus on conflicting rather than common interests. Democracies may find it easier to negotiate cooperative arrangements with other democracies. Political differences can lead to legacies of mistrust developing between countries.

CONCLUSION

Booming economy, high population, land locked location, vast area, remotely separated rural areas, which are not easily accessible, large reserves of oil, excellent sunshine, large mining sector and cattle farming on a large-scale are factors which are most influential to the total water scene in Sudan.

Water resources plans are developed to guide future decisions and are to be developed for each river basin and state, as well as for the country. The overall objective is to coordinate efforts and establish guidelines and priorities for water allocation and water pricing. The priorities established for water allocation would be used in critical drought conditions. The water quality classification of water bodies by different classes of use is the basis for truly integrating the quality and quality of water management. Water pricing is the single most controversial instrument of the law. The pricing system recognises the economic value of water, as stated in the principles of the policy, but is also the most difficult step to implement. It is expected that the pace of implementation will increase and the quality of work will improve in addition to building the capacity of the private and district staff in contracting procedures. The financial accountability is also easier and more transparent.

The communities should be fully utilised in any attempts to promote the local management of water supply and sanitation systems. A charge in water and sanitation sector approaches from supply-driven approach to demand-responsive approach call for full community participation.

The community should be defined in terms of their primary role as user/clients. Private-sector services are necessary because there are gaps, which exist as a result of the government not being able to provide water services due to limited financial resources and increase in population.

Table 7. Diverse water challenge

Country	Egypt	Sudan
Per capita annual water resources 2000 (m^3)	34	1187
Per capita annual withdrawal (m^3)	921	666
Per capita annual withdrawal for agriculture (m^3)	86	94

Source: World Resources Institute 1998 and 2000. World Resources 1998-1999. World Resources 2000-2001.

There is little notion of 'service, invoice and move on'. As a result, there are major problems looming with sustainability of completed projects. The factors affecting the eco-environmental changes are complex. There are interrelated and interact. The deterioration problems of water and sanitation have attracted some attention in recent years. There is an urgent need to study possible rehabilitation measures to ensure a sustainable and excellent water quality and improved sanitation.

APPENDIX 1. THE DESCRIPTIVE FORMULAES

Main criteria	Sub-criteria	Formulae
Food	Grain security	$\sum_{It=1}^{N} \frac{Grain\ production\ (t/y)}{\mathrm{Re}quired\ grain\ consumption\ (t/y)}$
Economic	Value added	$\sum_{i=1}^{N}$ [Return (SD) – Cost (SD)] Where N = number of sectors (agriculture, industrial, domestic water supply, navigation, and power generation)
Water	Water security	Water security = [total available water resources – Total water requirements] (10^6 m^3 / year)
	Water sustainability	$\sum_{i=}^{K} \frac{Volume\ of\ water\ used\ from\ resource}{Potential\ of\ resource}$ Where K = number of water resources (surface water, shallow groundwater, deep groundwater, desalination, etc.)
Socio-economic	National employment	$\sum_{i=1}^{K}$ Total employment of sector $_{I}$ Where K = number of sectors (agriculture, industrial, domestic water supply, power generation and navigation)
	New settlers	Total homes tender of agriculture sector
	Land *per capita*	$\frac{Total\ cultivated\ land}{Total\ population}(hectare\ /\ capita)$
	Water *per capita*	$\frac{Available\ water\ resources}{Total\ population}$ (m^3/(capita.y))
	Shisto paracities prevalence	Shisto paracities prevalence = number of shisto patients over the country
	Typhoid paracities prevalence	Number of typhoid patients over the country

Main criteria	Sub-criteria	Formulae
Environmental	Agro-chemical *per capita*	$\sum_{i=1}^{N} \frac{Pesticide\ used\ in\ (kg/y)}{Population}$ Where N = number of pesticide used all over the country
	Industrial effluent integrated water quality index	Integrated water quality index for sewage or industrial wastewater PTi = $\frac{APLPi}{SP/Pi}$ IWQI = $\frac{1}{K}\sqrt{\sum_{i=1}^{i=K} PI_i^2}$ Where: PTi = population index for parameter K APLPi = actual population loads of parameter K SPLPi = standard pollution load of parameter K $IWQI_{S/I}$ = Integrated water quality index (S= Sewage wastewater, I= industrial wastewater) K = number of proposed water quality parameters
	Domestic effluent integrated quality index	Same as above

REFERENCES

Amir, B. (1998). *Water development in Sudan.* Arab League. Khartoum: Sudan.

Anon. (1979). *Map of the world distribution regions. MAB Tech. Note 7.*

ASCE Task Committee on sustainability criteria. (1998). Sustainability criteria for water resource systems. Reston, Virginia, US: ASCE.

Brown, L. and Halweil, B. (1999). *Vital signs 1999.* New York, US.

FAO United Nations Food and Agriculture Organisation. (1999). The state of food in security in the world. Rome. Italy.

James, W. (1994). *Managing water as economic resources.* Overseas Development Institute (ODI). UK.

Noureddine, R. M. (1997). Conservation planning and management of limited water resources in arid and semi-arid areas. In: *Proceedings of the 9th Session of the Regional Commission on Land and Water Use in the Near East.* Rabat: Morocco.

Omer, A. M. (1995). Water resources in Sudan. *NETWAS* 2(7).

Omer, A. M. (2000). Water and environment in Sudan: the challenges of the new millennium. *NETWAS* 7(2): 1-3.

Overseas Development Administration (ODA). (1987). Sudan profile of agricultural potential. Survey. UK.

Salih, A. M. A. and Ali, A. A. G. (1992). Water scarcity and sustainable development. *Nature and Resources* 28(1).

Seckler, D. (1992). Private sector irrigation in Africa-Water resources and irrigation policy studies. Winrock International Institute for Agricultural Development.

UNESCO Working Group M.IV. (1999). Sustainability criteria for water resource systems. Cambridge, United Kingdom: Cambridge University Press.

In: Horizons in Earth Science Research. Volume 11
Editors: Benjamin Veress and Jozsi Szigethy
ISBN: 978-1-63321-296-1

Chapter 6

LASER RAMAN SPECTROSCOPY STUDY ON THE GROWTH ZONING OF FELDSPAR, ZIRCON, APATITE AND THEIR GEOLOGICAL MEANING

Yaoqi Zhou*, Wenyan Qin, Yuncui Zhang, Qian Liu, Wendong Liang, Huiwen Yue, Qi Sun and Han Zhang
School of Geosciences in China University of Petroleum,
Qingdao, Shandong, P.R. China

ABSTRACT

Zircon, apatite and feldspar are widely distributed in the crust, and they are common in rocks of various genetic types. Feldspar occurs as the main mineral while zircon and apatite are important accessory minerals of igneous rocks. Chemical composition and crystal structure of these minerals offer temperature, pressure and other formation environmental parameters, and they record the origin, migration and evolution of magma as well.

Ordering degree of alkali feldspars, zoning structure of plagioclase feldspars, zoning structure and elements composition of zircon, zoning structure and geochemical features of apatite were researched in this experiment with LRM technology, and the variation of Raman spectra was summarized and qualitative and semi-quantitative relationships were established. Combined these minerals' information with regional tectonic situation and thermal evolution history, the forming environment of magmatic crystallization and its geological significance were discussed.

* Email: Zhouyq@upc.edu.cn.

SECTION ONE: FELDSPAR'S LASER RAMAN SPECTROSCOPY AND GEOLOGICAL SIGNIFICANCE

Based on basic principle of Raman spectra and mineralogy, crystal chemical characteristics of feldspar, ordering degree of alkali feldspars and zoning structure of plagioclases were studied by means of laser Raman spectroscopy. The study found that the plagioclases show mainly two intense bands in the region 450-600 cm^{-1} while a triplet is generally observed in the alkali feldspar spectra. In addition, it is also possible to discriminate between the individual members in the alkali feldspar or plagioclase by comparing the number, frequencies and relative intensity of Raman spectra.

X-ray powder diffraction analysis of alkali feldspars can help get some parameters such as the ordering degree. Laser Raman spectra laws of alkali feldspar samples with increasing ordering degree were founded after testing: along with the increasing ordering degree, the number of spectra peaks increases and peaks becomes sharp in the low-frequency region (<400cm^{-1}); peaks at 480cm^{-1} and 510cm^{-1} gradually split into three peaks at 450cm^{-1}, 480cm^{-1} and 510cm^{-1}, shoulder peak at 480 cm^{-1} becomes apparent and sharp, peaks near 650cm^{-1}, 750 cm^{-1} and 815cm^{-1} are sharper. Overall, the Raman spectra peaks split and become narrow with the increase of ordering degree, this law can help to identify the ordering degree qualitatively. Hence, according to relationship between ordering degree and equilibrium temperature, equilibrium temperature of feldspars can be calculated roughly, it is between 575.7°C to 714.0°C.

Three types of zonings are identified through electron probe test of plagioclase zoning: normal zoning, reverse zoning and small-scale oscillatory zoning superimposed on partial large-scale zoning. Both normal and reverse zonings have little composition change, few zoning numbers and narrow width, which illustrates that the crystallization environment keeps relatively stable. Raman spectra characteristics can help identity normal and reverse zoning; it is normal zoning if intensity of characteristic peaks at 515cm^{-1}-500cm^{-1} decreases gradually from core to edge, while reverse zoning presents a converse trend. Besides, changes and offset direction of spectrum peak at 800cm^{-1} can also be used to distinguish zoning types. In addition, intensity of characteristic peak at 500cm^{-1}-515cm^{-1} does not decrease or increase continuously, it always shows a high - low - high - low or low - high - low - high repeated fluctuation changes and the number of changes represents zoning number. Therefore, combining zoning features of the experimental samples with previous analysis of the causes of zoning and the tectonic evolution of the Sulu orogenic belts, we deduce that the feldspar in the magmatic rocks experienced a stable-agitated-eruptive crystallization process.

The study of feldspar Raman spectra before 1990s is limited to research on end-member composition and ordering degree and so on, for example, V.P.Vodoyanova et al. (1984) studied Raman spectra of potash feldspar, D.W. Matson, S.K.Sharrna et al. (1986) studied Raman spectra of orthoclase-anorthite and nepheline-anorthite. T.P. Mernagh (1991) discussed difference of Raman spectra between alkali feldspar series and plagioclase series. These researches started distinguishing feldspar with Raman spectra.

For igneous rocks, zoning structure is widespread in plagioclase crystals, especially in neutral ones. The zoning of plagioclase reflected the environment change in the evolution of magma; it is closely related to magmatic crystallization rate and crystal formation conditions, which can be used to induce the evolution history of magmatic rocks. Therefore, the causes and geological meaning of plagioclase zoning became the focus of the study.

The purpose of the research is to study the composition of plagioclase zoning with laser Raman spectroscopy. Laser Raman is a new efficient and simple method because it's sensitive to zoning structure. Widespread attentions are drawn because determining ordering degree is an important way to induce the equilibrium temperature of feldspar. In spite of previous tests, the use of laser Raman spectroscopy still be immature in this respect. Lots of researches are needed to explain the relationship between Raman spectra and ordering degrees.

THE CHARACTERISTIC OF LASER RAMAN SPECTROSCOPY OF ALKALI FELDSPAR SERIES

Laser Raman spectroscopy has higher spatial resolution and this technology is lossless, which make it a complement for scanning electron microscope and electron probe analysis. Besides, it needs less or no sample preparation.

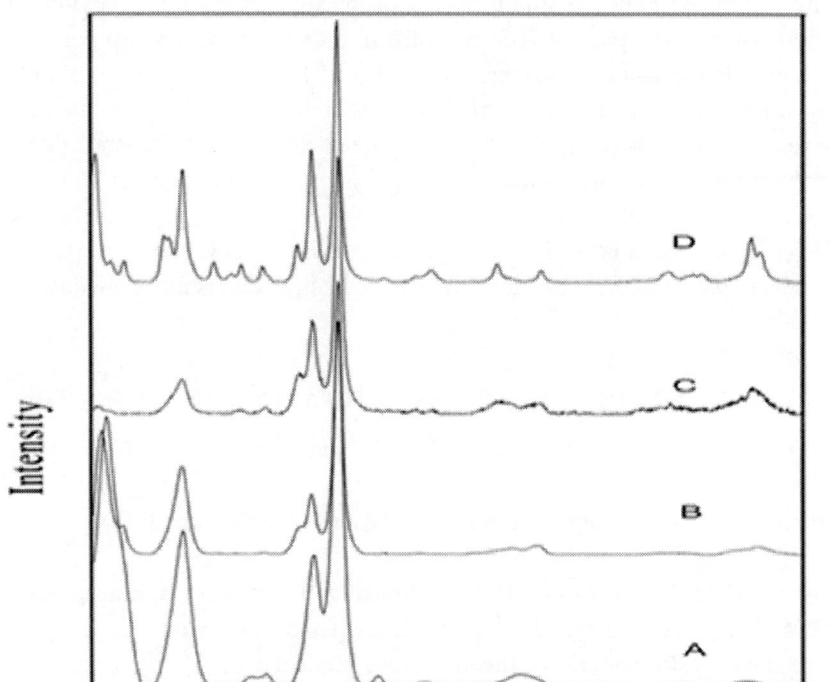

Figure 1. Raman spectra comparison of four kinds of alkali feldspar. (A: anorthoclase; B: sanidine; C: orthoclase; D:microcline).

The previous studies focused on end-member components of two group of feldspar, disorder of Si and Al, phase transformation among plagioclase. Only several studies have compared the Raman spectra of limited numbers of feldspar, this research seemed to be the first to compare Raman spectra of all species in two feldspar subfamily.

As can be seen from figure 1, the peak at 513cm^{-1} or 514cm^{-1} is the strongest band among the spectra. The other two strong bands can be seen at 450cm^{-1}～500cm^{-1}, they constitute the characteristic triplet together with the spectrum at 513cm^{-1} or 514cm^{-1}.

Conclusions can be drawn from Figure 1 as follows:

(1) If we have Raman spectroscopy of high quality, whether the shoulder peak splits or not can be used to distinguish anorthoclase and sanidine, it is anorthoclase if the shoulder peak splits, and the contrary is sanidine.
(2) Sanidine and orthoclase have similar Raman spectra; number and relative intensity of every weak bands are different. Besides, in high frequency region, spectra band at 995cm^{-1} can be seen in orthoclase but it does not appear in sanidine .
(3) Orthoclase and microcline are different in low frequency (<400cm^{-1}), intermediate frequency (400cm^{-1}～600cm^{-1}) and high frequency (>600cm^{-1}). Peak at 265cm^{-1} split into two relatively sharp peaks in spectra of microcline. The split of shoulder peak at 455cm^{-1} is more obvious in spectra of microcline; and a new band appeared at 1102cm^{-1} in high frequency in spectra of microcline.
(4) The difference between microcline and anorthoclase is more obvious. Compared with anorthoclase, peak at 265cm^{-1} splits into two relatively sharp peaks and peak at 455cm^{-1} also splits and a new band occurs at 1102cm^{-1} in the spectra of microcline.
(5) According to the order of anorthoclase→sanidine→orthoclase→ microcline, or the ascending order according to the Or content contained in mineral, there are some features in general as follows: the sharpness of spectrum peak at 265cm^{-1} gradually became small until divided into two relatively sharp peaks, the shoulder peak at 455cm^{-1} was gradually decomposed and became more obvious, the peaks near 1100cm^{-1} was gradually replaced by spectra band which is more obvious and sharp.

THE CHARACTERISTIC OF LASER RAMAN SPECTROSCOPY OF PLAGIOCLASE SERIES

Some conclusions are drawn according to the analysis of figure 2:

(1) For six kinds of plagioclases, the number and intensity of Raman spectra below 400cm^{-1} change slightly. This is because of the crystal directionality. But crystal directionality does not affect the high wave number area.
(2) In comparison with the intermediate oligoclase, andesine and labradorite, spectral band of albite is fainter or wider, and the band number is fewer.
(3) Raman spectra of oligoclase, andesine and labradorite are similar. The spectra strongly decline within 400cm^{-1}, overlapping and becoming much wider. Spectral peaks of albite at 764cm^{-1} and 813cm^{-1} are replaced by a wide spectrum. The stretching Raman band of albite above 1000cm^{-1} is also replaced by a wider, weaker band.

(4) Oligoclase, andesine and labradorite could be distinguished by the number of spectral peaks at $800cm^{-1}$ or the intensity and width of that at $1100cm^{-1}$.

(5) The Raman spectra of bytownite and anorthite are similar, both of their bands are very faint or wide besides of that at $485cm^{-1}$, $503cm^{-1}$ and $561cm^{-1}$.But more bands could be found in the Raman spectrum of anorthite because of its four different subcells.

(6) With the increase of An, albite turns to oligoclase, andesine, labradorite, bytownite and anorthite successively. Generally its Raman spectra are shown as below: spectral bands are faint, wide and fewer→ strong, sharp and more →faint, wide and fewer. The shoulder at $565cm^{-1}$ is obvious and sharp. The spectral peaks of albite at $764cm^{-1}$ and $813cm^{-1}$ are replaced by a wide spectrum.

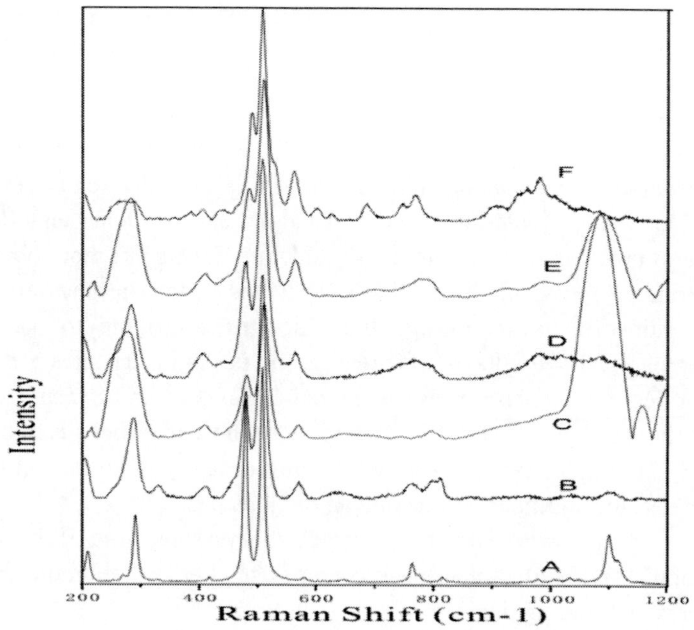

Figure 2. Raman spectra of six kinds of plagioclase (A: albite; B: oligoclase; C: andesine; D: labradorite; E: bytownite; F: anorthite).

According to the study on laser Raman spectra of feldspar minerals, we could identify feldspars by the characteristic spectral peak at $500cm^{-1}$-$515cm^{-1}$.The peak is the strongest in the spectrum and its Raman shift would change slightly in terms of different feldspars. Alkali feldspar and plagioclase could be distinguished by the position, number and relative intensity of Raman spectral peaks as well as their spectral characters. A triplet could be found in spectrum of alkali feldspar, two strong spectral bands at $450cm^{-1}$-$600cm^{-1}$ in plagioclase. Besides, series of alkali feldspars and plagioclases could be distinguished by the position, number, and relative intensity of Raman peaks.

LASER RAMAN SPECTRA STUDY ON THE ORDERING DEGREE OF ALKALI FELDSPAR

The features of Raman spectra with respect to three vibration modes in feldspar crystal are obvious. Combined with the Raman spectra study on structural units of Silicon-oxygen tetrahedron, we compared features of Raman spectra changes, and then qualitative judgment of ordering degrees was made by Raman spectra. Besides, the equilibrium temperature of feldspar could be inferred by the study on its ordering degrees.

The conclusions are as below:

(1) Lattice vibration of minerals and vibrations between K^+ and oxygen contribute to the spectral peaks in low frequency area below 400cm^{-1}.There is an increase in the number of peaks with the raising ordering degrees, and the shape of peaks become sharper. So ordering degrees could be roughly judged by the number and sharpness of spectral peaks in low frequency area.

(2) The bending vibration of Si/Al-O-Si bridge contributes to the spectral peak in intermediate frequency area at 400cm^{-1}- 600cm^{-1}.The reflection peaks at 510cm^{-1} and 480cm^{-1} gradually divide into three peaks at 450cm^{-1}, 480cm^{-1} and 480cm^{-1} with the increase of ordering degrees and the shoulder at 450 cm^{-1} is more obvious and sharp. So ordering degrees could be qualitatively judged by the sharpness of spectral peaks.

(3) The vibration of oxygen bridge bond doesn't contribute to the peaks in high frequency area above 600cm^{-1}. Generally, the peaks in that area are faint. With the ordering degrees decrease, peaks at 650cm-1, 750cm^{-1} and 815cm^{-1} become smaller and smaller, and the first peak disappears at the end. There is a obvious peak at 1000cm^{-1} in samples A5 and A6 with highest ordering degrees. But at 1100cm^{-1}, an envelope occurs and finally turns into an obvious peak.

(4) On the whole, with ordering degree increasing, the full width at half maximum(FWHM) of Raman peak becomes smaller, which could make qualitative judgments on ordering degrees.

RECOVER EQUILIBRIUM TEMPERATURE OF FELDSPAR CRYSTALS

Temperature is the main factor of feldspar ordering degrees, so ordering degrees could reflect temperature change during the crystallizing of feldspars. One purpose of this experiment is to analyze the formations of various feldspars with different ordering degrees and restore their equilibrium temperature by studying alkali feldspar's laser Raman spectra. Their temperatures could be estimated by the formula T(°C)=856.2594-395.1256Sm. Although the equilibrium temperature could be identified by the ordering degree and structural parameters of feldspars, it is not the same with crystallization temperature. The reason is that the structural state of feldspar decreases and then become stable as the reduction of temperature. So the deduced temperature is the stabilized one.

Besides, according to the latest achievements, temperature is not the key factor of ordering degrees. For feldspar, its composition and duration of temperature change is more critical than crystallization temperature. The condition with high temperature and low rate of temperature decrease tends to make feldspar be orderly state because of more Al-Si exchange.

LASER RAMAN SPECTRAL ANALYSIS OF PLAGIOCLASE'S ZONING STRUCTURE AND RECOVER FORMATION ENVIRONMENT

Plagioclase's zoning draws the attention of researchers because it reflects environmental change during magma crystallization and it can be used to deduce that process. And it is also relevant with the rate and condition of crystallization.

The conclusions are as below:

(1) Three types of zonings are found: normal zoning, reverse zoning and small-scale oscillatory zoning superimposed on partial large-scale zoning.
(2) Normal zoning is basic in inner part and is acid in outer part. The reason is that during the dramatic temperature decrease of magma crystallization, early formed basic plagioclase crystals can not wait to react with magmas and some acid plagioclases around basic ones because of their lower fusion point which is only 1118°C.
(3) The pressure change is the main factor forming reverse zoning. The crystallized plagioclase changes from acid to basic as water pressure increases.
(4) Supersaturation of magma diffusion and temperature change accounts for normal-scale oscillatory zoning, while "supersaturation-nucleation-depletion" mechanism accounts for small-scale one.
(5) Normal and reverse zonings all have features of small composition changes, fewer numbers and narrow width, which reflects that the crystallization environment is relatively stable.
(6) In comparison of normal zoning, reverse one is of larger numbers, width as well as rate of An changes from core to edge. These features reflect that the environment where reverse zoning formed is relatively oscillatory.

LASER RAMAN SPECTRA MICRO-ZONE ANALYSIS OF PLAGIOCLASE'S ZONING STRUCTURE

Laser Raman Spectroscopy tends to be another efficient way of testing plagioclase's zoning structure because it is sensitive to the change in the structure. So we analyzed the Raman spectra of sample zonings, tried to find the rules of spectra changes during zoning changes and then classified different types of zonings according to Raman spectra.

The conclusions are as below:

(1) Low frequency area below 400cm^{-1}: the change of spectral peaks in that area can not distinguish normal zoning and reverse zoning because both zonings becomes more obvious and sharper from core to edge.
(2) Intermediate frequency area at 400cm^{-1}-600cm^{-1}: Different zonings cannot be distinguished at 478cm^{-1}-488cm^{-1} with the same reason above. But at 500cm^{-1}-515cm^{-1}, featured peaks of normal structure gradually decline, while reverse one are opposite. Featured peaks at 500cm^{-1} -515cm^{-1} show fluctuation of high value→ low value→ high value or low value→ high value→ high value. Times of fluctuation demonstrate numbers of zonings.
(3) High frequency area above 600cm^{-1}: From core to edge, Spectral peaks of normal zoning at 800cm^{-1} move to low frequency area, they are more apparent. Their shapes are sharper. While peak value of reverse zoning declines, the peaks is not obvious and turn into a big envelope moving toward high frequency area. Spectral peaks at 1100cm^{-1} can not identify different zonings because the shape of both peaks become sharper from core to edge.

CONCLUSION

Conclusion are drew based on field sampling, laboratory analysis, as well as the previous achievements of Raman spectra, ordering degrees and zoning structures.

(1) Specified Raman spectrum at 500cm^{-1} -515cm^{-1} could identify feldspars. There is a triplet in alkali feldspar's spectrum while there are two strong spectra at 450cm^{-1} -600cm^{-1} in feldspar's spectrum. Besides, the position, number and relative intensity of peaks could also distinguish the two kinds of feldspars.
(2) The links between the ordering degrees and Raman spectra of feldspars are as below: a) numbers and sharpness of spectral peaks in low frequency area below 400cm^{-1} increase as feldspar's ordering degrees increase. b) In intermediate frequency area at 400cm^{-1} -600cm^{-1}, reflection peaks at 480cm^{-1} -510cm^{-1} split into three peaks near 450cm^{-1}, 480cm^{-1} and 510cm^{-1} respectively, and the peak near 480cm^{-1} is more obvious and sharper. c) In high frequency area above 600cm^{-1}, with the decrease of ordering degrees, peaks near 650cm^{-1}, 750cm^{-1} and 815cm^{-1} becomes smaller and smaller, and finally the first peak disappears. d) On the whole, Raman peaks will split and narrow their FWHM as ordering degrees increase, which could make qualitative judgment to samples' ordering degrees.
(3) According to the relationship between the equilibrium temperature and ordering degree of X-ray proposed by Ma Hongwen(1988), T(°C)=856.2594-395.1256Sm, it can be roughly estimated that the equilibrium temperature of feldspar ranges from 575.7°C to 714.0°C. However, the structural state of feldspar decreases with decreasing temperature and will not be retained until it reaches certain stability. Therefore, the temperature we estimated here is merely the temperature of steadily stated feldspar. The initial crystallization temperature of feldspar might be higher.

(4) The zoning shown in the samples of the studied area covers both normal-zoning and reversed-zoning, some contains small-scale fluctuant zoning and oscillatory zoning which piled inside the zoning. The normal zoning is small in number, with almost unchanged compositions and narrower width, reflecting that it crystalized in a relatively static environment. The difference between normal zoning and reverse zoning is that the latter has more rings, along with wider width and a greatly changed An content on the core and edge. This reflects that the reversed-zoning was formed from relatively turbulent origin magma. In short, the original magma which generated the crystal had been experienced a stable - volatile - extrude process, and environment was relatively stable.

(5) The low frequency region below $400cm^{-1}$, shoulder peak at $478cm^{-1} \sim 488cm^{-1}$, and spectral peak at $1100cm^{-1}$ cannot serve as the basis for judging the normal and reversed zoning, for these two have similar characteristics in those areas. It's a normal zoning that shows a gradual downward trend of its characteristic peaks at $500cm^{-1}$~$515cm^{-1}$(or vice versa). However, the characteristic peaks there does not rise or decrease continuously, they repeatedly fluctuates and forms several high-low-high-low (or vice versa) rotations. The number of characteristic peaks changes represents the number of the rings. Secondly, the spectral peak around $800cm^{-1}$ becomes more obvious and acute and has a tendency of moving towards the low frequency area, opposing with the reversed-zoning, which tend to fade as it moves towards the high frequency area and would form a big envelope at the end.

(6) Combining the normal, reverse and oscillatory zonings of the samples, previous causing analysis of the zonings, and tectonic evolution of the Sulu orogenic belt, I inferred the following crystallization process of feldspar in igneous rocks. The environment of primitive magma in early cretaceous was relatively stable at first, without intensive variation of the physical and chemical conditions, resulting in the formation of plagioclase crystals with uniform composition instead of zoning. After that, because of the sharp decline in the temperature of the upward migrated magma, the earlier formed basic plagioclase crystals had no sufficient time to blend and react with the surrounding magma, and then some acidic plagioclase grew around the earlier formed basic plagioclase. Thus normal zoning was formed. From inside to outside, its ingredient changes from basic to acidic. From the seldom changed composition, the small ring numbers, and the narrow width of the normal-zoning, we can see that the crystal crystallized in a relatively stable static environment, whose magma temperature does not drop quickly. After that, the magma continues to move upward, causing the changes of total pressure, water and volatile matters, forming the reverse zoning in a relatively high crystallization temperature. In the process of the crystallization of normal and reversed zoning, the small-scale oscillatory zoning was formed, due to many insignificant changes of temperature, pressure and other external conditions. After the plagioclase zoning crystals is formed, the original magma will continue to travel upward and erupt violently when it reaches the ground, therefore the original magma who crystallized these igneous rocks has experienced steady - turbulence - eruption process.

SECTION 2. LASER RAMAN SPECTROSCOPY AND MICRO-STRUCTURE OF DIFFERENT GENESIS OF ZIRCON IN SULU OROGEN

Cathodoluminescence image, electron probe and laser Raman spectroscopy were used to analyze zircon samples from Rongcheng (Weihai), Mountain Laoshan, Yakou and Lingshan Island (Jiaonan). Analysis is based on systematic mineralogy, petrology, internal structure of samples. This research shows that the internal structure and element characteristics and other information can help identify the genesis of zircon: (1) Cathodoluminescence image shows the oscillatory zoning of the magma genesis zircon, furthermore the increasing content of Hf and decreasing Zr/Hf ratio from the core to the edge of the crystal or from the inner side to the outside of the zone; (2) Metamorphic genesis particles shows irregular morphology, complicated and diversified internal structure. Zr content increases while Hf decreases from core to the edge of the zircon, which leads to the decreasing Zr/Hf ratio.

We identified zircons from different origins and examined them using Micro-Raman analysis. From the core to the edge of the magmatic zircon, the characteristic peak of Raman curve gradually declines, and the ratio between relative intensity and half-width height increases with the increasing metamorphic intensity; from the core to the edge of the metamorphic zircon, the characteristic peak of Raman curve intensity increases or basically remain unchanged. The stronger the metamorphism is, the higher the ratio between relative intensity and half-width height would be, and the half high width will increase generally. Combining Sulu orogen tectonic setting and zircon geochronology data analysis, we found that the zircon formed under high to ultra-high pressure metamorphic rocks in the Neoproterozoic , whose Δ439 Raman curve is mainly distributed between a high value area of 500-700; the zircon of Laoshan Yakou volcanic tuff , 118.9Ma, has a Δ439 value distributed between 300-500, which is in the medium value area; the zircon found in the thick rhyolitic volcanic tuff Lingshan Island Laohuzui region, 119.2Ma, has a Δ 439 value in 400 or less, and some even as small as 200, which is obviously in a low value area.

The geological history of sulu orogenic belt in east Shandong province experienced many complex tectonic, magmatic activities which resulted in complex internal structure and elements composition of zircon in the geological body. The feature difference of zircons which can be recognized by its laser Raman curves, provides some reference to the interpretation of geological history background and formation age of geological body containing zircon with various genesis.

Laser Raman microprobe as a new technology plays an important role in the research of micro-area structure of minerals, combined with CL images, it can provide relevant information of mineral's metamict degree and internal structure, and offer fast, convenient inferences to the mineral's composition change and genesis feature of different part. The changes of characteristic peak values at 1008cm-1, 439cm-1 and 357cm-1 and FWHM in Raman spectra reflect the degree of metamict. Meanwhile, the metamict degree and change of U, Th in zircon linked intrinsically within the same geological history. Therefore, it is considered the differences of peak values and FWHM in different regions of zircon reflect the changes of U, Th content in these areas.

The study of the micro-area structure characteristic of zircon and the distribution and migration characteristic of U, Th and other elements by the systematic research, has an important significance for the explanation of the formation environment and growth history of zircon. Meanwhile, it can provide some reference to the interpretation of tectonic thermal history and formation age of geological body containing zircon.

There are mainly five characteristic peak values of the laser Raman curve of zircon, which can be grouped into three classes, with its major wave number: 202-225 cm^{-1}, 1008cm^{-1}(mode B_{1g}), 357cm^{-1}(mode E_g)and 439cm^{-1}, 974cm^{-1}(mode A_{1g}). With strong relative intensity, the phonon vibration modes at 357cm^{-1} (E_g), 439cm^{-1} (A_{1g}) and 1008cm^{-1}(B_{1g}) can be easily measured, and they correspond to the internal vibrations of silica tetrahedron of zircon. The characteristic peak at 202-225cm^{-1} corresponds to the translation and rotation of the external lattice of zircon crystals. The Raman spectra peaks at 357cm^{-1} and 439cm^{-1} are most sensitive to the changes of crystalline state. The E_g and A_{1g} Raman spectra peaks can be observed in different directions with intensity changes. The metamictization of zircon can make the bond length and bong angle of silica tetrahedron irregular, causing peaks shift and half-width increases. Among all the previous studies, the most effective index of the metamict degree is the change of the peak value of inverse stretching vibrational spectra B_{1g} and half-width in silica tetrahedron.

We studied the systematic mineralogical characteristics, petrography, internal structure, major elements, trace elements, and laser Raman spectral features of zircon, by analyzing the rock samples from Weihai Rongcheng, Laoshan Yakou and Jiaonan Lingshandao. Together with the investigation of tectonic evolution background, a detailed study of the internal structure, geochemical and laser Raman features of the sample and a combination analysis of the regional tectonic settings in eastern Sulu orogen Jiaonan - Weihai orogenic, we concluded the following conclusions:

(1) Different geological genesis of zircon leads to its different internal structure. In cathodoluminescence images, most magmatic zircons crystals are euhedral - subhedral, in which the magmatic oscillatory zoning is characteristically formed. Metamorphic zircon crystals varies in shape, from anhedral to euhedral. Its internal structure varies and its shape and internal structure changes with changing metamorphic conditions.

(2) With the electronic probe, major elements and trace elements in different causative zircon have different variation trend. In the magmatic zircon, from the core to edge in the crystals(the band is not obvious in the crystal) and in each band of the crystal from the inner side to the outside, the content of the HfO_2 increases, the value of ZrO_2 and the rate of ZrO_2 / HfO_2 decrease; in the metamorphic zircon, from the center to the edge crystal, the content of ZrO_2 increases, the level of HfO_2 drops, and the rate of ZrO_2 / HfO_2 increases. That is opposite to the magmatic zircon.

(3) Laser Raman microprobe is a good way to detect the geological genesis of zircon. From the core to the edge of magmatic zircon, the characteristic peak of Raman curve gradually declines, and the ratio between relative intensity and half-width increases with the increasing metamorphic intensity. From the core to the edge of metamorphic zircon, the intensity of characteristic peak increases or basically remains unchanged. The stronger the metamorphism is, the higher the ratio between relative intensity and half-width would be, and the half-width will increase generally.

(4) The eastern Shandong region in Sulu orogenic belt has experienced complex tectonic and magmatic activities in the geological history, causing complex internal structure characteristics of zircon in geological body. Using laser Raman curve, we can better

understand the difference of internal structure of zircon and its origion. In the direction of Rongcheng-Yakou-Lingshandao, the stratigraphic age gradually becomes newer, the value of Raman characteristic peak at Δ439 in zircon has a declining trend, the metamorphic intensity changes from strong to weak.

SECTION 3. LASER RAMAN CHARACTERISTIC OF ZONING WITHIN APATITE

As a common accessory mineral in various types of rocks, apatite is commonly used in studies of isotope tracer and low temperature geochronology due to its stability over a wide range of temperature and pressure.

Changes in composition, temperature and pressure of surrounding melt result in the chemical zoning within apatite. Variation of growth environment can be recorded by the regular change in content of trace elements from core to edge of apatite. In this experiment, laser Raman spectroscopy was applied to analyze zoning within apatite separated from sedimentary rocks and metamorphic rocks, according to characteristic peak's variation in laser Raman spectra of different zones within apatite and the known thermal history of apatite, the features and rules of different zones within apatite were inferred, which could help explore new methods applied on the low temperature geochronology of apatite.

Zoning within apatite is most readily observed using cathodoluminescence(CL) imaging. Apatites mostly are yellow-green, green or brown lavender in the CL images. In this research, zoning was found in the apatite separated from eclogites and sandstone samples and it was classified into three types by the genesis. Zoning within apatite is correlated with the distribution of elements.

Electron microprobe was applied to analyze the changes of element content in different locations of apatite and element zoning profile can be obtained. It is happened only sometimes, not universal, that wave crest or trough in element zoning profiles of apatite coincides with the light and shade zones within apatite.

Laser Raman spectroscopy was applied to analyze apatite separated from Sulu eclogites and Songliao basin sandstones, detecting different locations within the apatite from core to edge. The obtained information about shifts, intensity, FWHM of characteristic peaks indicates the former one is much more complex.

The characteristic peaks of different zones within apatite do not shift obviously, but peak intensities change little and show a decreased tendency, which can be attributed to the high temperature and pressure that apatite have gone through during growth.

The main peak intensity of inner zones within apatite is stronger than that of outer zones. The strengthen of peak intensity can be attributed to the high temperature and pressure the apatites have gone through during growth, depending on the known thermal history of apatites and laser Raman spectra of different zones within apatite.

As a common accessory mineral in various types of rocks, apatite is commonly used in studies of isotope tracer and low temperature geochronology due to low closure temperature of apatite fission track and stability of apatite over a wide range of temperature and pressure. Recently, various analysis technologies have been used by scholars to research the distribution and diffusion of trace elements and rare earth element, reveal mantle melt and magmatic differentiation, identify genetic type of granite and so on.

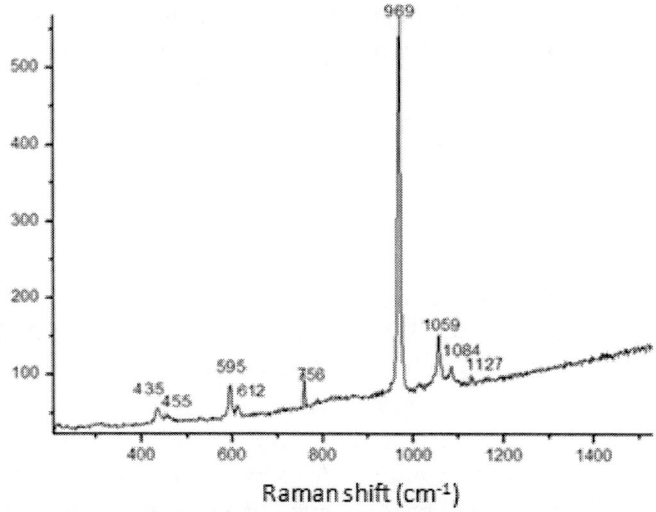

Figure 3. Laser Raman spectra of apatite.

Cathodoluminescence imaging can reveal the crystal structure and growth environment of apatite. Some scholars have researched zoning within ziron, rutile and some other minerals, in this paper we try to apply the relationship between temperature and pressure changes in mineral growth with zoning within those minerals to zoning apatite. Analysis on apatite using laser Raman spectroscopy is still in exploration. Compared with infrared spectrum, laser Raman spectrum is more sensitive to crystal structure and give more narrow band, the changes in microstructure and bonding features of crystal can be obtained by analyze the variation on laser Raman spectra. In laser Raman spectrum of apatite (figure 3), nine characteristic peaks can be identified in 200-1600cm^{-1} region: peak at 1127cm^{-1} (weak),1084cm^{-1} (weak),1059cm^{-1} (strong), 969cm^{-1} (strongest), 756cm^{-1} (strong), 612cm^{-1} (weak), 595cm^{-1} (strong), 455cm^{-1} (weak) and 435cm^{-1} (strong), sometimes, peaks can be detected near 288cm^{-1} and 205cm^{-1}. Anionic complexes in apatite, $[PO_4]^{3-}$, have four shift peaks: peak v1 at 962cm^{-1}-965cm^{-1} is caused by asymmetric stretching vibrations; peak v2 (relative strong peak) at 419cm^{-1}-431cm^{-1} results from bending vibration; peak v3 at 1040cm^{-1}-1049cm^{-1} is caused by asymmetrical stretching vibration; peak v4(weak absorption band) at 575cm^{-1}-593cm^{-1} result from asymmetric bending vibration. Strong absorption band at 1061cm^{-1}-1088cm^{-1} can be identified in laser Raman spectra of apatite in which $[CO_3]^{2-}$ replace partial $[PO_4]^{3-}$, this band indicates $[CO_3]^{2-}$ occupy lattice positions in apatite, its appearance cannot be attributed to simple mixture of carbonate into apatite. Stretching vibration band of $[OH]^-$ in apatite is around 3568 cm^{-1} , which is lower than that of $[OH]^-$ in typical hydroxide (3630 cm^{-1}-3700cm^{-1}). The lower stretching frequencies of $[OH]^-$ in apatite may be attribute to bond length between $[OH]^-$ and other ions and bond length of $[OH]^-$ itself. Vibration of $[F]^-$ is at 4136cm^{-1}-4142cm^{-1}, belonging to stable strong band. Yu Liu et al. researched on the vibration mode outside the lattice(variation vibration outside the tetrahedron) of the manmade fluorapatite single crystal and found that frequency of external vibration mode was low(＜400cm^{-1}) and easy to be covered by Rayleigh line [13].

Xianyong Liu et al. have researched on apatite of different genetic types, variation on laser Raman peak of ionic group in different subspecies of apatite separated from phosphorite have been analyzed. The v4 vibration band of $[PO_4]^{3-}$ in different subspecies shifts largely, the other three bands show minor shift; vibration band of $[OH]^-$ shifts largely, which attribute to the content of $[OH]^-$; vibration band of $[F]^-$ is relatively stable, minor shifts exist in different subspecies.

Conclusions can be drawn from this experiment:

(1) Zoning within apatite can be divided into three types: 1) Oscillatory zoning exist in the core, mantle and rim of apatite. Boundary between core and mantle and boundary between mantle and rim are distinguished by overall light and shade degree of oscillatory zoning, zoning is center based, the boundaries between different zones are always parallel to the crystal growth face, the outlines of zoning boundary are always straight, the crystalline form of apatite develop well. 2) The core is an uniform ensemble, oscillatory zoning always develops in the mantle and rim. Boundary between core and mantle is obvious, while there is a gradual transition from mantle to rim and do not exist obvious boundary. The core generally has a regular outline, oscillatory zoning is parallel to the crystal growth face and the shape of it is regular. Apatite with those characteristics may crystallize in early period and continue growing in later period, it may suspend in the fluid during growth and excitation elements which control the luminescent effect of apatite can only diffuse limitedly (T.J.Dempster et al.). 3) There is almost no oscillatory zoning existing in the core, mantle and rim, those parts can be distinguished by different luminescent colors, boundaries are obvious but not regular, sometimes, they can be embayment.

(2) Element content in different location of apatite varies. It can be found that only sometimes, not universal, that wave crest or trough in element zoning profiles apatite coincides with light and shade zones within apatite. Generally, cathodoluminescence characteristics coincides with trace element characteristics, but because there are still uncertainties existing in the excitation scheme of cathodoluminescence, the cathodoluminescence characteristics cannot identify all luminescence stimulated by ions, but they can give us partial qualitative inference. For the time being, the simple and effective quantitative relation between trace element content and cathodoluminescence intensity cannot be obtained. In this experiment, only partial rare earth elements and trace elements were obtained, so the correspondence cannot be amply demonstrated.

(3) Cathodoluminescence technique can help us detect zoning within apatite that cannot be seen under general microscope. Zoning is not directly correlated to distribution of F, Cl, P, Si, La in apatite, which illustrates the zoning shown in CL image is stimulated by some other rare earth elements and trace elements besides those elements.

(4) Laser Raman characteristics of apatite separated from Sulu eclogites are much more complex than that of apatite separated from Songliao basin sandstone. The former one just show the strong peak in 964cm^{-1} of PO_4, while the characteristic peak of SO,

CO can be detected in the later one. This illustrates the two have big difference in crystal structure and elementary composition and some other aspects.

(5) It can be qualitative inferred that the variation of characteristic peak intensity can reflect the thermal history which apatite have gone through.

SECTION ONE: FELDSPAR'S LASER RAMAN SPECTROSCOPY AND GEOLOGICAL MEANING

1.1. INTRODUCTION

1.1.1. Purpose

Feldspar is widely distributed in the crust, and common in rocks of various genetic types. The chemical composition and crystal structure of feldspar record important information on magmatic origin, evolution and conditions of magmatism etc [1-2].

Two main feldspar series are alkali feldspar and plagioclase. According to Stewart(1974) and Ye(1984),Temperature is the main factor than affects the ordering degrees of alkali feldspars. Equilibrium temperature could be calculated by Al^{3+}occupancy in T1 site because of the functional relationship between them [3-4]. Plagioclase's composition and crystal structure record the temperature, pressure as well as other vital information during crystallization and its zoning structures reflects of magmatic activity process.

Laser Raman Microprobe (LRM) has been applied on geology as a micro-region analysis technique since 1980. Soon, it came to be an important technique because of its high definition and high speed. In microstructure researches, LRM is used to determine the structure of matters because Δv indirectly reflects the molecular skeleton structure such as all kinds of functional groups.

1.1.2. Research Status

Mysen (1981) believes that the number of bridge oxygen in silicon-oxygen tetrahedron reflects difference of structural units. The units could be classified to 5 types Q0, Q1, Q2, Q3 and Q4, the figure 1.4 represents the number of bridge oxygen. After that some people believe that good relationships between structural unit type unit and the specific Raman peaks are existed.

Before 1990s, laser Raman spectral study on feldspar limited to endmember component or ordering degrees. T. P. Mernagh (1991) analyzed the Raman spectral differences on all kinds of feldspars in plagioclase series and alkali feldspar series [5].

Xiejun (2008) finds the spectral peaks in low and intermediate frequency region splits with the increasing ordering degree of alkali feldspar. He also discusses the relationship between Si(Al)-O_{nb} symmetric stretching vibrations in high frequency region and Si(Al)-O_b-Si symmetrical bending vibration in intermediate frequency region [6-7].

1.1.3. Research Contents

Basic principle of laser Raman spectra.

Crystal-chemical features of feldspars.

Laser Raman spectral analyze on feldspar series.

Laser Raman spectral study on ordering degrees of alkali feldspars and zoning structure of plagioclases.

Recover equilibrium temperature of feldspar crystals and formation environment of zoning structures.

1.2. OVERVIEW ON LASER RAMAN SPECTROSCOPY

1.2.1. Brief Introduction to Raman Spectra

Raman effect is the inelastic scattering of light, which means light is scattered to all directions when monochromatic light irradiating the sample. And the frequency of some spectral in the scattered light changes ($\nu=\nu_0\pm\Delta\nu$). This effect was found by Raman and Krishman in 1928 [8].

In scattered light, scattering line whose frequency doesn't change is Rayleigh line ($\nu=\nu_0$). Scattering line with higher frequency is Stokes line($\nu=\nu_0-\Delta\nu$), on the contrary, it is Anti-Stokes line($\nu=\nu_0+\Delta\nu$)(Figure 1.2.1). We put Stokes line and Anti-Stokes line together as Raman line, all the Raman lines constitute a Raman spectrum.

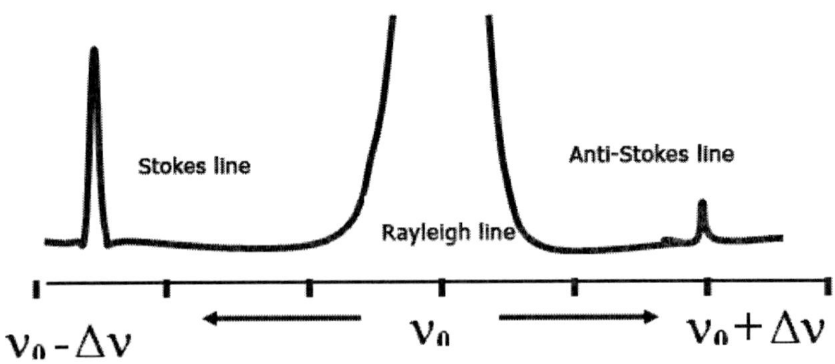

Figure 1.2.1. Schematic diagram of three scattering lines.

1.2.2. Forming Mechanism

The Raman effect occurs when light impinges upon a molecule and interacts with the electron cloud and the bonds of that molecule. For the spontaneous Raman effect, which is a form of light scattering, a photon excites the molecule from the ground state to a virtual energy state. When the molecule relaxes it emits a photon and it returns to a different rotational or vibrational state. The difference in energy between the original state and this new state leads to a shift in the emitted photon's frequency away from the excitation wavelength.

If the final vibrational state of the molecule is more energetic than the initial state, the emitted photon will be shifted to a lower frequency for the total energy of the system to remain balanced. This shift in frequency is designated as a Stokes shift.

If the final vibrational state is less energetic than the initial state, then the emitted photon will be shifted to a higher frequency, and this is designated as an anti-Stokes shift. Raman scattering is an example of inelastic scattering because of the energy transfer between the photons and the molecules during their interaction.

A change in the molecular polarization potential—or amount of deformation of the electron cloud—with respect to the vibrational coordinate is required for a molecule to exhibit a Raman effect. The amount of the polarizability change will determine the Raman scattering intensity. The pattern of shifted frequencies is determined by the rotational and vibrational states of the sample.(Figure 1.2.2).

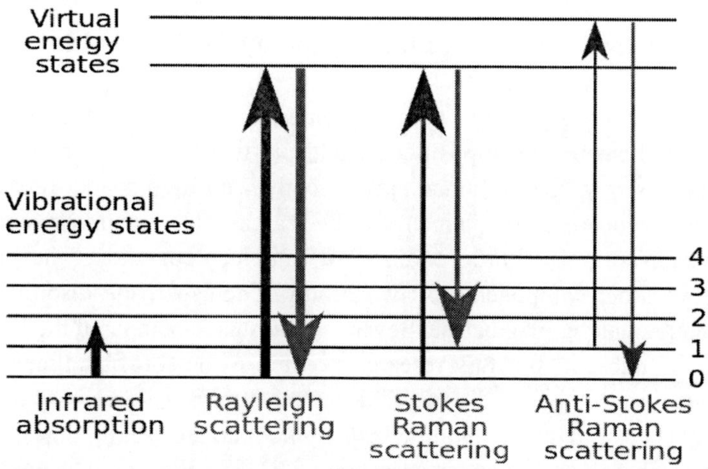

Figure 1.2.2. Schematic diagram of Raman scattering energy level.

1.2.3. Parameters

Raman shift Δv: It is the most important parameter, which reflects all kinds of molecular vibration mode of the sample. Δv is determined by molecular structure instead of light source [9]. Raman intensity I: It is determined by the intensity and angel of incident light as well as numbers of molecules in sample.

Full width at half maximum(FWHM) of Raman peak: It moves inversely to ordering degrees.

Symmetry degree of spectral peaks: Many peaks merge as the ordering degrees increasing.

1.2.4. Laser Raman Spectroscopy

Before laser was invented, mercury arc lamp was used as light source, Raman effect wasn't obvious. After 1960 when laser was applied to Raman spectrometer, micro-zone analysis were realized and soon laser Raman spectroscopy came into being [10].

1. 3. RESEARCH ABOUT THE CHARACTERISTICS OF LASER RAMAN

1.3.1. Spectrum of Feldspar Series

Feldspar belongs to framework silicate mineral, most of them can be expressed by chemical composition of $K[AlSi_3O_8]$(Or)-$Na[AlSi_3O_8]$(Ab)- $Ca[AlSi_3O_8]$(An).

Complete isomorphism series at high temperatures and incomplete isomorphism series at low temperatures of Or and Ab constitute alkali feldspar subfamily [11]. Complete isomorphism series of Ab and An constitute plagioclase subfamily.

The alkali feldspar and plagioclase series is mainly divided by the composition of feldspar.

1.3.2. Characteristics of Laser Raman Spectra of Alkali Feldspar Series

The purpose of this study was to define whether the Raman spectra sensitive enough to various structure and chemical composition of feldspar to allow us distinguish these feldspar with Raman spectroscopy. Laser Raman spectroscopy with high spatial resolution and non-destructive feature, which is a complement for the scanning electron microscope and electron microprobe, has advantages of little or no need sample preparation. The previous studies focused on end-member components of two group of feldspar, the disorder of Si and Al, phase transformation among plagioclase. Several studies have compared the Raman spectra of limited number of feldspar, but this research seemed to be the first time to compare the Raman spectra of various types of the two feldspar subfamily.

The Raman spectra data of alkali feldspar and plagioclase in this experiment were obtained from database of Raman spectroscopy-RRUFF. We tried to find each kind of feldspar's Raman spectral data with same area and same experimental conditions, and then mapped them with software of Origin75 and CorelDrawX3.

As can be seen from Figure 1.3.1, the peak at 513cm^{-1} or 514cm^{-1} is the strongest band among the spectra. The other two strong bands can be seen at 450cm^{-1}～500cm^{-1}, they constitute the characteristic triplet together with the spectrum at 513cm^{-1} or 514cm^{-1} which present in the Raman spectra of most of alkali feldspar. Conclusions can be drawn by Figure 1.3.1 as follows:

(1) Anorthoclase and sanidine can be distinguished through the comparison that whether the shoulder was decomposed or not, if we have Raman spectroscopy of high quality. It would be anorthoclase if the shoulder was not decomposed, and the contrary is sanidine.

(2) Sanidine and orthoclase have similar Raman spectra picture, the difference is the number and relative intensity of every weak bands. Besides, the spectra band at 995cm^{-1} can be seen in orthoclase but not appear in spectra of sanidine in high frequency region.

(3) Orthoclase and microcline are differences in low frequency of below 400cm^{-1}, intermediate frequency of 400cm^{-1}～600cm^{-1} and high frequency of above 600cm^{-1}. The spectrum peak at 265cm^{-1} was divided into two relatively sharp peaks in spectra

of microcline. The division at 455cm^{-1} is more obvious in spectra of microcline; and a new band appeared at 1102cm^{-1} in high frequency in spectra of microcline.

(4) The difference between microcline and anorthoclase is more obvious. Compared with anorthoclase, microcline has two relatively sharp peaks at 265cm^{-1} and 455cm^{-1}; and a new band at 1102cm^{-1}.

(5) According to the order of anorthoclase→sanidine→orthoclase→ microcline, or the ascending order according to the Or content contained in mineral, there are some features in general as follows: the sharpness of spectrum peak at 265cm^{-1} gradually became small until divided into two relatively sharp peaks, the shoulder peak at 455cm^{-1} was gradually decomposed and became more obvious, the peaks near 1100cm^{-1} was gradually replaced by spectra band which is more obvious and sharp.

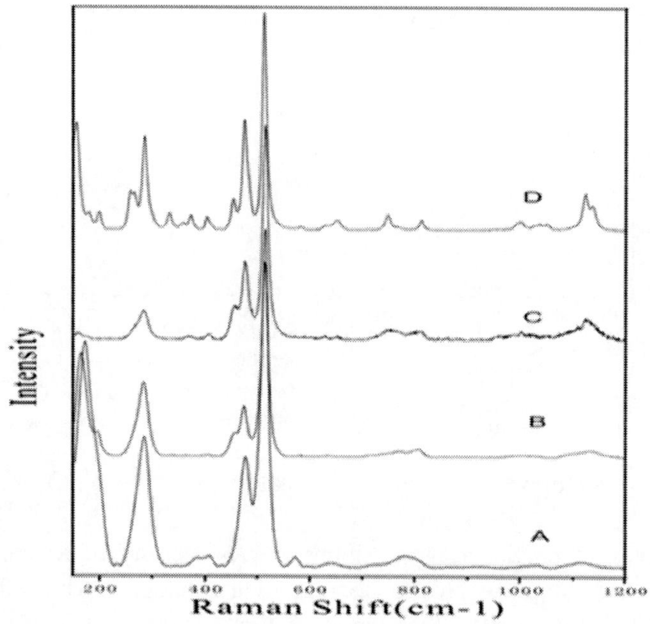

Figure 1.3.1. Raman spectra comparison of four kinds of alkali feldspar. (A: anorthoclase; B: sanidine; C: orthoclase; D:microcline).

1.3.3. Characteristics of Laser Raman Spectra of Plagioclase Series

There are many measuring methods for plagioclase ingredient, but all were insufficient.

However, Laser Raman spectroscopy can determine accurately the structural changes caused by compositional changes, which may be a high precision and effective method for the plagioclase component measurement.

From the figure we can see, just as the research finding of T.P.Mernagh [5], the spectra peaks concentrated in these three regions if we classified the figure into three part with the point of 400cm^{-1}and 600cm^{-1} ,that is low, medium and high frequency region. According to the crystal directionality, the number and intensity of the observed Raman spectra peaks under a slight change in region below 400cm^{-1}, while hardly affected in the higher frequency.

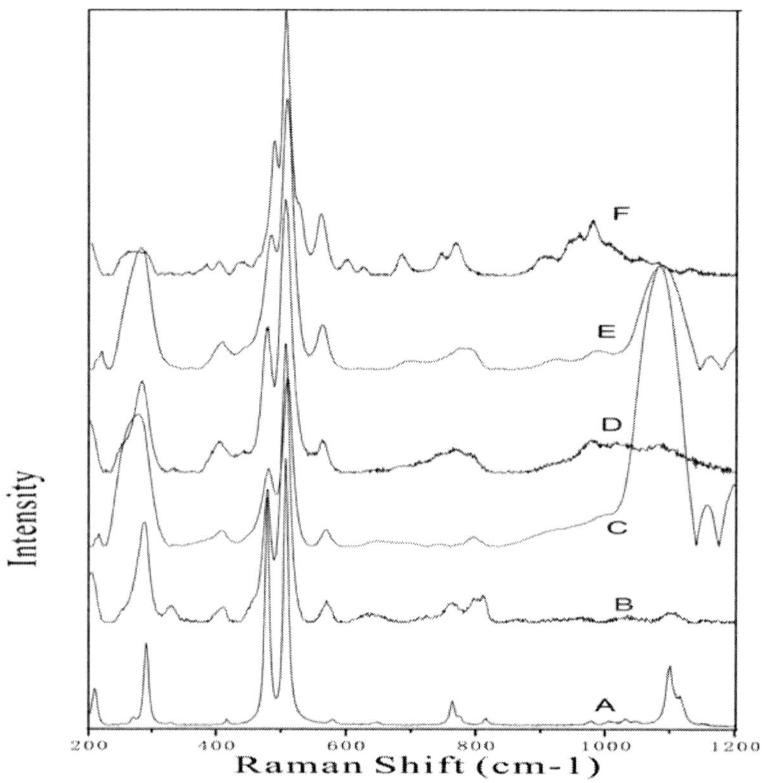

Figure 1.3.2. Raman spectra of six kinds of plagioclase. (A: albite; B: oligoclase; C: andesine; D: labradorite; E: bytownite; F: anorthite).

Some conclusions are drawn according to the analysis of Figure 1.3.2:

(1) For six kinds of plagioclases, the number and intensity of Raman spectra below $400cm^{-1}$ change slightly. This is because of the crystal directionality. But crystal directionality does not affect the high wave number area.
(2) In comparison with the intermediate oligoclase, andesine and labradorite, spectral band of albite is more faint or wider, and the band number is fewer.
(3) Raman spectra of oligoclase, andesine and labradorite are similar. The spectra strongly decline within $400cm^{-1}$, overlapping and becoming much wider. Spectral peaks of albite at $764cm^{-1}$ and $813cm^{-1}$ are replaced by a wide spectrum. The stretching Raman band of albite above $1000cm^{-1}$ is also replaced by a wider, weaker band.
(4) Oligoclase, andesine and labradorite could be distinguished by the number of spectral peaks at $800cm^{-1}$ or the intensity and width of that at $1100cm^{-1}$.
(5) The Raman spectra of bytownite and anorthite are similar, both of their bands are very faint or wide besides of that at $485cm^{-1}$, $503cm^{-1}$ and $561cm^{-1}$.But more bands could be found in the Raman spectrum of anorthite because of its four different subcells.
(6) With the increase of An, albite turns to oligoclase, andesine, labradorite, bytownite and anorthite successively. Generally its Raman spectra are shown as below: spectral

bands are faint, wide and fewer→ strong, sharp and more →faint, wide and fewer. The shoulder at 565cm^{-1} is obvious and sharp. The spectral peaks of albite at 764cm^{-1} and 813cm^{-1} are replaced by a wide spectrum.

According to the study on laser Raman spectra of feldspar minerals, we could identify feldspars by the characteristic spectral peak at 500cm^{-1}-515cm^{1}.The peak is the strongest in the spectrum and its Raman shift would change slightly in terms of different feldspars.

Alkali feldspar and plagioclase could be distinguished by the position, number and relative intensity of Raman spectral peaks as well as their spectral characters.

A triplet could be found in spectrum of alkali feldspar, two strong spectral bands at 450cm^{-1}-600cm^{-1} in plagioclase. Besides, series of alkali feldspars and plagioclases could be distinguished by the position, number, and relative intensity of Raman peaks.

1.4. Laser Raman Spectra Analysis of the Order Degree of Alkali Feldspar and Recover of Equilibrium Temperature

The order degree of feldspar is that the degree of Si replaced by Al and alternative law in the basic structural unit of feldspar crystals. The crystals' order degree will be higher if this alternative is regular and numerous. Different order degree reflects the changes of physical and chemical conditions during the magmatic crystallization process. Order degree is a dimensionless quantity, which is usually expressed by S=0~1, Where S = 0 indicates completely disordered, S = 1 means fully ordered [11].

The parameters such as feldspar component, order degree and equilibrium can be used to analyze the rock's formation temperature and study the diagenesis and mineralization [12].

1.4.1. Factors Affecting the Order Degree of Feldspar

Research found that, influencing factors of feldspar's order degree mainly are forming temperature, cooling rate, the total composition of the rock, pressure and activity of the water, the depth and geological age of rock formations and so on, during such, the temperature is the most important factor [7]. It should be noted that the previous study overemphasized the role of temperature, which is one-sided in complex geological processes. The composition of feldspar and change rate of temperature plays an important role than original crystal temperature in deciding order degree of feldspar. Despite the high temperature state of feldspar, Al and Si will be full exchange and form ordered structure if the cooling rate is slow.

Scholars now held different views on the relationship between these affecting factors and the order degree, especially, feldspar forms under interaction of various conditions, it will be one-sided without comprehensive study.

1.4.2. Common Methods of Determining Order Degree

From the beginning of the thirties, X-ray diffraction method is widely used in the field of geology. After that, a number of measurement ways for the order degree have appeared. And now X-ray diffraction, infrared spectroscopy and optical methods are used more commonly [13].

1.4.2.1. X-Ray Diffraction

The basis of determining order degree of feldspar is that, some value of the diffraction angle 2θ changes when the structure of feldspar changes. So we can obtain the order degree via relating charts or calculating if the 2θ has been got. Now the common methods are triclinicity and three peaks method.

Triclinicity method: triclinicity (Δ) is the deviation degree that crystal axis angle of triclinic alkali feldspar to monoclinic alkali feldspar crystal axis α, γ. The angular distance of these two peaks changes with the deviation degree of angle α, γ, so Smith proposed using reflection peaks (131) and ($1\bar{3}1$) to determine triclinicity. The two peaks overlap for monoclinic feldspar, however, the angular distance of two peaks is largest for maximum microcline, so we can get the triclinicity via Coefficient conversion (Δ)=12.5×[d(131)-d($1\bar{3}1$)] (according to Smith) orΔ=7.8×[d(130)-d($1\bar{3}0$)] (according to Mackenzie) [14]。

Calculate order degree with triclinicity easy to operate and have high precision, but its sensitivity was largely influenced by twinned, while the phenomenon of twins in feldspar crystals are very common. Besides, the two peaks (131) and ($1\bar{3}1$) overlap when the order degree of potassium feldspar between 0～0.4, so we cannot use this formula to calculate triclinicity. Three peaks method: Wright (1968) has used value 2θof peaks 060、 $\bar{2}04$ and $20\bar{1}$ to determine the composition and structure of the feldspar [3], which provides a new way for the study of feldspar's composition and structure. Ragland improved this method and calculated order degree with formula Sm= $(9.063+2\theta_{060}-2\theta_{\bar{2}04})/0.340$ [15].

According to other scholars' study, feldspar and albite is isomorphism minerals under high temperature conditions. Bowen and Tuttle (1950) proposed using d（$\bar{2}01$） to determine the composition of synthetic alkali feldspar under certain conditions [16]. Orville（1967） [17]、Jones （1969） [18] proposed using Or mol% molecules percent rather than weight percent to determine the composition of alkali feldspar.

Using the following relationships when the structure state has been known (Wright, 1968) [3]:

High sanidine - high albite series:

$$Or\% = 2030.05 - 92.18 \times 2\theta\ (\bar{2}01)$$

Orthoclase series:

$$Or\% = 1930.77 - 87.69 \times 2\theta\ (\bar{2}01)$$

Maximum microcline series:

$$Or\% = 2031.77 - 92.19 \times 2\theta\ (\bar{2}01)$$

1.4.2.2. Infrared Spectroscopy

Infrared spectroscopy is a new way to determine order degree of feldspar, sensitive and effective, especially reliable for the determination of plagioclase number. On the following basis, the changes of Si/Al order degree will cause the changes of structure state and infrared spectra has obvious response to the symmetry degree of mineral structure, so we can use it to determine order degree of feldspar.

1.4.2.3. Optical Method

Optical method using the conoscopic method to determine the alkali feldspar's 2V and then determine the order degree, the accuracy of 2V is ± 0.5°. Optical methods has been used less because of its high requirements of sample and complex experimental conditions, here is the formula:

$$St = 0.025\ (2V - 44^{\circ})$$

1.4.3. Determine Order Degree of Alkali Feldspar with X-Ray Powder Diffractometry

1.4.3.1. Preparation of the Experimental Samples

For this experiment, we explored the mountain Xiaozhu of Huangdao area and collected 8 samples, ultimately six of them were picked out as good representations. Regional geological map of Huangdao area are shown in Figure 1.4.1, the photo of each sample are shown in figure 1.4.2.

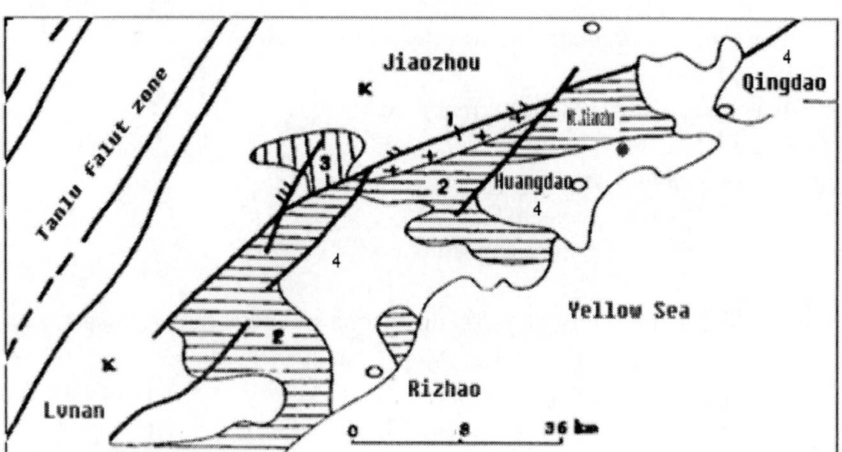

Figure 1.4.1. Geologic map of Huangdao and surrounding areas. (region 1～3 are metamorphic rocks, region 4 is igneous rocks).

Figure 1.4.2. Field photos of six experimental samples (Y1: Quartz monzonite; Y2:.Flesh pink granite; Y3: Offwhite granite; Y4: Granodiorite; Y5: Quartz syenite; Y6: Quartz syenite).

Treating the samples before the experiment, that is grinding potassium feldspar powder. The procedure as follows: crushing the whole-rock samples to get mixed particles in 0.2mm ~ 0.4mm size, removing iron by magnetic separation with electromagnetic separator, separating quartz grains from the feldspars by grains float method, then separating plagioclase from the potassium feldspar by heavy-liquid separation, and getting the pure potassium feldspar particles. Finally grinding these particles into 200 mesh in the agate mortar and getting the potassium feldspar powder.

1.4.3.2. Test and Analysis of the Samples

The experiment is measured in X-ray diffraction laboratory of State Key Laboratory of Heavy Oil in China University of Petroleum (East China), test by Wang Huaiping.

The X-ray diffraction chamber was produced by the PANalytical B.V., instrument model is X'Pert PRO MPD, and the experimental conditions were as follows:

Anode target type: CuKα, operating wavelength: 1.54060 (Å);
Scanning speed: 0.30 ° / s;
Acceleration voltage: 40KV, tube current 80mA;
Scan range 2θ = 5 ° ~ 60 °;
Scan mode: continuous scanning;
Step length: 0.017 °, time per step: 8.3978s;
Emission slit opening is 1 °, diffusion slit is 2 °.

We got six figure of X-ray powder diffraction of potassium feldspar samples, six corresponding peak list and a phase analysis table from this experiment.

The X-ray powder diffraction results of the sample Y1, Y2, Y3, Y4, Y5 and Y6 shows in figure 1.4.3, figure 1.4.4, figure 1.4.5, figure 1.4.6, figure 1.4.7 and figure 1.4.8. Phase identification results are shown in Table 4.1. Standard card information of each phase is shown in Table 1.4.1.

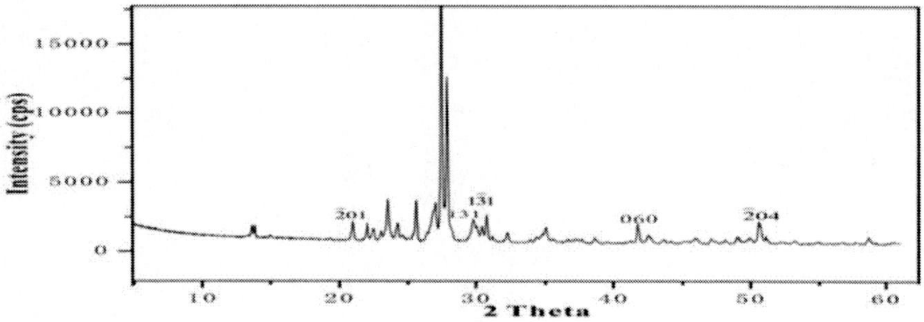

Figur 1.4.3. X-ray powder diffraction of sample Y1.

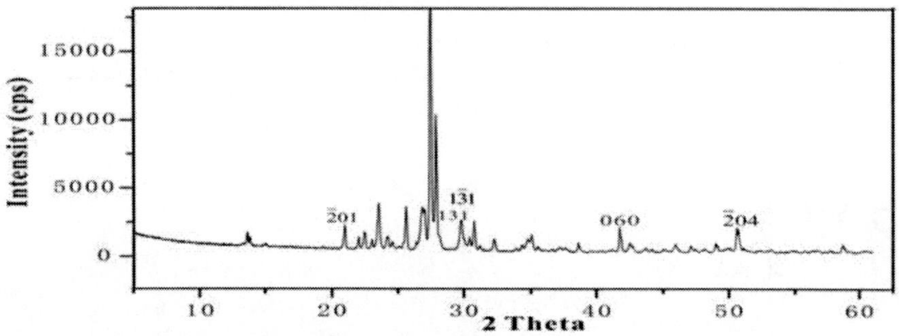

Figure 1.4.4. X-ray powder diffraction of sample Y2.

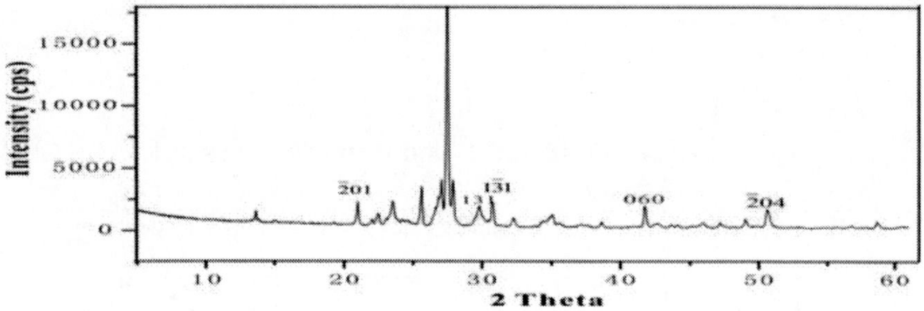

Figure 1.4.5. X-ray powder diffraction of sample Y3.

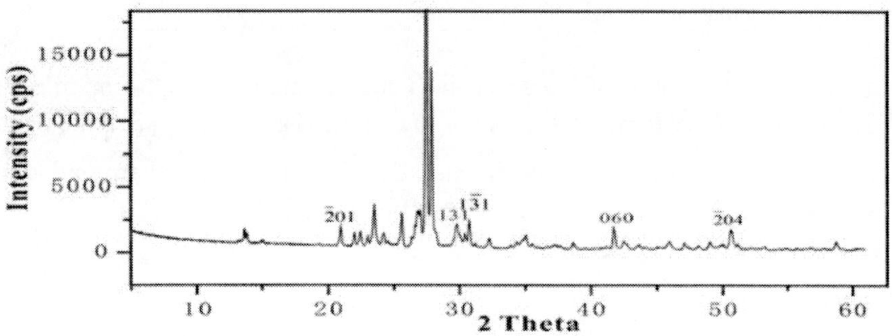

Figure 1.4.6. X-ray powder diffraction of sample Y4.

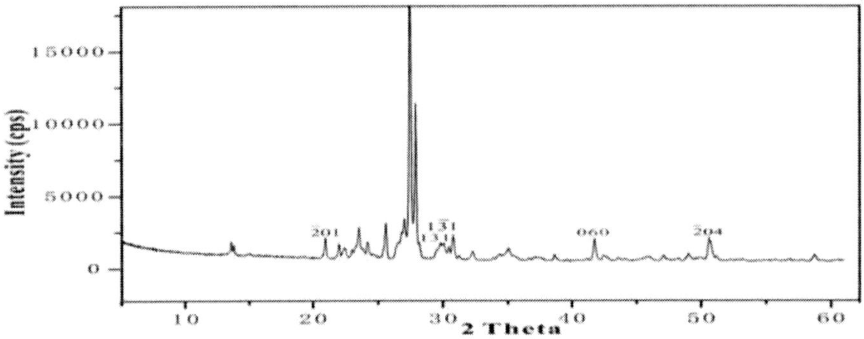

Figure 1.4.7. X-ray powder diffraction of sample Y5.

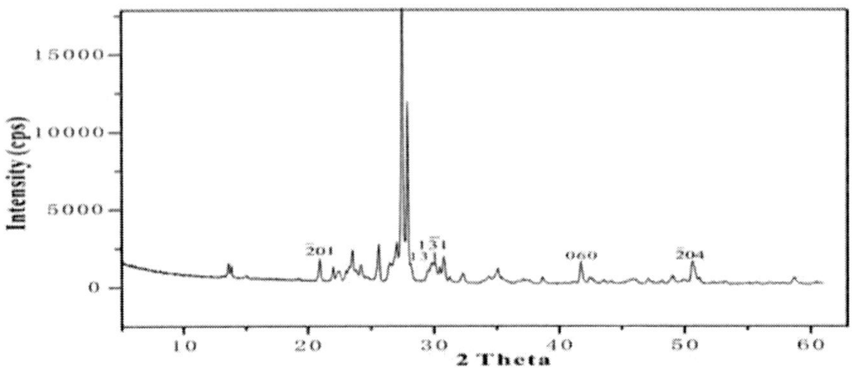

Figure 1.4.8. X-ray powder diffraction of sample Y6.

Table 1.4.1. Phase identification results of six samples

Sample code	main phase	secondary phase	PDF card of main phase
Y1	middle microcline	albite	00-019-0932
Y2	middle microcline	order albite	00-009-0466
Y3	orthoclase	low temperature albite	01-086-0438
Y4	middle microcline	albite	00-019-0932
Y5	middle microcline	order albite	00-019-0932
Y6	middle microcline	order albite	00-019-0932

We can see from the table1.4.1 that all the samples' main phase are middle microcline besides Y3, also contain seldom order albite or low albite. From the figure 1.4.3~figure 1.4.8 we can see that, $\bar{2}01$ is the characteristic peak of albite, it is very sharp and has a larger intensity, which is consistent with the phase identification results.

Table 1.4.2. Standard card information of six samples' phase

Sample code	PDF card number	d(A)	h	k	l	2θ(°)
Y1	00-019-0932	4.22794	$\bar{2}$	0	1	21.0126
		2.99449	1	3	1	29.8378
		2.96491	1	$\bar{3}$	1	30.1425
		2.16454	0	6	0	41.6938
		1.80135	$\bar{2}$	0	4	50.6336
Y2	00-009-0466	4.23477	$\bar{2}$	0	1	20.9783
		2.99302	1	3	1	29.8529
		2.96921	1	$\bar{3}$	1	30.0979
		2.16368	0	6	0	41.7111
		1.80280	$\bar{2}$	0	4	50.5899
Y3	01-086-0438	4.23338	$\bar{2}$	0	1	20.9853
		2.99686	1	3	1	29.8137
		2.96812	1	$\bar{3}$	1	30.0111
		2.16294	0	6	0	41.8346
		1.80276	$\bar{2}$	0	4	50.5914
Y4	00-019-0932	4.23720	$\bar{2}$	0	1	20.9661
		2.99690	1	3	1	29.8133
		2.97206	1	$\bar{3}$	1	30.0684
		2.16492	0	6	0	41.6861
		1.80335	$\bar{2}$	0	4	50.5735
Y5	00-019-0932	4.23485	$\bar{2}$	0	1	20.9779
		2.99748	1	3	1	29.8074
		2.96873	1	$\bar{3}$	1	30.1028
		2.16316	0	6	0	41.7578
		1.80475	$\bar{2}$	0	4	50.5764
Y6	00-019-0932	4.23375	$\bar{2}$	0	1	20.9834
		2.99625	1	3	1	29.8199
		2.96902	1	$\bar{3}$	1	30.0998
		2.16377	0	6	0	41.7094
		1.80359	$\bar{2}$	0	4	50.5664

Table 1.4.3. The parameters derived from X-ray powder diffraction method

Sample code	Sm	triclinicity	structural parameters	potassium feldspar number
Y1	0.36	0.3698	0.9826	94.6184
Y2	0.54	0.2976	1.4233	97.7805
Y3	0.48	0.3790	1.2371	97.1352
Y4	0.54	0.3105	1.3783	98.9052
Y5	0.71	0.3594	1.8357	97.8174
Y6	0.60	0.3403	1.5868	97.3104

To calculate the order degree of alkali feldspar with the above formula, we still need these five parameters $2\theta_{060}$、 $2\theta_{\bar{2}04}$、 d（131）、d（$1\bar{3}1$）、$2\theta_{\bar{2}01}$ which can obtain from the table1.4.2. So we can calculate the six samples' order degree Sm, triclinicityΔ, structural parameters η, potassium feldspar number Or as follows [15, 19-21]:

Order degree Sm=$(9.063+2\theta_{060}-2\theta_{\bar{2}04})/0.340$

Triclinicity$\Delta=12.5\times[d(131)-d(1\bar{3}1)]$

Structural parameters $\eta=6.68\times2\theta_{060}-7.44\times2\theta_{\bar{2}04}+99.182$

Potassium feldspar number Or$=2031.77-92.19\times2\theta_{\bar{2}01}$

Arrange these parameters in table 1.4.3.

1.4.4. Laser Raman Spectra Study of Order Degree of Alkali Feldspar

The three vibration modes of feldspar crystals have obviously corresponding Raman spectral features, combined the previous study of the Raman spectra of silicon-oxygen tetrahedron structural unit, we compared carefully the Raman spectra variation with a larger order degree, and then anti-showed that determine of the order degree qualitatively with Raman spectral features. Besides the order degree is a reflection of crystallize environment and equilibrium temperature and other conditions of the mineral, so it can be deduced the equilibrium temperature of feldspar through the study of the order degree.

1.4.4.1. Experimental Conditions

The experiment was completed in Key Laboratory of Tectonics and Lithosphere Dynamics of China University of Petroleum (East China). The instrument—LabRam-010 Raman spectrometer (figure 1.4.9)— was produced in France J.Y company, with characteristics of fast test speed and high sensitivity. The experimental conditions were as follows:

Laser wavelength: 514.53nm;
Laser average power of about: 100mW;
Pulse frequency: 16.6Hz;
Confocal aperture: 1000μm;
Silt width: 400μm;
Number of scans: 1;
Integration time: 10s;
Range of scan wave number: $150cm^{-1} \sim 1200cm^{-1}$.

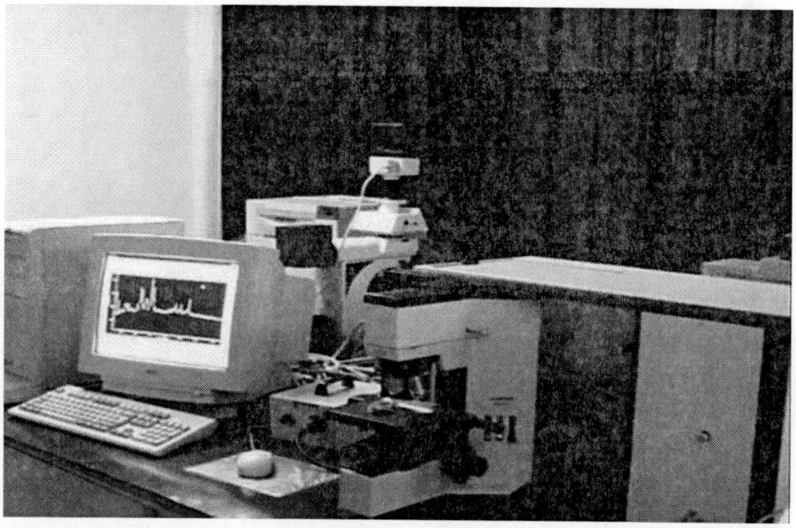

Figure 1.4.9. The used Raman spectrometer of the experiment.

The samples for the experiment without special preparation, and the alkali feldspar samples were potassium feldspar powder. Take appropriate samples to be tested, and place directly under the Raman spectrometer to determine. Because of the anisotropic characteristics of the heterogeneity of mineral, the Raman spectrum may be different in various directions, so we need choose the better orientation to make the Raman characteristic peaks of silicon-oxygen tetrahedron obviously.

But note that, no matter what changes the orientation is, as long as the silicon-oxygen tetrahedron characteristic peaks clearly manifested, then the other Raman peaks' relative position will not change.

1.4.4.2. The Analysis of the Laser Raman Spectra of the Sample

The six Raman spectrum pictures of the six samples obtained from this experiment, shown in figure 1.4.10 after CorelDrawX3 software processing, a list of the main spectra peaks' Raman shift (see table 1.4.4).

Combined this experiment, we analyzed the spectra characteristic of the six feldspar samples shown in figure 1.4.10, found that their number and the sharpness of the peaks differ, and the Raman spectra exist some regular changes with the change of order degree.

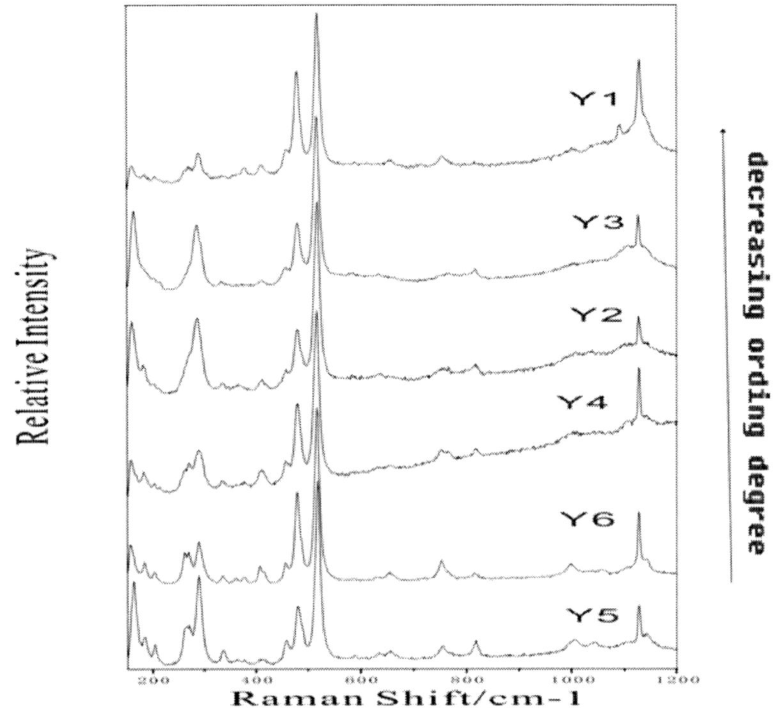

Figure 1.4.10. Raman spectra of six samples.

Table 1.4.4. Characteristic peaks of Raman spectra of six samples

Sample code	Raman shift of characteristic peaks /cm^{-1}		
	<400 Low frequency region	400~600 Middle frequency region	600~1200 High frequency region
Y1	174、281、298、384	416、455、475、513	761、814、1122
Y2	158、180、285、334、365	409、459、480、516	757、814、1128
Y3	189、211、256、293、333	415、460、482、513	766、819、1102
Y4	158、181、201、270、287、333	408、458、480、513	753、819、1128
Y5	161、182、202、267、288、335	410、459、480、516	654、753、815、1006、1128
Y6	156、181、201、265、288、334、374	408、447、479、515	652、751、814、998、1128

The order degree of Y5 is highest and it also has the largest number of Raman spectra peaks, the low-frequency region has five peaks and the FWHM is small, the three group high peaks may comes from the spectral splitting on 516cm^{-1} and 480cm^{-1} [22-26]. With the smaller of order degree, the number of Raman spectra peaks gradually reduced and the FWHM become larger. The order degree of Y1 is lowest and its peaks' number is smallest, the FWHM is significantly greater than the other samples.

The conclusions are as below:

(1) Lattice vibration of minerals and vibrations between K^+ and oxygen contribute to the spectral peaks in low frequency area below 400cm^{-1}.There is an increase in the number of peaks with the raising ordering degrees, and the shape of peaks become sharper. So ordering degrees could be roughly judged by the number and sharpness of spectral peaks in low frequency area.

(2) The bending vibration of Si/Al-O-Si bridge contributes to the spectral peak in intermediate frequency area at 400cm^{-1}- 600cm^{-1}.The reflection peaks at 510cm^{-1} and 480cm^{-1} gradually divide into three peaks at 450cm^{-1}, 480cm^{-1} and 480cm^{-1} with the increase of ordering degrees and the shoulder at 450 cm^{-1} is more obvious and sharp. So ordering degrees could be qualitatively judged by the sharpness of spectral peaks.
(3) The vibration of oxygen bridge bond doesn't contribute to the peaks in high frequency area above 600cm^{-1}. Generally, the peaks in that area are faint. With the ordering degrees decrease, peaks at 650cm-1, 750cm^{-1} and 815cm^{-1} become smaller and smaller, and the first peak disappears at the end. There is a obvious peak at 1000cm^{-1} in samples A5 and A6 with highest ordering degrees. But at 1100cm^{-1}, an envelope occurs and finally turns into an obvious peak.
(4) On the whole, with ordering degree increasing, the full width at half maximum(FWHM) of Raman peak becomes smaller, which could make qualitative judgments on ordering degrees.

1.4.5. Recover Equilibrium Temperature of Feldspar Crystals

Temperature is the main factor of feldspar order degree, so order degree could reflect temperature change during the crystallizing of feldspars. One purpose of this experiment is to analyze the formations of various feldspars with different order degrees and restore their equilibrium temperature by studying alkali feldspar's Raman spectra [27-28].

We can study from the previous research that the temperature is a important factor affecting the order degree of feldspar.

Generally, the order degree of high temperature feldspar is lower than that of low temperature feldspar, because that the Al and Si inside the feldspar crystals strenuous exercise when the outside temperature is high, so if the temperature suddenly reduced, some crystal feature under high temperature too late to change it to be preserved which resulted in low order degree. On the contrary, if the outside temperature is low or the temperature slowly decreased even if it is high, these crystals' order degree is high because of the fully exchange between Al and Si.

Ma Hongwen (1988) determined the order degree of Monoclinic feldspar derived from the granitic rocks of Yulong porphyry copper belt with X-ray and infrared spectroscopy, and thought that occupancy of Al in alkali feldspar is a function of the equilibrium temperature, the regression equation between order degree and equilibrium temperature [28]:

$$T\ (°C) = 856.2594 - 395.1256Sm$$

This formula can be used to roughly estimate the equilibrium temperature of feldspar, so we can calculate the equilibrium temperature of the Y1～Y6 according to the relationship, the calculate data are shown in table1.4.5. Visible, the feldspar samples' equilibrium temperature of this region between 575.7 °C～714.0°C.

Besides, we can draw the equilibrium temperature by structural parameter ηbecause of the good positive correlation between order degree and structural parameter.

Table 1.4.5. Equilibrium temperature of samples Y1～Y6

sample code	Sm	triclinicity	structural parameters	equilibrium temperature
Y1	0.36	0.3698	0.9826	714.0142
Y2	0.54	0.2976	1.4233	642.8916
Y3	0.48	0.3790	1.2371	666.5991
Y4	0.54	0.3105	1.3783	642.8916
Y5	0.71	0.3594	1.8357	575.7202
Y6	0.60	0.3403	1.5868	619.1840

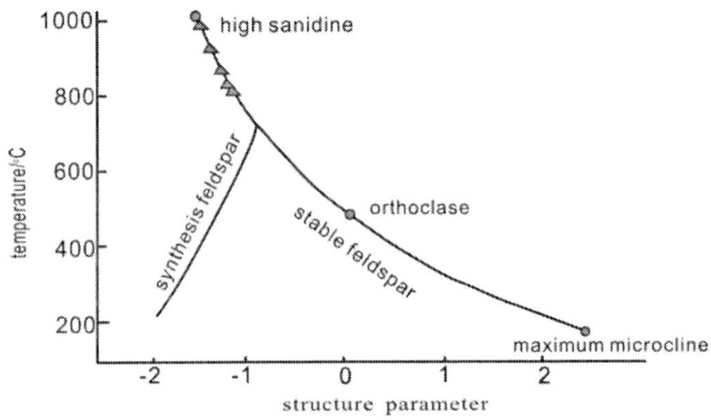

Figure 1.4.11. Relationship between equilibrium temperature of feldspar crystal and structural parameterη.

Structural parameter (η) $=6.68\times2\theta_{060}-7.44\times2\theta\,\bar{2}_{04}+99.182$, its relationship with equilibrium temperature are shown in figure 1.4.11. We put the calculated structural parameter $\eta$of the six samples into the figure 1.4.11, and came to approximate value with the above equilibrium temperature.

Although the equilibrium temperature could be identified by the ordering degree and structural parameters of feldspars, it is not the same with crystallization temperature. The reason is that the structural state of feldspar decreases and then become stable as the reduction of temperature. So the deduced temperature is the stabilized one. Besides, according to the latest achievements, temperature is not the key factor of ordering degrees. For feldspar, its composition and duration of temperature change is more critical than crystallization temperature.

1.5. Laser Raman Spectral Analysis of Plagioclase's Zoning Structure and Recover Formation Environment

Zoning structures means the ring extinction in orthogonal polarizing microscope because the changing light body of chemical composition of plagioclase.

For igneous rocks, zoning structure is widespread in plagioclase crystals, especially in neutral ones. Plagioclase's zoning draws the attention of researchers because it reflects environmental change during magma crystallization and it can be used to deduce that process. And it is also relevant with the rate and condition of crystallization.

1.5.1. Classification of Plagioclase's Zoning Structure

According to the composition, the zonings of plagioclase are often classified into three types: normal zoning, reverse zoning and oscillatory zoning. From inner parts to outer parts, normal zoning turns form basic to acid, reverse zoning turns oppositely.

Extinctions are existed under orthogonal polarizing microscope. Discontinuous zoning presents Discontinuous extinctions, and vise versa. Oscillatory zonings are composed of basic and acid zonings alternatively, like sedimentary rhythm in the strata [29].

Homma(1932) detailed classified plagioclase's zonings according to the changes of An in them(figure 1.5.1) [30]. In figure 1.5.1, I-N (Normal) represents non-rhythmic normal zonings, I-E(Even) represents non-rhythmic even zonings, I-R(Reverse) represents non-rhythmic reverse zonings; Ⅱ-N represents wavy rhythmic normal zonings Ⅱ-E represents wavy rhythmic even zonings, Ⅱ-R represents wavy rhythmic reverse zonings; Ⅲ-N represents positive rhythmic normal zonings, Ⅲ-E represents positive rhythmic even zonings, Ⅲ-R represents positive rhythmic reverse zonings;

Ⅳ-N represents negative rhythmic normal zonings, Ⅳ-E represents negative rhythmic even zonings, Ⅳ-R represents negative rhythmic reverse zonings;

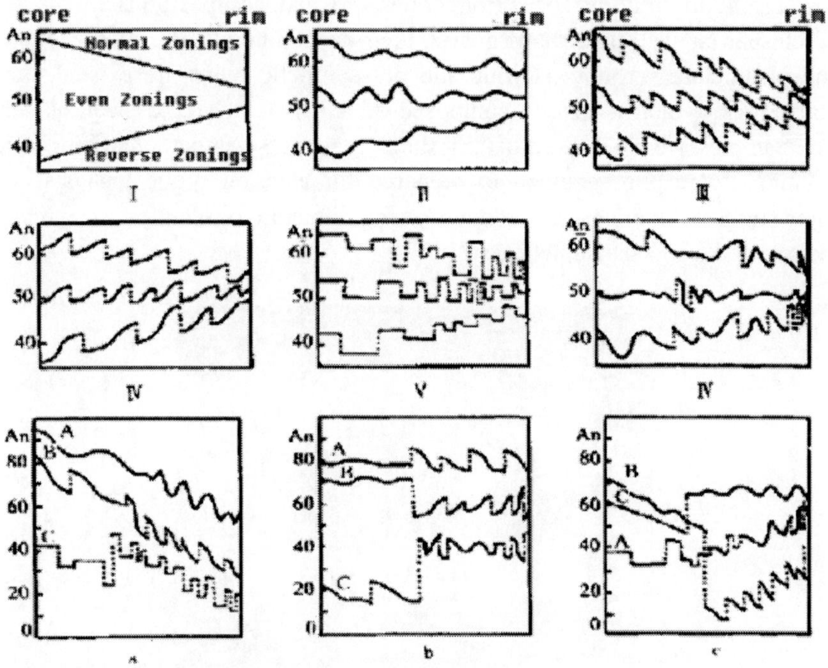

Figure 1.5.1. Classification of plagioclase's zonings (Homma, 1932) [30].

V-N represents flat rhythmic normal zonings, V-E represents flat rhythmic even zonings, V-R represents flat rhythmic reverse zonings; VI-N represents irregular rhythmic normal zonings, VI-E represents irregular rhythmic even zonings, VI-R represents irregular rhythmic reverse zonings; a represents composite structure zonings, b represents hybrid structure zonings, c represents complex structure zonings.

1.5.2. Causes of Plagioclase's Zoning Structure

Normal zoning is basic in inner part and is acid in outer part. The reason is that during the dramatic temperature decrease of magma crystallization, early formed basic plagioclase crystals cannot wait to react with magmas and some acid plagioclases around basic ones because of their lower fusion point which is only 1118°C. The pressure change is the main factor forming reverse zoning. The crystallized plagioclase changes from acid to basic as water pressure increases. Supersaturation of magma diffusion and temperature change accounts for normal-scale oscillatory zoning, while "supersaturation-nucleation-depletion" mechanism accounts for small-scale one.

1.5.3. Electron Probe Analyze on the Composition of Plagioclase's Zoning Structure

1.5.3.1. Test Sample Preparation

All the plagioclase samples were acquired from Mt. Xiaozhu, Huangdao district. 6 samples were collected to do the test. Sample H1, H2, H3, H4 and H5 are granodiorite, they were acquired from different distances from crater, the main compositions are plagioclase and potassium feldspar, as well as 25% of quartz, their dark minerals are mainly amphibole and biotite. The plagioclase's zoning is wide and polysynthetic twins are existed. Sample 6 is normal quartz rock, which is mainly composed of Alkali feldspars, as well as some acidic plagioclase. Samples were made into thin slices to obverse under polarizing microscope. Obvious zonings were photographed to prepared for electron probe analysis and Raman spectroscopy study on zonings. 37 zonings were found in 6 samples, one representative zoning was selected on each sample(figure 1.5.2).

H1 (-)

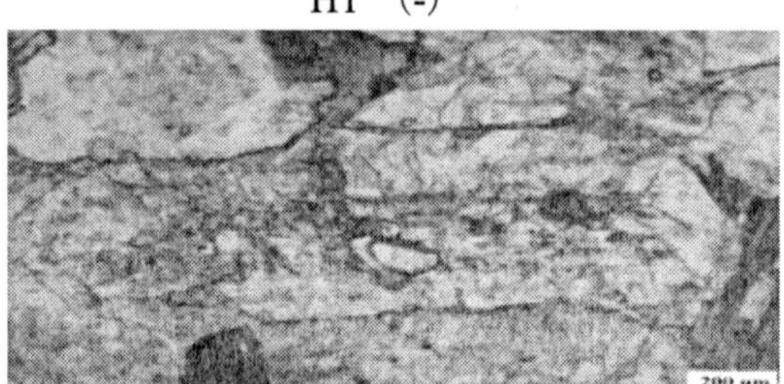

H2 (-)

Figure 1.5.2. (Continued).

H3 (-)

H4 (+)

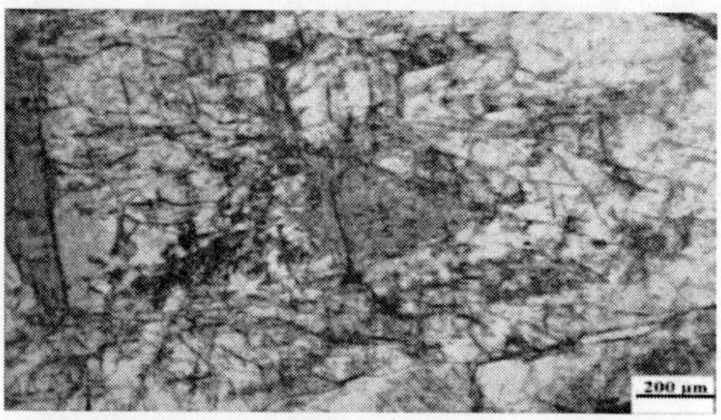

H5 (-)

H6 (+)

Figure 1.5.2. Features of feldspar's zonings under microscope.

1.5.3.2. Electron Probe Analysis on Sample Composition

This test was done by the electron probe instrument in the electron probe Lab, Materials Science and Engineering College, Shandong University of Science and Technology. The instrument mode is JXA-8230. Energy Dispersive Spectroscopy's mode is 350. Experimental conditions are listed as below.

Accelerating voltage: 150KV;
Probe current: 1×10^{-8}A;
Beam diameter: 5μm;

Mineral standards: Mineral standards of American National Standards Institute Albite and amphibole standard were used to test SiO_2 and other elements respectively.

Element Ca's relative content in different points on the scan line was tested. Distribution curve of Ca's content is similar to that of An because The contents of Ca and An are positively proportioned.

In order to easily calculate the An's content of each point on the zonings, two test points were selected to the margin or core of the zonings.

Results of the test are 6 An distribution curves (from Figure 1.5.3 to Figure 1.5.8) in line scan sections corresponding to 6 samples and 1 Chemical composition analysis table of rock (Table 1.5.1)

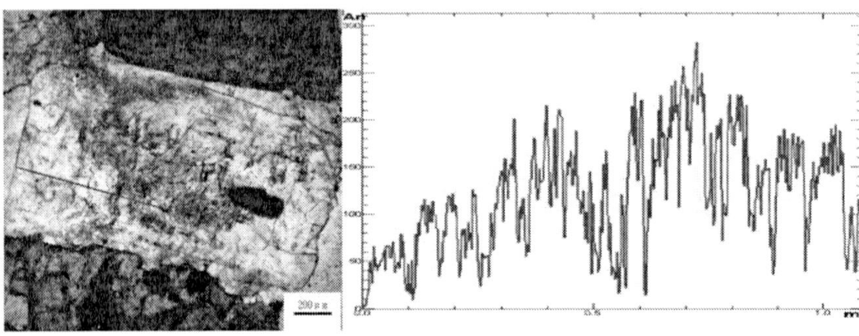

Figure 1.5.3. An's distribution curve in plagioclase zoning's line scan section of sample H1.

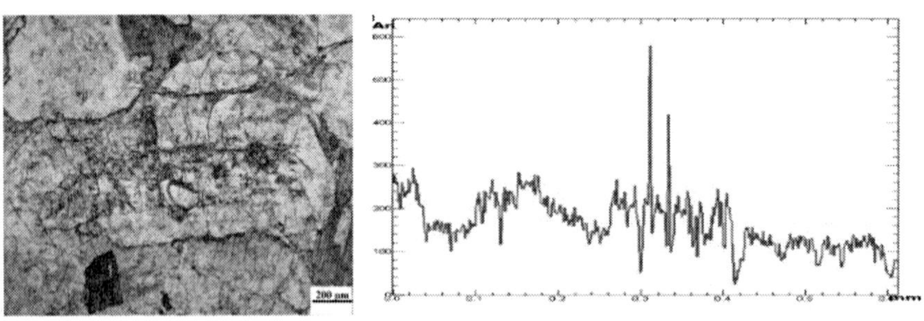

Figure 1.5.4. An's distribution curve in plagioclase zoning's line scan section of sample H2.

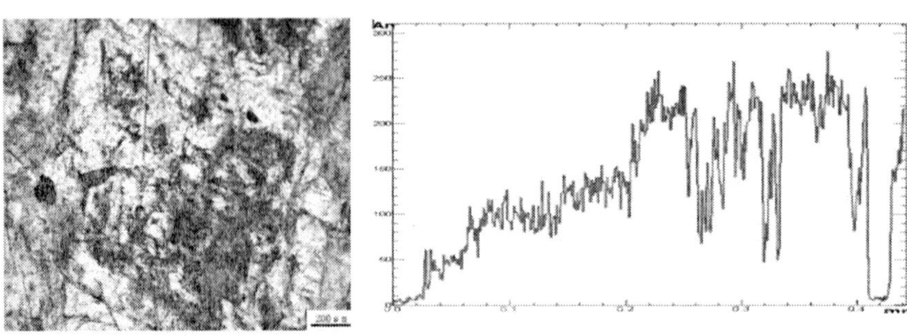

Figure 1.5.5. An's distribution curve in plagioclase zoning's line scan section of sample H3.

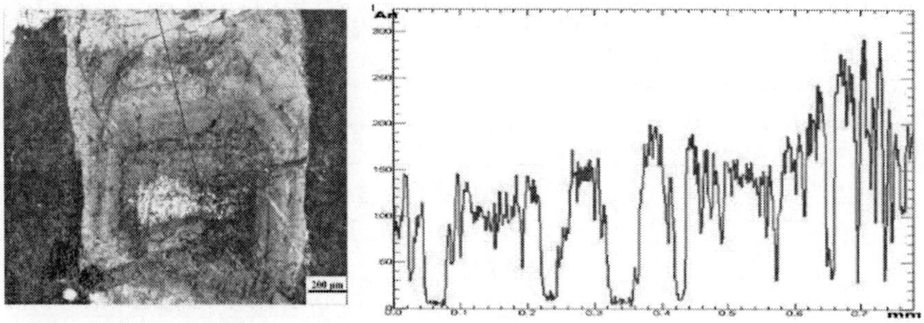

Figure 1.5.6. An's distribution curve in plagioclase zoning's line scan section of sample H4.

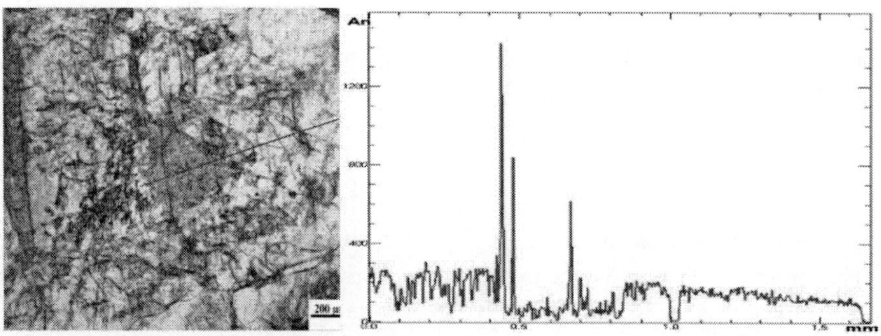

Figure 1.5.7. An's distribution curve in plagioclase zoning's line scan section of sample H5.

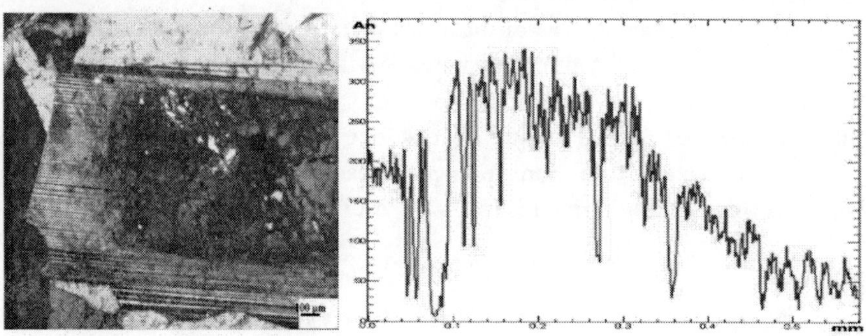

Figure 1.5.8. An's distribution curve in plagioclase zoning's line scan section of sample H6.

Table 1-5-1. Data of electron probe analysis on the 6 samples

sample	H1		H2		H3		H4		H5		H6	
composition	A	B	A	B	A	B	A	B	A	B	A	B
Na_2O	3.55	5.24	6.63	6.94	6.37	7.00	8.10	7.76	7.17	7.32	7.85	7.99
SiO_2	67.31	66.35	56.83	57.94	55.53	57.44	60.93	60.47	58.46	59.34	60.47	60.77
Al_2O_3	16.39	20.40	25.83	25.71	26.08	24.92	23.83	24.30	25.31	25.23	23.99	24.47

Table 1-5-1. (Continued)

sample	H1		H2		H3		H4		H5		H6	
composition	A	B	A	B	A	B	A	B	A	B	A	B
CaO	2.97	5.90	8.71	8.09	9.19	7.45	5.78	6.60	7.67	7.23	6.47	6.20
K_2O	6.81	0.19	0.26	0.26	0.24	0.29	0.38	0.34	0.28	0.32	0.53	0.53
FeO	0.07	0.05	0.13	0.14	0.14	0.12	0.27	0.15	0.15	0.18	0.25	0.24
total	97.12	98.15	98.40	99.10	97.57	97.23	99.31	99.64	99.07	99.65	99.58	99.23

The conclusions are as below:

(1) Three types of zonings are found: normal zoning, reverse zoning and small-scale oscillatory zoning superimposed on partial large-scale zoning.
(2) Normal zoning is basic in inner part and is acid in outer part. The reason is that during the dramatic temperature decrease of magma, early formed basic plagioclase crystals can not wait to react with magmas and some acid plagioclases around basic ones because of their lower fusion point which is only 1118°C.
(3) The pressure change is the main factor forming reverse zoning. The crystallized plagioclase changes from acid to basic as water pressure increases.
(4) Supersaturation of magma diffusion and temperature change accounts for normal-scale oscillatory zoning, while "supersaturation-nucleation-depletion" mechanism accounts for small-scale one.
(5) Normal and reverse zonings all have features of small composition changes, fewer numbers and narrow width, which reflects that the crystallization environment is relatively stable.
(6) In comparison of normal zoning, reverse one is of larger numbers, width as well as rate of An changes from core to edge. These features reflect that the environment where reverse zoning formed is relatively oscillatory.

1.5.4. Laser Raman Spectra Micro-Zone Analysis of Plagioclase's Zoning Structure

Laser Raman Spectroscopy tends to be another efficient way of testing plagioclase's zoning structure because it is sensitive to the change in the structure. So we analyzed the Raman spectra of sample zonings, tried to find the rules of spectra changes during zoning changes and then classified different types of zonings according to Raman spectra.

The test was done in the Laboratory of Geochemistry and Lithosphere Dynamics, China University of Petroleum (East China). The instrument was Raman spectrometer LamRam-010 produced by France company J.Y. The instrument uses confocal microscopy system, features high sensitivity and test speed.

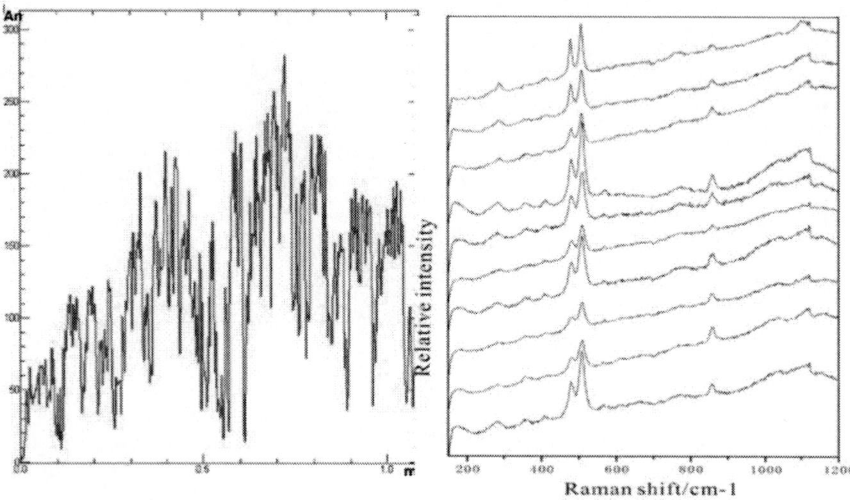

Figure 1.5.9. Comparison of laser Raman spectra and An variation in sample H1.

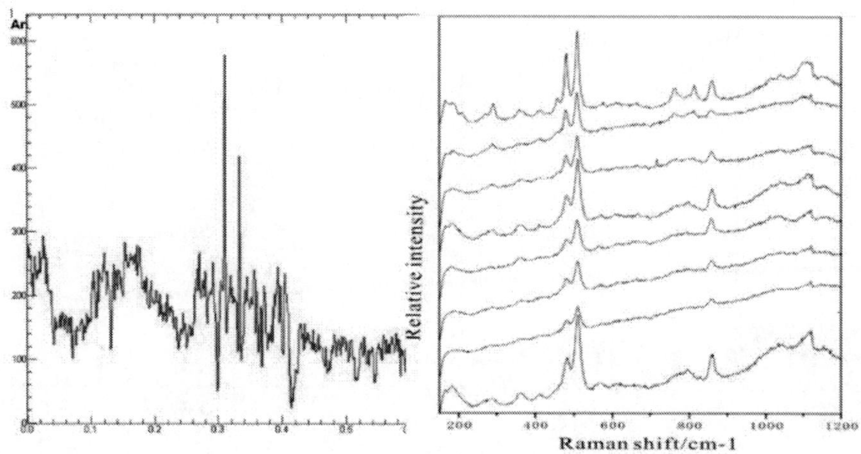

Figure 1.5.10. Comparison of laser Raman spectra and An variation. in sample H2.

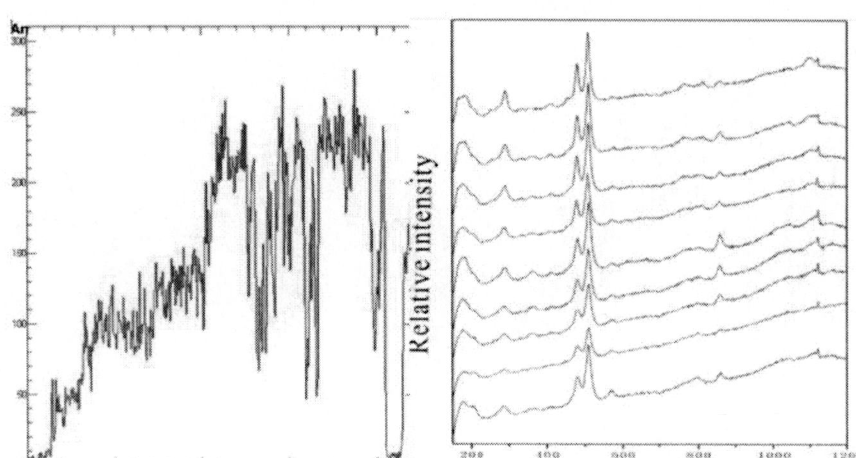

Figure 1.5.11. Comparison of laser Raman spectra and An variation in sample H3.

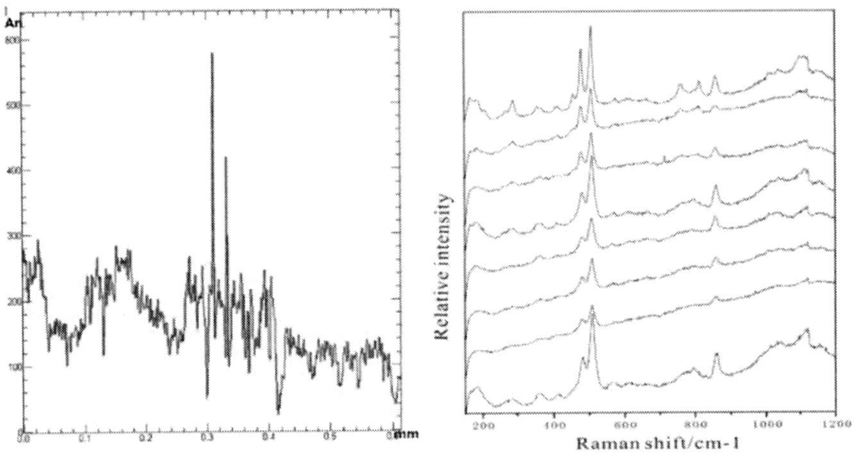

Figure 1.5.12. Comparison of laser Raman spectra and An variation in sample H4.

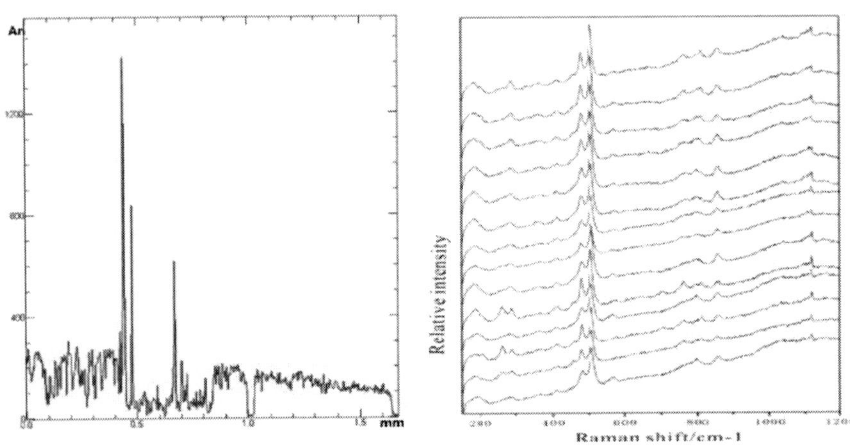

Figure 1.5.13. Comparison of laser Raman spectra and An variation in sample H5.

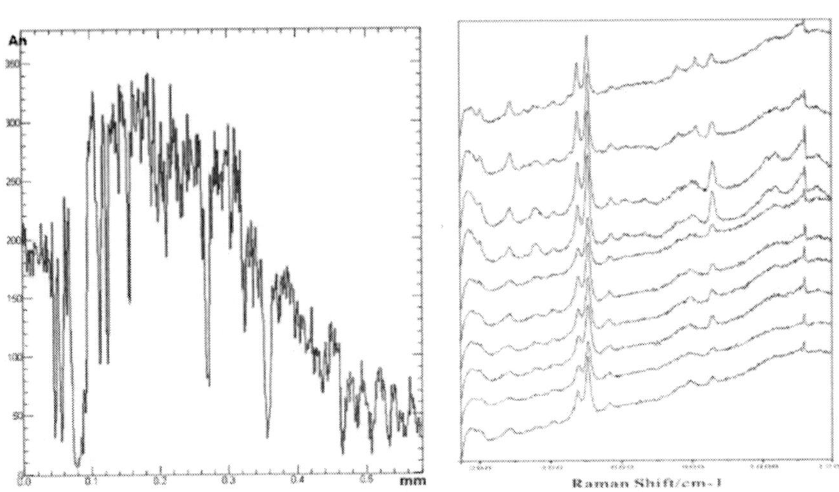

Figure 1.5.14. Comparison of laser Raman spectra and An variation in sample H5.

The samples are also H1~H6. Plagioclase zonings in the samples were found and marked under microscopes, and furthermore were photographed. After the detailed study of zonings, points were made from inner parts to outer parts under laser Raman spectrometer. To make them comparative, the line of making points is the same with electron probe scan line

Data of the test were 6 Raman spectra (from Figure 1.5.9 to Figure 1.5.14) corresponding to 6 samples respectively. All of the spectra were processed by CorelDrawX3. From figure 1.5.9~figure 1.5.14, feldspar spectra could be divided to three areas: low frequency area below 400cm-1, intermediate frequency area at $400cm^{-1}$-$600cm$-^{1}, high frequency area above $600cm^{-1}$. Spectral peaks in low frequency area results from vibrations between positive ion and oxygen as well as lattice vibration of mineral frameworks. The peak is significantly affected crystal orientation. Peak in intermediate frequency area is the most obvious one, which results from bridging oxygen bending vibrations of Si/Al-O-Si. Peak in low frequency area is normally faint, which results from non-bridging oxygen symmetric stretching vibrations of Si/Al-O.

The conclusions are as below:

(1) Low frequency area below $400cm^{-1}$: the change of spectral peaks in that area cannot distinguish normal zoning and reverse zoning because both zonings becomes more obvious and sharper from core to edge.
(2) Intermediate frequency area at $400cm^{-1}$-$600cm^{-1}$: Different zonings cannot be distinguished at $478cm^{-1}$-$488cm^{-1}$ with the same reason above. But at $500cm^{-1}$-$515cm^{-1}$, featured peaks of normal structure gradually decline, while reverse one are opposite. Featured peaks at $500cm^{-1}$ -$515cm^{-1}$ show fluctuation of high value→ low value→ high value or low value→ high value→ high value. Times of fluctuation demonstrate numbers of zonings.
(3) High frequency area above $600cm^{-1}$: From core to edge, Spectral peaks of normal zoning at $800cm^{-1}$ move to low frequency area, they are more apparent. Their shapes are sharper. While peak value of reverse zoning declines, the peaks is not obvious and turn into a big envelope moving toward high frequency area. Spectral peaks at 1100cm-1 cannot identify different zonings because the shape of both peaks become sharper from core to edge.

1.5.5. To Recover Formation Environment of Feldspar Crystallization

Tested sampled were all acquired from Mt. Xiaozhu, Huangdao district which belongs to Sulu orogen in regional structure. The orogen were formed from the collision between North China plate and Yangtze plate at Triassic to the late Mesozoic. And after that, on major tectonic movement was experienced.

Rocks in Sulu orogen are mainly metamorphic and igneous, as well as a small amount of sediment rocks. Mesozoic igneous rocks which are mainly basic intrusive rocks and different types of granite are widely spread in that area.

After the collision, large amount of igneous rocks were formed at late Triassic (225Ma~205Ma), late Jurassic (170Ma~140Ma) and early Cretaceous (170Ma~140Ma). The subducted thickening lithosphere in South China is the the main source rocks which erupts in orogenic stage. Igneous rocks in early Cretaceous are the most developed ones and could be classified to two stages. The early stage rocks formed at 145Ma～130Ma, features gneissosity. That means tectonic deformation occurs and orogon was still active. The late stage rocks formed at 130Ma～120Ma doesn't feature tectonic deformation, which means oregon was stable.

According to the tectonic evolution in Sulu orogen and previous analysis on the causes of zonings, as well as the samples feature normal zonings and reverse zonings, the crystallization of feldspar are inferred as below:

After collisions of Sulu orogen was completed, magma in that area is highly active in early Cretaceous. The reason is that the thickened lithosphere in the orogen were structurally collapsed and partial melted, materials in the upper mantle went into asthenosphere, and then asthenosphere magma intruded to lithosphere. In that period, many minerals like apatite and pyroxene are preferentially crystallized due to the stable conditions of original magma, without any intensive physical or chemical changes. In plutonic crystallization conditions, previously crystallized basic plagioclase forms uniform crystals which are almost identical to the original magma instead of zonings because there is sufficient time for them to react with surrounding magma.

With the dramatic temperature decrease during magma moves upward , early formed basic plagioclase crystals cannot wait to react with the surrounding magmas and some acid plagioclases crystallize around basic ones because of their lower fusion point which is only 1118°C, and then normal zoning which is basic in inner part and acid in outer part comes into being. Normal zonings have features of small composition changes, fewer numbers and narrow width, which reflects that the crystallization environment is relatively stable and the temperature decrease of magma is not too fast. When magma continually move upward, the decreasing magma pressure results to the decrease on water in magmas. As the dissolved water could decrease magma crystallization temperature, when water becomes less, crystallization temperature becomes higher. Finally reverse zoning with higher crystallization temperature come into being.

In the process of the crystallization of normal and reversed zoning, the small-scale oscillatory zoning was formed, due to the many insignificant changes of temperature, pressure and other external conditions. After the plagioclase zoning crystals is formed, the original magma will continue to travel upward and erupt violently when it reaches the ground, therefore the original magma who crystallized these igneous rocks has experienced steady - turbulence - eruption process.

CONCLUSION

Conclusion are drew based on field sampling, laboratory analysis, as well as the previous achievements of Raman spectra, ordering degrees and zoning structures.

(1) Specified Raman spectrum at 500cm^{-1} -515cm^{-1} could identify feldspars. There is a triplet in alkali feldspar's spectrum while there are two strong spectra at 450cm^{-1} -600cm^{-1} in feldspar's spectrum. Besides, the position, number and relative intensity of peaks could also distinguish the two kinds of feldspars.

(2) The links between the ordering degrees and Raman spectra of feldspars are as below: a) numbers and sharpness of spectral peaks in low frequency area below $400cm^{-1}$ increase as feldspar's ordering degrees increase. b) In intermediate frequency area at $400cm^{-1}$ -$600cm^{-1}$, reflection peaks at $480cm^{-1}$ -$510cm^{-1}$ split into three peaks near $450cm^{-1}$, $480cm^{-1}$ and $510cm^{-1}$ respectively, and the peak near $480cm^{-1}$ is more obvious and sharper. c) In high frequency area above $600cm^{-1}$, with the decrease of ordering degrees, peaks near $650cm^{-1}$, $750cm^{-1}$ and $815cm^{-1}$ becomes smaller and smaller, and finally the first peak disappears. d) On the whole, Raman peaks will split and narrow their FWHM as ordering degrees increase, which could make qualitative judgment to samples' ordering degrees.

(3) According to the relationship between the equilibrium temperature and ordering degree of X-ray proposed by Ma Hongwen(1988), T(°C)=856.2594-395.1256Sm, it can be roughly estimated that the equilibrium temperature of feldspar ranges from 575.7°C to 714.0°C. However, the structural state of feldspar decreases with decreasing temperature and will not be retained until it reaches certain stability. Therefore, the temperature we estimated here is merely the temperature of steadily stated feldspar. The initial crystallization temperature of feldspar might be higher.

(4) The zoning shown in the samples of the studied area covers both normal-zoning and reversed-zoning, some contains small-scale fluctuant zoning and oscillatory zoning which piled inside the zoning. The normal zoning is small in number, with almost unchanged compositions and narrower width, reflecting that it crystalized in a relatively static environment. The difference between normal zoning and reverse zoning is that the latter has more rings, along with wider width and a greatly changed An content on the core and edge. This reflects that the reversed-zoning was formed from relatively turbulent origin magma. In short, the original magma which generated the crystal had been experienced a stable - volatile - extrude process, and environment was relatively stable.

(5) The low frequency region below $400cm^{-1}$, shoulder peak at $478cm^{-1}$ ~ $488cm^{-1}$, and spectral peak at $1100cm^{-1}$ cannot serve as the basis for judging the normal and reversed zoning, for these two have similar characteristics in those areas. It's a normal zoning that shows a gradual downward trend of its characteristic peaks at $500cm^{-1}$~$515cm^{-1}$(or vice versa). However, the characteristic peaks there does not rise or decrease continuously, they repeatedly fluctuates and forms several high-low-high-low (or vice versa) rotations. The number of characteristic peaks changes represents the number of the rings. Secondly, the spectral peak around $800cm^{-1}$ becomes more obvious and acute and has a tendency of moving towards the low frequency area, opposing with the reversed-zoning, which tend to fade as it moves towards the high frequency area and would form a big envelope at the end.

(6) Combining the normal, reverse and oscillatory zonings of the samples, previous causing analysis of the zonings, and tectonic evolution of the Sulu orogenic belt, I inferred the following crystallization process of feldspar in igneous rocks. The environment of primitive magma in early cretaceous was relatively stable at first, without intensive variation of the physical and chemical conditions, resulting in the formation of plagioclase crystals with uniform composition instead of zoning. After that, because of the sharp decline in the temperature of the upward migrated magma, the earlier formed basic plagioclase crystals had no sufficient time to blend and react

with the surrounding magma, and then some acidic plagioclase grew around the earlier formed basic plagioclase. Thus normal zoning was formed. From inside to outside, its ingredient changes from basic to acidic. From the seldom changed composition, the small ring numbers, and the narrow width of the normal-zoning, we can see that the crystal crystallized in a relatively stable static environment, whose magma temperature does not drop quickly. After that, the magma continues to move upward, causing the changes of total pressure, water and volatile matters, forming the reverse zoning in a relatively high crystallization temperature. In the process of the crystallization of normal and reversed zoning, the small-scale oscillatory zoning was formed, due to many insignificant changes of temperature, pressure and other external conditions. After the plagioclase zoning crystals is formed, the original magma will continue to travel upward and erupt violently when it reaches the ground, therefore the original magma who crystallized these igneous rocks has experienced steady - turbulence - eruption process.

SECTION 2. LASER RAMAN SPECTROSCOPY AND MICRO-STRUCTURE OF DIFFERENT GENESIS OF ZIRCON IN SULU OROGEN

1.1. INTRODUCTION

Zircon is one of the most common and widely used accessory minerals in geology, and can be found in all three types of rocks. Because of its high physical and chemical stability, zircon will hardly occur alteration, even through geological processes of weathering, transport, erosion, etc., which means that zircon can save a large amount of geological information. What's more, Zircon's crystal lattice is capable of accommodating a wide range of U, Th, Hf, Ti and REE and other trace elements, which makes zircon a preferred target of U-Pb isotopic dating.

As a new technology, Laser Raman microprobe plays an important role in the study of the micro-structure of minerals, combining with the zircon cathodoluminescence image. It provides information related to the decrystallization and internal structure of minerals, and provide reference of compositional changes, genetic characteristics of different parts of minerals, etc.

Systematic study of zircon Raman spectral features helps further understand the structural characteristics of micro-zircon as well as its distribution and migration characteristics of U, Th and other elements in the environment, which zircon grown in. Thereby, we can better interpret the tectonic thermal history and the forming age of a zircon-contained geological body.

Rubatto D, etc. (2000) [31] noted that the magmatic and metamorphic zircon have distinct Th, U elements and Th / U ratios. Vavra et al (1999) [32] found that there is a positive correlation between w (U) value and growth rate of metamorphic zircons, and a negative correlation between Th / U and the growth rate, by studying the w (Th), w (U) and Th / U characteristics of hyperplastic metamorphic zircon in Ivrea, indicating that to some extent, the growth rate affected the composition of elements in zircon. Belousova et al (2002) [33] found a generally growing trend of trace elements in magmatic zircons from ultrabasic→mafic→granite rocks, using statistical analyzsis of magmatic zircons' trace element characteristics (figure 2.1.1).

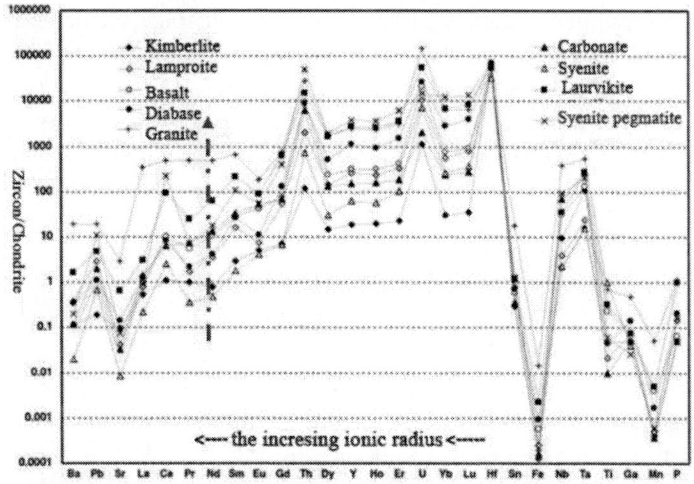

Figure 2.1.1. Mean chondrite-normalized trace element diagram of different types of magmatic zircons [33].

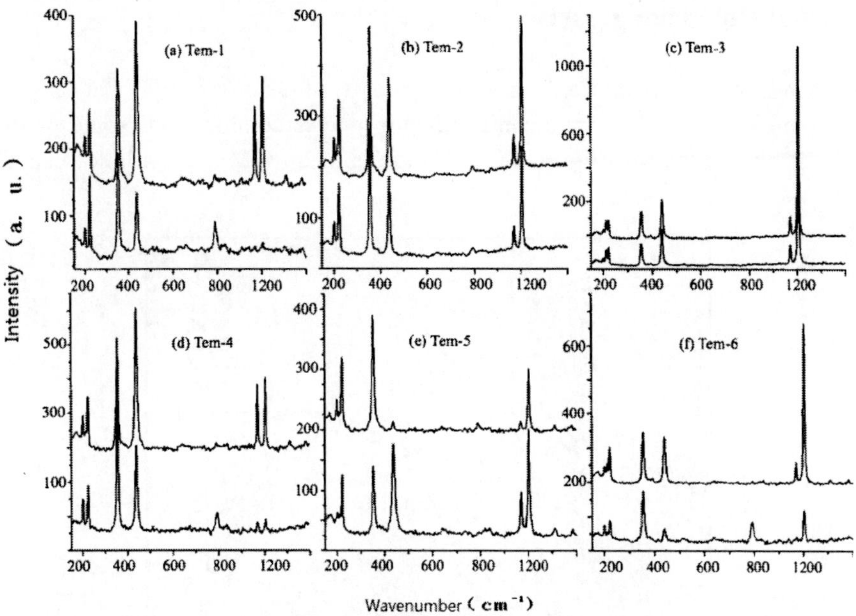

Figure 2.1.2. Raman spectra of Australian TEMORA1 standard zircon. (0., 45., 90 and 135. Incident azimuth) (Wang Yinxi et al, 2007) [34].

Wang Yinxi et al (2007) [34] analyzed the laser Raman spectra of eight geological standard zircons (TEMORA 1), which were generated in incident azimuth 0◦ , 45◦ , 90◦ and 135◦ , and found that the standard zircon TEM Raman spectra (figure 2.1.2) are in good agreement with the theoretical values. Different U, Th distribution trends in zircon will lead to value changes of the Raman peaks. This variation can therefore help determine the origin of zircon.

In this paper, we studied zircons from Shandong Rongcheng, Weihai, and Qingdao Mountain Laoshan & Ling Shan Island in Sulu orogen. Based on the previous studies of zircon, we focus on zircons' internal structure, major and trace elements distribution, and micro-Raman spectral characteristics, and summarized the variation law of element content and structure changes of zircons formed due to different causes and the law of intensity of characteristic peak and FWHM change in Raman spectra of zircon at different evolutionary stages, thereby providing a reference for the subsequent study of the thermal history of Sulu orogen.

2.2. Samples

Samples used in this paper are mainly from three places: Rongcheng Weihai, Qingdao Mountain Laoshan Yakou, and Qingdao Jiaonan Lingshan Island, which are located in Jiaonan - Weihai orogen, the east side of Sulu orogen.

2.2.1. Weihai Rongcheng gneiss

Rongcheng, weihai is located in the northeast of Jiao nan-Wei hai origen, where developed the district diorite - monzonite granitic gneiss combination (some people call it Rongcheng gneiss units [35]), figure 2.2.1. Its age is 682-787Ma, according to single zircon dating.

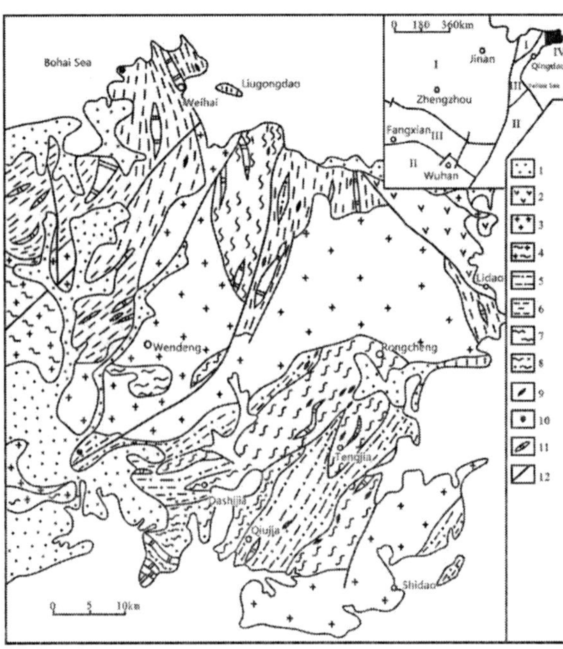

Figure 2.2.1. Rongcheng geological map (Songming Chun et al., 1995 [35]).

1 Quaternary; 2 Mesozoic volcanic rocks; 3 Yanshanian intrusive rocks; 4 INDOSINIAN gneissic granite; 5 Qiu gneiss; 6 Weihai gneiss; 7 Teng gneissic; 8 Dashijia gneiss; 9 eclogite; 10 granulites; 11 crustal xenoliths (early Proterozoic Jingshan Group); 12 break; I. North China Block; II. Yangtze block; III Qinling - Dabie - Jiaonan metamorphic complex distribution; green shaded area is the sampling location.

In this paper, samples are shown in table2.2.1, and are divided into Qiu gneiss by geography. Photographs of rock hand specimens are shown in figure 2.2.2. The microscopic characteristics of granitic gneiss and gneissic mixed granite are shown in figure 2.2.3 and figure 2.2.4.

Table 2.2.1. Sample information of Weihai Rongcheng

No.	Lithology	GPS
wh-006	Flesh red mixed granite	36°59'58"N 122°17'16"E
wh-010	Quartz vein inclusions	36°59'58"N 122°17'16"E
wh-014	Gray gneissic mixed granite	36°59'58"N 122°17'16"E
wh-015	Gray granite	36°59'58"N 122°17'16"E

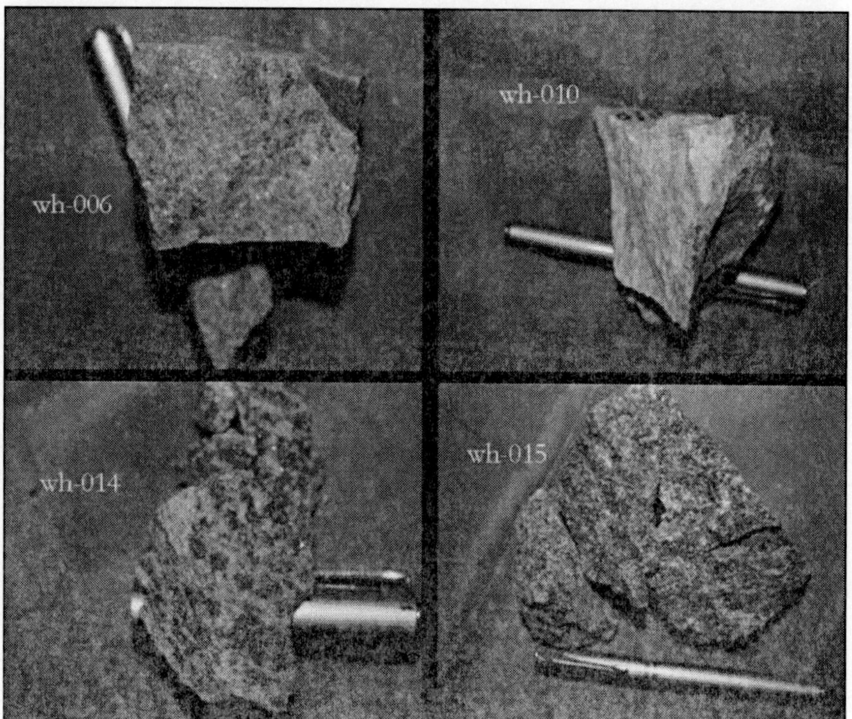

Figure 2.2.2. photographs of Weihai Rongcheng hand specimens.

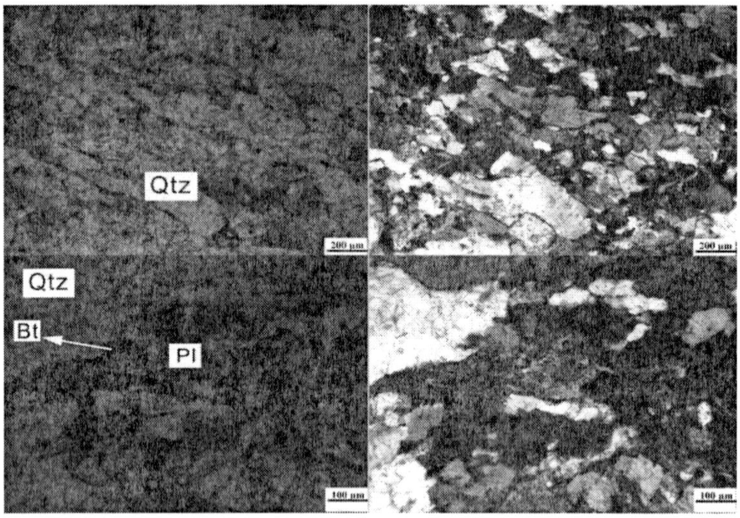

upper left-50 times polarized light, upper right-50 times orthogonal light; single polarized left 100 times, 100 times lower right orthogonal light, the main mineral component of quartz (Qtz), plagioclase (Pl) suffer kaolinised presented the surface is not smooth morphology, biotite (Bt) was directed or semi-directed filling the gap in the light minerals.

Figure 2.2.3. microscope photos of granitic gneiss.

2.2.2. Tuff of Mountain Laoshan Yakou

Mountain Laoshan Yakou area is located in the southeastern margin of the Laoshan Granite body, figure 2.2.5. Our study used zircon from tuff layers in Yingfang section, a set of thin interbedded chert and tuff, located found along an easy road. This maybe the same formation as Fu Yongtao et al (2010) have previously reported a set of andesitic volcanic clastic sedimentary rocks in Yakou-Ba Xiandun path [36]. Using glassy tuff samples LA-ICP-MS dating, we found the zircon U-Pb age is (118.9 ± 3.3) Ma, which belongs to the Qingshan Cretaceous volcanic tuff [37] .

microscopic photograph of granitic gneiss hand specimens (below left-50 times single-polarized light , right-50 times orthogonal light;mainly quartz (Qtz), dark minerals such as biotite (Bt) and Hui Stone (Cpx) filling the gaps between quartz).

Figure 2.2.4. microscope photos of mixed gneissic granite.

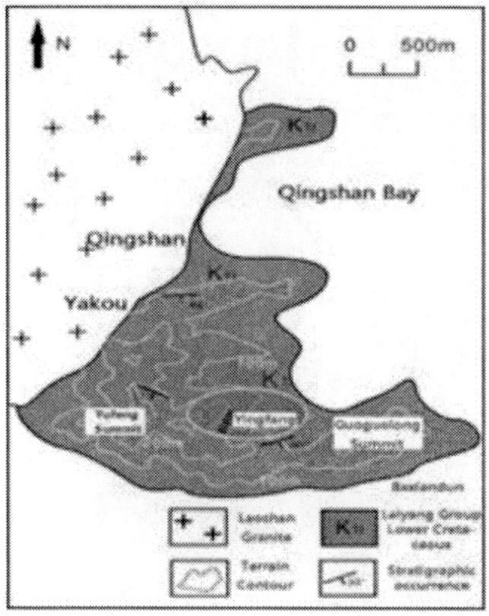

Figure 2.2.5. Simplified geological map of Laoshan Yakou (modified from Fu Yongtao, etc., 2010) [36].

2.2.3. Tuff of Jiaonan-Lingshan Islands

Lingshan Island is located in the southeast of the Yellow Sea, about 10km from Mount Dazhu, Qingdao Jiaonan. Table 2.2.3 has detailed sample information.

Table 2.2.3. Sample data of Lingshan Island

1	LHZ-002	Lao Huzui tuff	N : 35°44.596 ; E : 120°09.802 ; H : 100m ; Δ : 15m
2	LHZ-YXZ	Lao Huzui Andesite	
3	YJD-12	Yang Jiaodong tuff	
4	CCN-02	Chuan Chang tuff	N:35°45`914``, E:120°09`476``

We use samples taken from a large segment of tuff layers in Laohuzui region. Tuff strikes NNW, with a wide distribution, thinned towards northwest due to fault offset (figure 2.2.6).

Under the microscope, the sample is glassy rhyolitic tuff (figure 2.2.7), with a typical tuff structures. Debris are mainly rigid vitric glass, irregular, angular shaped, messy arrangement, either colorless, or gray, yellow, and brown. Their granularities are at 100μm or less, mostly non-deformed, and are situ stacked.

There may also be normal calcite, quartz, clay minerals deposition, mixed with the glass tuff.

According to Laohuzui region tuff samples' LA-ICP-MS dating results, zircons' ages concentrated in the early Cretaceous 108 ~ 126Ma, 206Pb/238U obtained a weighted mean age of (119.2 ± 2.2) Ma [39].

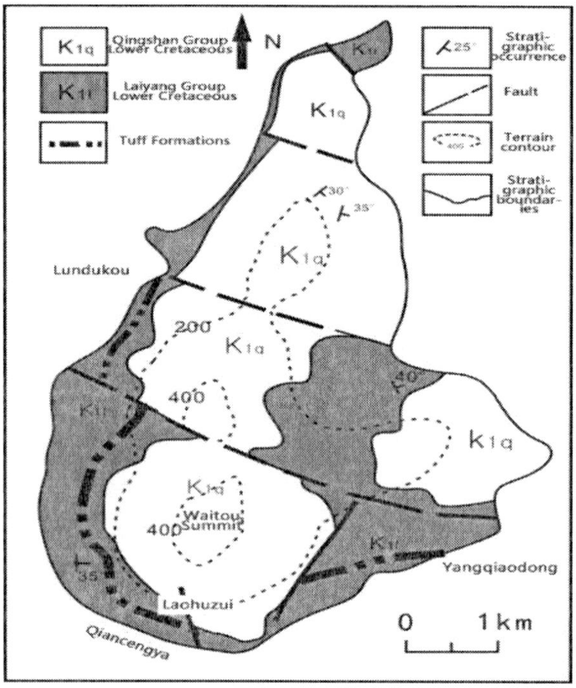

Figure 2.2.6. Geological sketch map of Lingshan Island (according to Luan Guangzhong, 2010) [38].

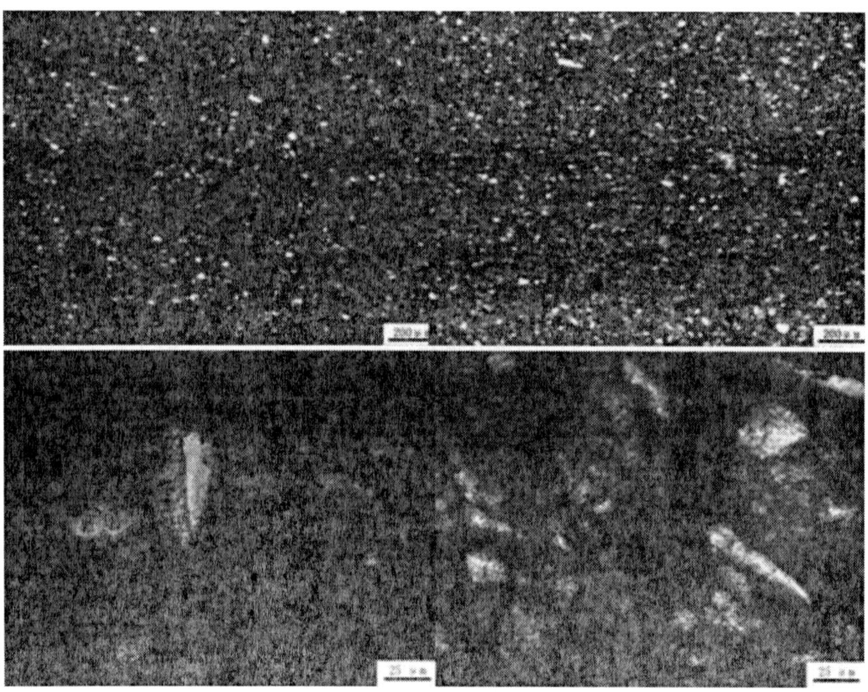

Left-two taken under monolithic light irradiation, right-two taken under orthogonal light irradiation, lower left-quartzite crumbs, lower right-debris and caulking, etc.

Figure 2.2.7. Microscopic photos of Laohuzui tuff.

2.3. INTERNAL STRUCTURE OF ZIRCON

2.3.1. Tuff of Mountain Laoshan Yakou

The euhedral - subhedral crystallized tuff zircons contain a small amount of mineral crystals (figure 2.3.1).

Their internal structures including obvious bilayer structure, typical magmatic oscillatory zoning, and fan-shaped sub-zone. Cores of zircons are black, with typical magmatic zoning.

However, edges are gray white – white on CL images, with very uneven width between each ring, ranging from a few microns to a hundred microns. CL images of the zircon rims displays, though not rule magmatic zone, but according to the study of mineral inclusions, our samples are not within the scope of metamorphic zircon. Sample rims also caused by magmatic crystallization, and may have been subjected to high pressure metamorphism or late fluid action effects which caused incomplete recrystallization and blurred their magmatic origin.

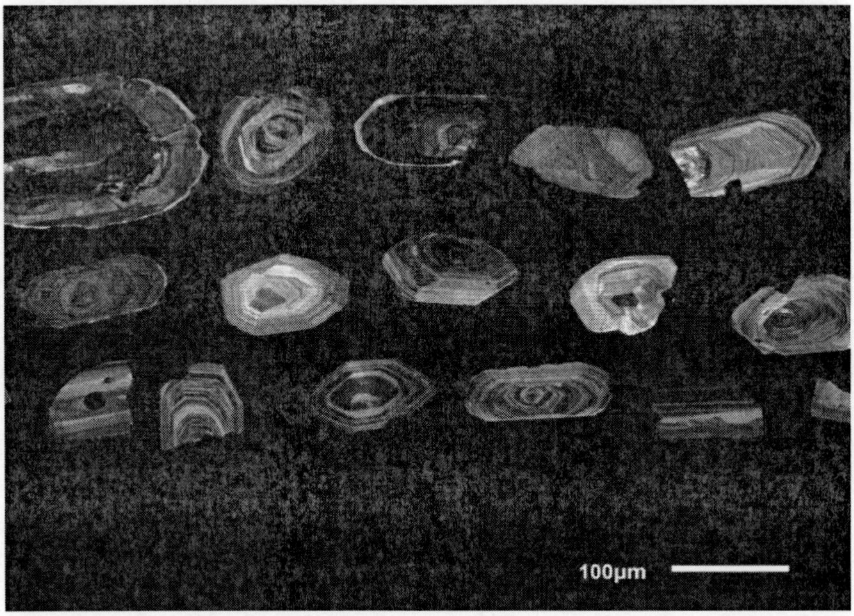

Figure 2.3.1. internal structure of tuff zircons of Laoshan Yakou.

2.3.2. Tuff of Jiaonan -Lingshan Island

Four samples were collected from Lingshan island: LHZ-002 (Laohuzui sectional tuff), LHZ-YXZ (Laohuzui sectional andesite), YJD-12 (Yangjiaodong sectional top tuff samples), CCN-02 (South Shipyard sectional tuff).

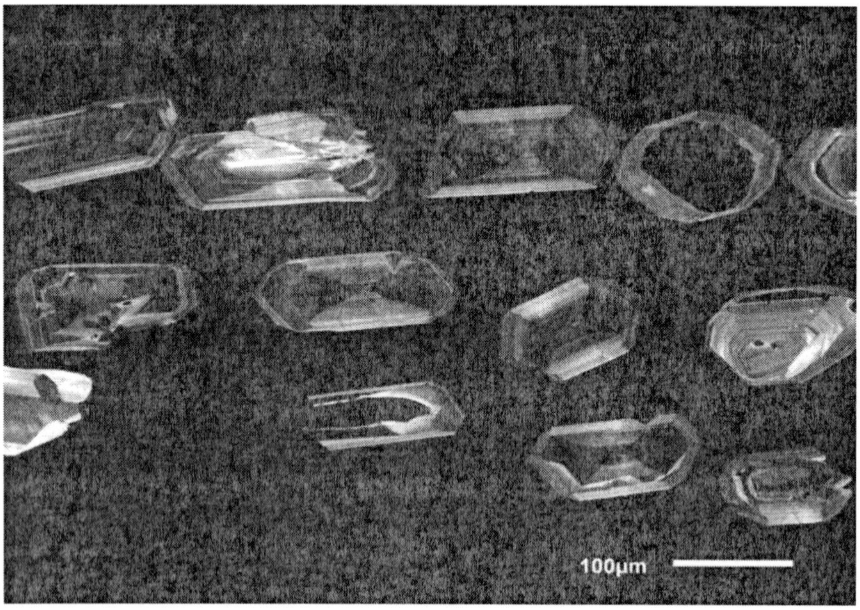

Figure 2.3.2. CL image of samples from Lingshan island.

All of which are tuff rocks, some are well-developed crystalline particles, elongated columnar, half cone, or oval-shaped, pale yellow – transparent; others are damaged due to mechanical crushing. Grains are mainly in the size of 100-200μm, whose aspect ratio ranges from 2:1 to 3:1, some up to 5:1. The CL images (figure 2.3.2) show rhythmic magmatic zoning, which indicate certain magmatic crystallization.

2.3.3. Weihai Rongcheng Gneiss

Samples from this area are mainly gray white - gray red - pale red mixed gneissic granite, with a complex granoblastic structure, containing abundant zircons, which mainly hosted in quartz, feldspar and other minerals, only a few are distributed among rock particles. Zircons' diameters are roughly between 0.1-0.3mm, they are mainly smooth irregular columnar or short columnar. Part of the zircon's core and its intermediate section have many UHP mineral inclusions Coe + Rt + Ttn, while the edge mainly composes of Qtz and other low-voltage minerals, which indicates these zircons began to grow in the UHP peak stage, then formed the core, the intermediate section and the edge during the retrograde process.

CL images (Figure 2.3.3) shows that the majorities of zircon crystals are irregular-shaped, and have typical inherited magmatic crystallization cores, whose rings are practically cut by the latters.

There are also some experienced post-magmatic tectonic thermal event reconstructed zircons, with blurred rings, Qtz+Ap+ within their cores, and Coe&Coe+Phe+Ap in the mantle area, which shows a significant regeneration structure in the CL images.

As for the margin, we can clearly see a retrograde edge in the CL images, for the mineral inclusions are not found in this part.

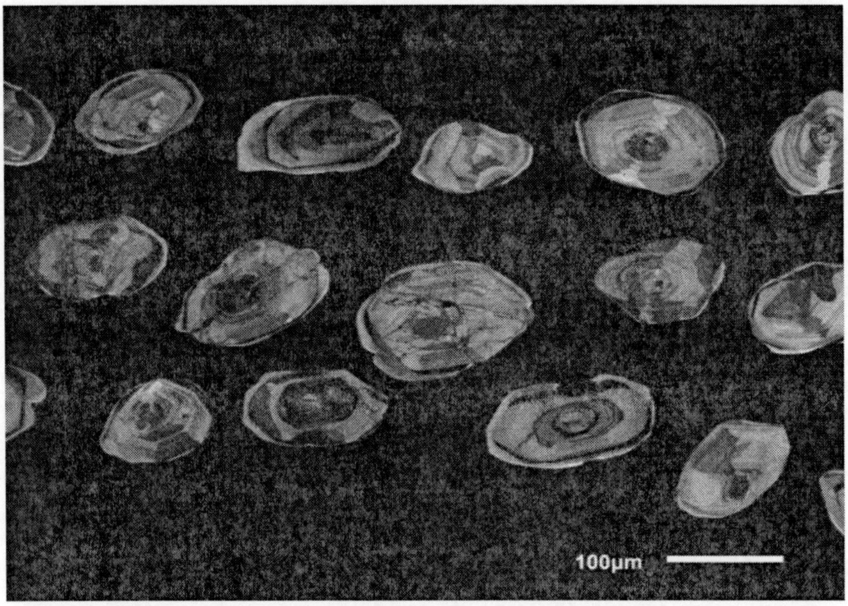

Figure 2.3.3. CL images of metamorphic zircons from Weihai Rongcheng.

2.4. Characteristics of Element Compositions of Zircon under the Electron Probe

The article mainly use the electron probe microanalyses techniques to research the element geochemistry characteristics of zircon. Because of the limitation of the sensitivity of the electronic probes experiments, element types that can be detected are mainly Zr, Si, Hf, Y.

Electron probes experiments of samples were carried out in electron probe laboratory of College of Material Science and Engineering, Shandong University of Science and Technology. This experiment was supervised by Hao Qingkun teacher and operated by the author himself. The adopted Instrument model is JXA-8230, whose DES is 350 and the Instrument is produced by Japanese electronics company. Specific experimental conditions are as follows:

Acceleration voltage: 15KV
Beam current: 1×10-8A
Beam diameter: 5μm

2.4.1. Electron Probe Microanalysis of Zircon of Tuff in Mountain Laoshan-Yakou Area

From the cathodoluminescence images of zircon(figure 2.4.1), we can see that the internal structure of LSYK-3 zircon mainly has fan-shaped zoning and fine vibration zone. In the process of the electronic probes experiments, we did the quantitative analysis of element content of zircon vertically (bright area) and horizontally(dark area), then we made the element content changing trend chart of Zr, Si, Hf,Y, getting the data and image like table 2.4.1 and figure 2.4.2.

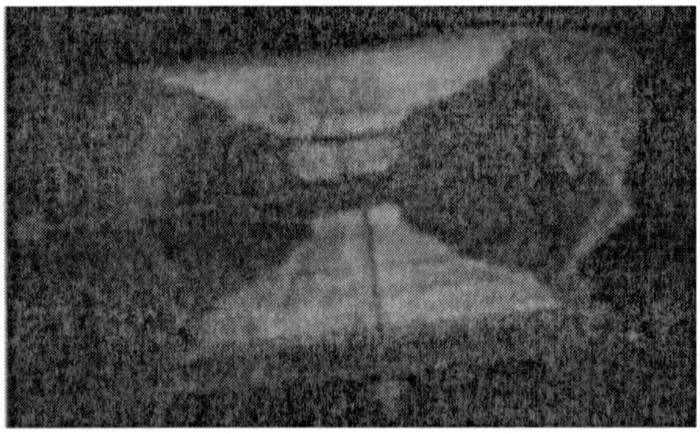

Figure 2.4.1.The sketch maps of electron microprobe analysis of LSYK-3 sample.

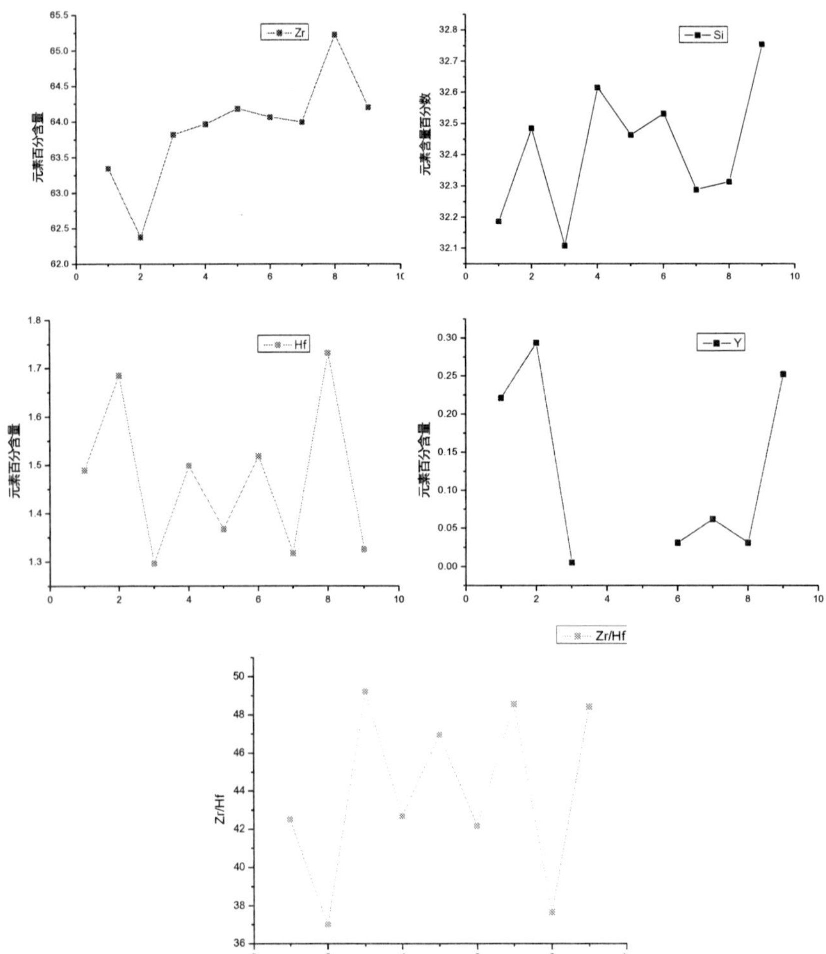

Figure 2.4.2. Trend of the element content change of Zr, Hf, Si, Y from the core to the edge in LSYK-3 vertically.

Table 2.4.1. Results of electron probe quantitative analysis of major elements in LSYK-3 zircon vertically (bright band)

Element	1	2	3	4	5	6	7	8	9	Minimum	Maximum	Average
SiO2	32.186	32.485	32.108	32.615	32.463	32.531	32.288	32.313	32.754	32.108	32.754	32.416
ZrO2	63.345	62.377	63.822	63.967	64.183	64.066	63.996	65.225	64.203	62.377	65.225	63.909
HfO2	1.490	1.685	1.297	1.499	1.367	1.519	1.318	1.732	1.326	1.297	1.732	1.470
Y2O3	0.221	0.294	0.005	-	-	0.031	0.062	0.031	0.252	-	0.294	0.100
Total	97.242	96.841	97.232	98.081	98.013	98.147	97.664	99.301	98.535	96.841	99.301	97.895

From table 2.4.1 and Figure 2.4.2, we can see that in the fan-shaped bright part of zircon which has fan-shaped band and zonal structure, the element content of Zr shows a gradual increasing trend from the core to the edge. The element content of Si increases slightly, but the total content stables at about 32.4%.

The mean value of element content of Hf is 1.47% with its characteristic of reciprocating change from low to high. Totally, the element content in dark band is low but high in bright band.

The element content of Y is low and we can only collect the parts whose content is above hundred thousandth. The variation of the element content has no very obvious regularity. In addition, based on the experimental data, we calculated the ratio of Zr/Hf and drew the trendline that value changes with the location, finding that the ratio of Zr/Hf has the very neat law of cyclical changes from high to low and the ratio has the trend of increase from the center to the edge.

We did the quantitative analysis of Zr, Si, Hf, Y in the area of fan-shaped dark band of zircon using the same test methods. The results are shown in table 2.4.2 in figure 2.4.3.

Table 2.4.2. Results of electron probe quantitative analysis of major elements in LSYK-3 zircon horizontally (dark band)

Element	1	2	3	4	5	6	7	8	Minimum	Maximum	Average
SiO2	33.102	32.589	32.866	32.525	32.803	33.149	33.596	33.579	32.525	33.596	33.026
ZrO2	64.116	64.097	62.921	63.152	63.612	63.342	61.501	61.601	61.501	64.116	63.043
HfO2	1.814	1.381	1.588	1.675	1.718	2.079	1.848	1.503	1.381	2.079	1.701
Y2O3	0.154	0.133	0.098	0.087	0.108	-	0.181	0.289	-	0.289	0.131
Total	99.186	98.200	97.473	97.439	98.241	98.570	97.126	96.972	96.972	99.186	97.901

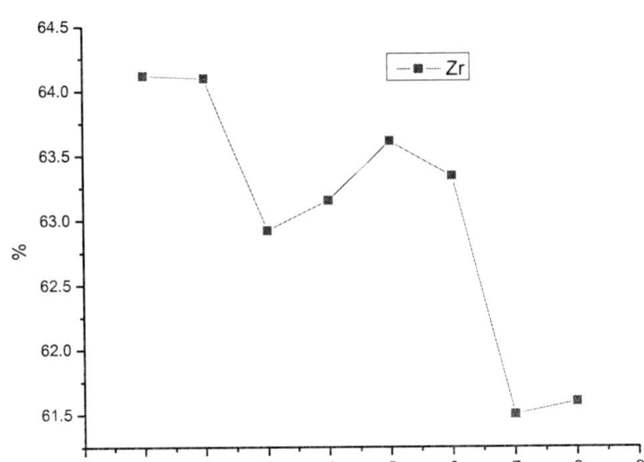

Figure 2.4.3. (Continued).

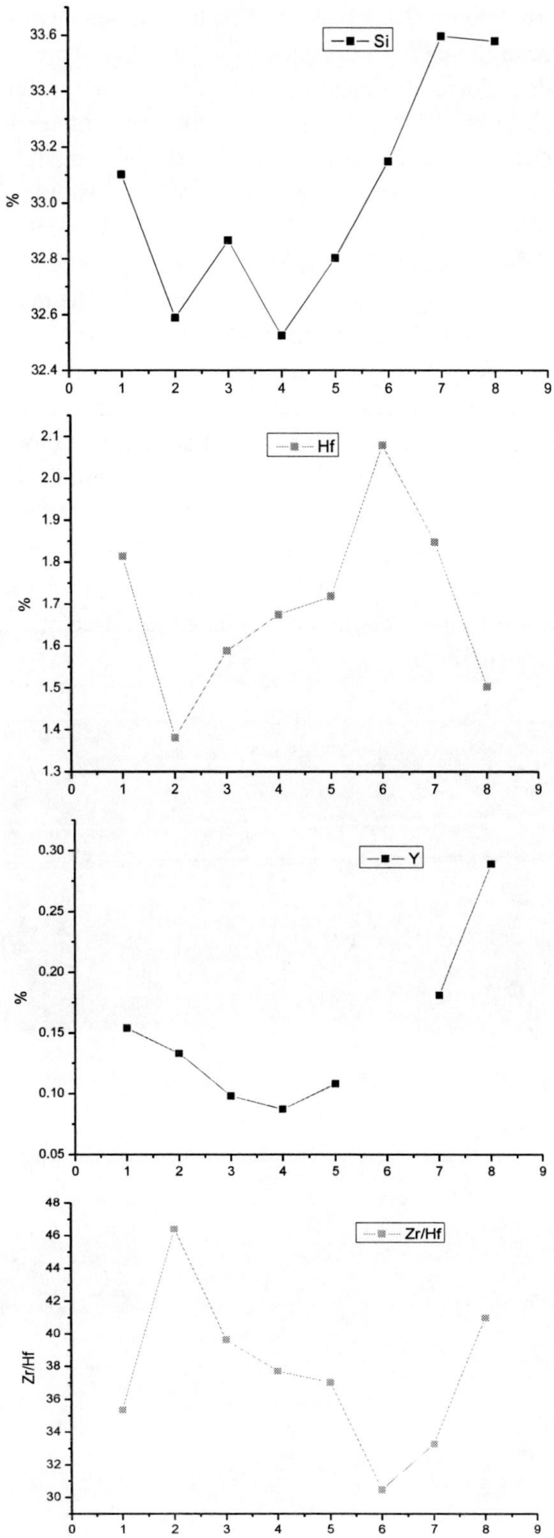

Figure 2.4.3. Trend of the element content change of Zr, Hf, Si, Y from the core to the edge in LSYK-3 zircon horizontally.

From table 2.4.2 and figure 2.4.3, we can see that the element content of Zr changes obviously from the core to the edge. It decreases first and then increases and decreases again when it reaches the edge. For the microstructural study of zircon, the light intensity of the band changes from dark to bright then to dark from the core to the edge in the CL image of zircon, which means that the element content of Zr decreases gradually in the dark area of microstructure and changes from low to high in the bright area and decreases again in the dark area of the edge. The element content of Si tends to be on the rise and the mean value of it is about 33.026%. The element content of Hr has the similar change trend with Zr. Althought Y is rare, it shows roughly down first and then up. The ratio of Zr/Hf has obvious change rule that the element content increases first from the core to the edge, then decreases in the middle part of a relatively bright area and then increases again in the edge of dark area.

Based on the above analysis, it can be seen that the change rule differs in different areas of the fan-shaped zoned zircon by comparing horizontal element content change with vertical change of the zircon grain. The element content of vertical part in the bright area is stable and it has the small reciprocating change rules with the different locations in different vibration band. But for the fan-shaped dark grey area, the variation of the element content has the obvious three-stage pattern, especially the change rule of Zr、 Hf and the ratio of Zr/Hf. So it can be discovered that the element content of Zr changes differently in the dark and bright area in the fan-shaped zoned zircon.

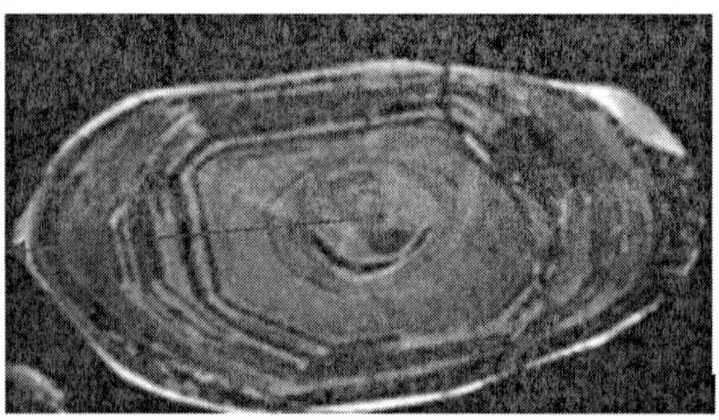

Figure 2.4.4. Test point direction of electron probe analysis of LSYK-6 zircon.

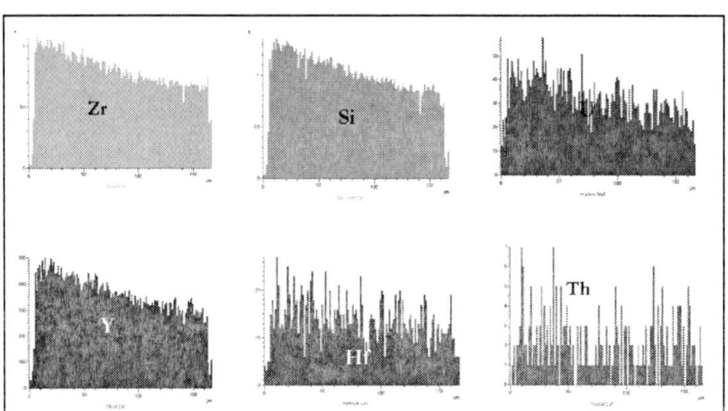

Figure 2.4.5. (Continued).

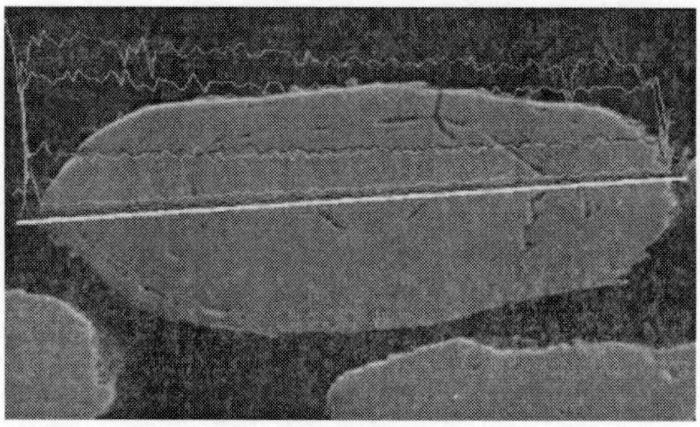

Figure 2.4.5. Scans of energy spectrum analysis of LSYK-6 zircon.

It decreases overall from inside to outside in the bright area but changes more complexly in the dark area. According to the CL image of zircon, we can speculate that the differences of the elements content change in the mineral have some kinds of correlations with the structure of the CL image and the structural differences of zircon is the external expression of element distribution. The degree of bright and dark of the CL image of zircon is correlative with the content of U and Th in an opposite way. Essentially, structural alternation is the direct expression of element types and quantity variance of mineral. We think that the element content of Zr decreases gradually from the core to the edge in the part of zircon which contains high amounts of U and Th and vice versa.

Besides LSYK-3, we choose LSYK-6 to make the electronic probe test analysis (figure 2.4.4) and do the energy spectrum analysis and quantitative analysis.

The results of the current study are as follows(figure 2.4.5, figure 2.4.6 and table 2.4.3). The elements in zircon are mainly Zr, Si, Hf, U and Th from the energy-spectrum scanning test. According to the energy spectrum curve, we can see the rough trend of variation of different elements. From the center to the edge, the element content increases slightly.

Combined with the image of internal structure and the data of electron probe quantitative analysis, we can see that the element content of zircon and the internal structure have the similar corresponding relation. The internal structure can be divided into three parts from CL image and the changing curve of the element content of Zr, Hf, Zr/Hf also has the obvious three-stage.

2.4.2. Electronic Probe Analysis on Zircons of Tuff in Lingshan Island, Jiaonan

In this paper, we analyse the zircons with obvious characteristics in Laohuzui area by energy spectrum analysis combined with quantitative analysis of elements. Though the zircons show obvious fan-shaped zoning rather than clear oscillatory zoning (figure 2.4.7), we can still find that crystalline phase can be divided into three phases. The research was carried on in two directions: the dark areas and the bight areas.

Table 2.4.3. Results of electron probe quantitative analysis of major elements in LSYK-6 zircon

Element	1	2	3	4	5	6	7	8	9	10	11	Minimum	Maximum	Average
SiO2	32.552	32.652	32.539	32.047	32.502	32.350	32.133	32.522	32.725	33.029	33.227	32.047	33.227	32.556
ZrO2	64.963	63.758	63.549	64.215	64.830	64.722	63.320	61.621	64.070	63.848	63.787	61.621	64.963	64.011
HfO2	1.503	1.706	1.458	1.695	1.261	1.466	1.566	1.880	1.728	1.887	1.294	1.261	1.887	1.588
Y2O3	-	0.015	-	-	-	-	-	0.201	-	-	-	-	0.201	0.015

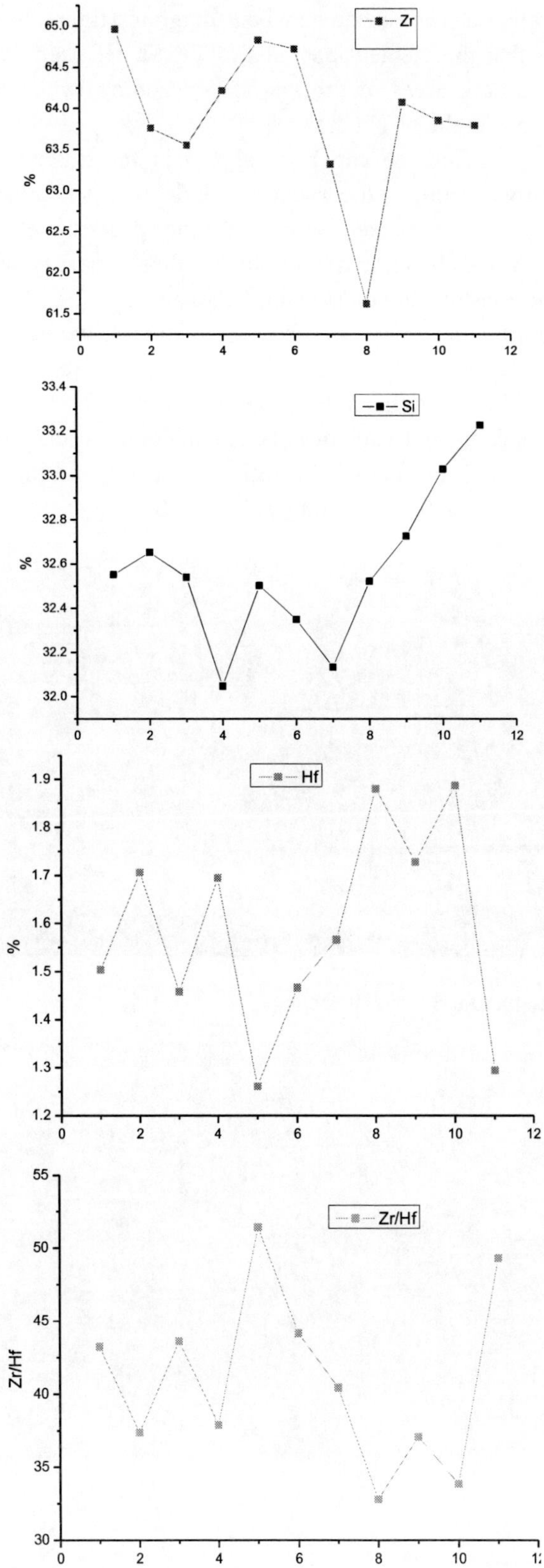

Figure 2.4.6. Trend chart of the content change of major elements in LSYK-6 zircon.

The electron microprobe spectroscopy analysis diagrams(figure 2.4.8 and figure 2.4.9) of LHZ - 002-1 show us that the element contents of Zr, Si, Hf keep increasing steadily from core to the edge in the dark areas of the fan-shaped zoning, while this changing trend of elements is not so obvious in the bright areas.

The reasons for the difference can be analysed in the internal structure images: the phenomenon of oscillatory zoning in the transverse dark areas is more obvious than that in the vertical bright areas, resulting a clearer increasing trend. Also, the element contents of Zr in the bright areas has a reverse changing trend to that in the dark areas, which is consistent with the characteristics of zircons in Yakou, Moutain Laoshan.

From the internal structure image of zircon LHZ-002-3, we can find a clear magmatic oscillatory zoning at the edge while an unclear zoning in the core. But the zoning in the core is more uniform and brighter than that at the edge.

All of these result increasing trends of the element contents of Zr, Si, Hf from core to the edge. The Zr/Hf ratio shows an opposite changing tendency, and the inflection point of the elements appears at the edge (table2.4.4 and figure 2.4.10).

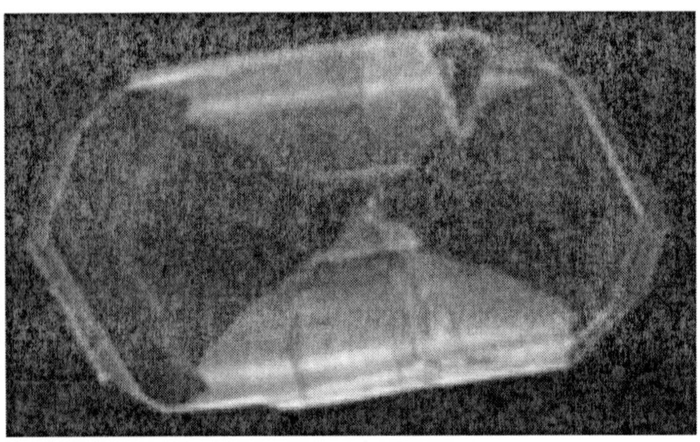

Figure 2.4.7. CL photograph of LHZ-002-1 zircon.

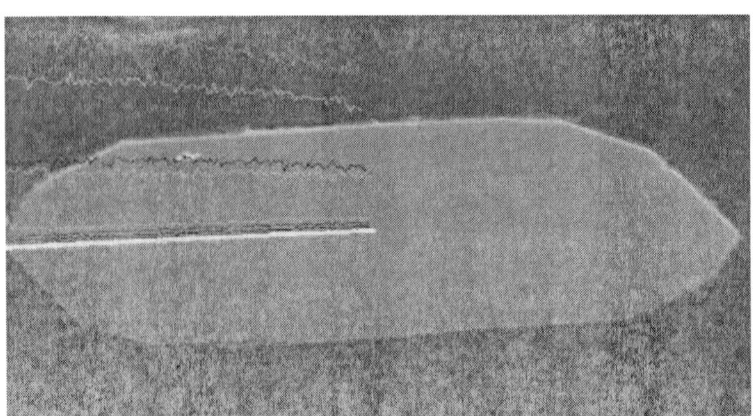

Figure 2.4.8. (Continued).

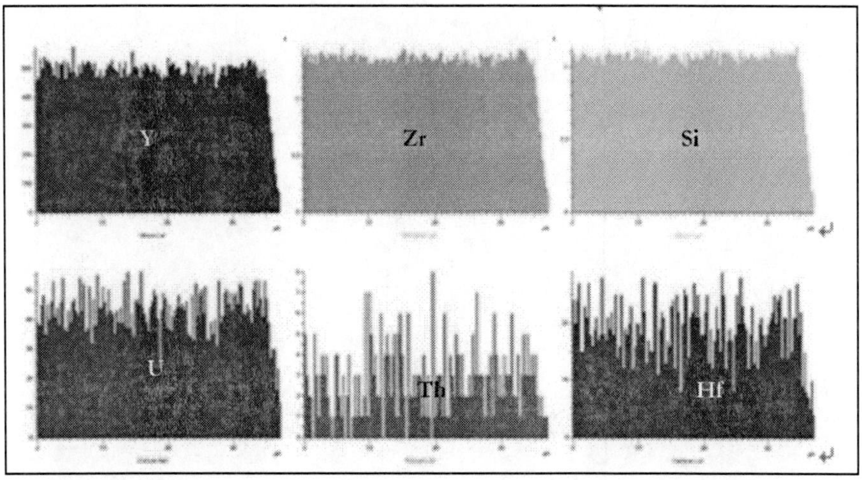

Figure 2.4.8. Transverse scans spectroscopy analysis of LHZ-002-1 zircon.

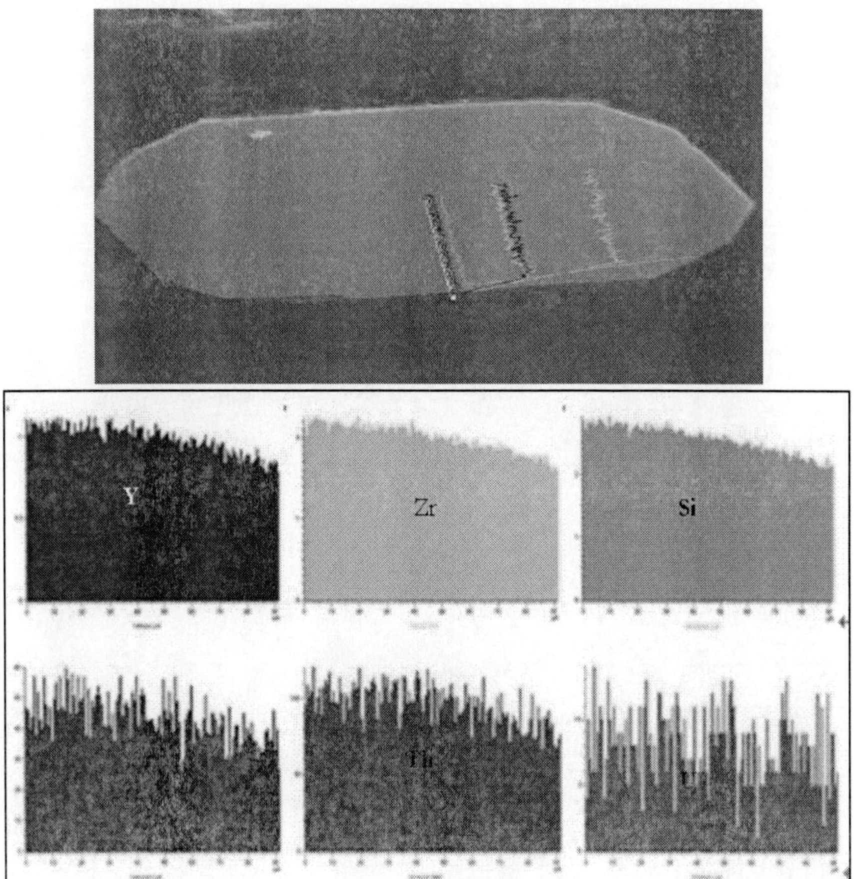

Figure 2.4.9. Vertical scans spectroscopy analysis of LHZ-002-1 zircon.

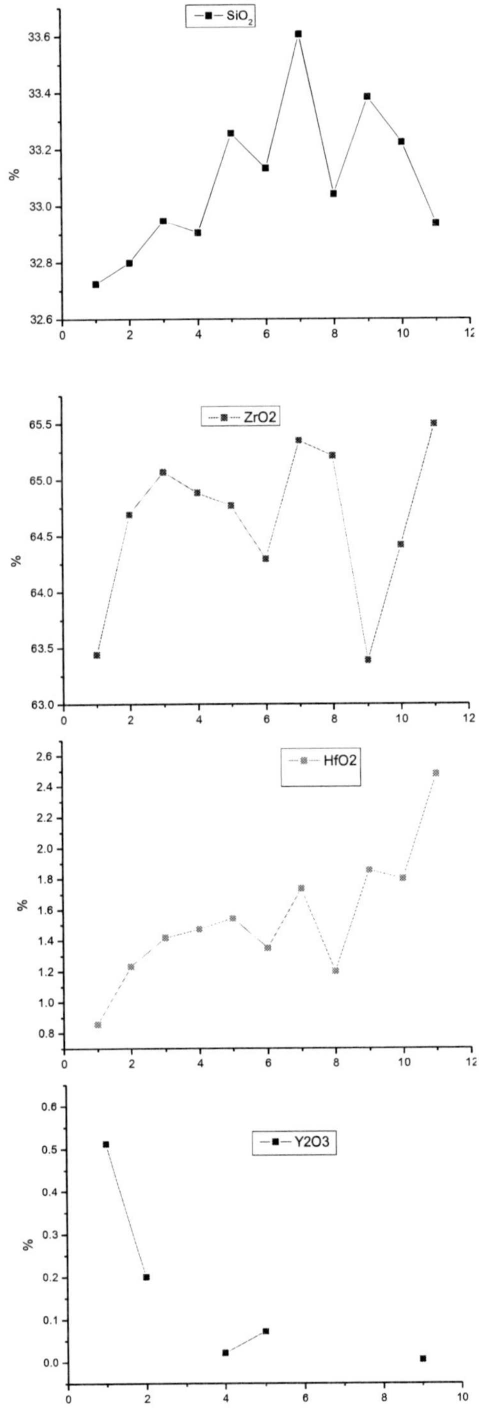

Figure 2.4.10. (Continued).

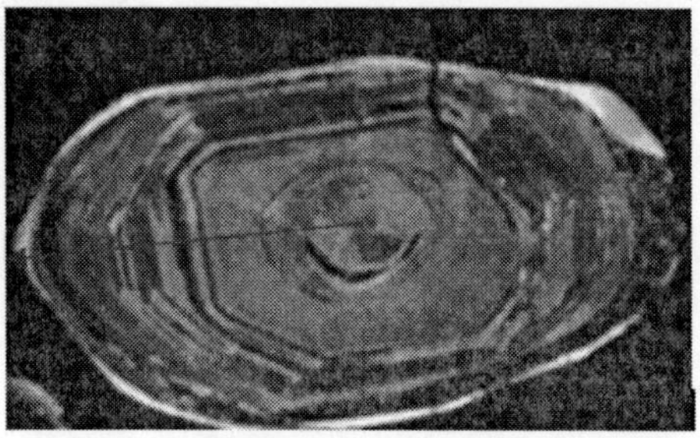

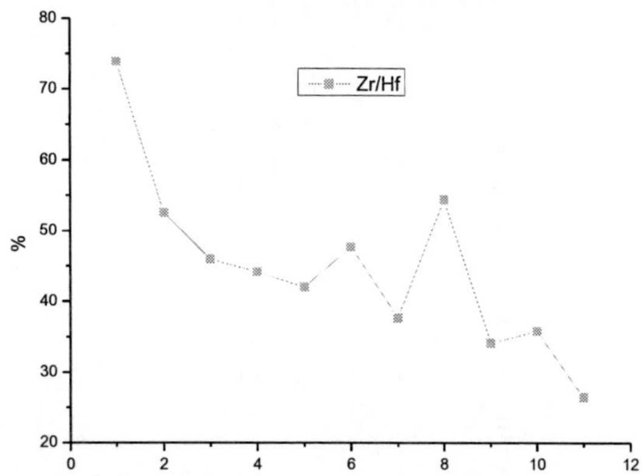

Figure 2.4.10. The changing trend of major elements in zircon LHZ-002-3.

2.4.3. Electronic Probe Analysis on Zircons of Gneiss in Rongcheng, Weihai

The spectral analysis in zircon Wh-015-2 (figure 2.4.11)and linear scan of elements content in zircon Wh-015-4 (figure 2.4.12)show us that the element contents of Zr, Si, Hf, Y of zircons in Rongcheng, Weihai decrease from core to the edge, which is reverse to the changing trend of that in Yakou, Moutain Laoshan and Lingshan Island, Jiaonan. The internal structure of zircon Wh-015-4 can be divided into three parts distinctly. In the linear scan of elements content in zircon Wh-015-4, the element contents of Zr, Si keep steady overall despite some initial fluctuation, while the element contents of Hf, Y show a gradual downward trend. More changing trends will be discussed in quantitative tests.

From two zircon quantitative analysis data and images (table 2.4.5, table 2-4-6 and figure 2.4.13, figure 2.4.14), we can be find that metamorphic zircon grains from the dark core to the bright edge, the content of Zr, Si, Hf increases gradually, Zr/Hf value in the place where color changes from dull to bright also increased obviously, and the bright part of Zr/Hf ratio presents the regularly changing. In contrast to the changing rule of the magma genesis of zircon, the dark band ratio is low, bright band with high ratio.

Table 2.4.4. The quantitative analysis results of major elements in zircon LHZ-002-3

Element	1	2	3	4	5	6	7	8	9	10	11	Minimum	Maximum	Average
SiO2	32.725	32.799	32.948	32.905	33.256	33.133	33.607	33.041	33.383	33.224	32.937	32.725	33.607	33.070
ZrO2	63.443	64.690	65.070	64.886	64.769	64.293	65.349	65.212	63.388	64.415	65.494	63.443	65.494	64.675
HfO2	0.858	1.231	1.416	1.471	1.542	1.349	1.735	1.198	1.856	1.799	2.479	0.858	2.479	1.511
Y2O3	0.512	0.200	-	0.021	0.072	-	-	-	0.005	-	-	-	0.512	0.075
Total	97.538	98.920	99.434	99.283	99.639	98.775	100.691	99.451	98.632	99.438	100.910	97.538	100.910	99.363

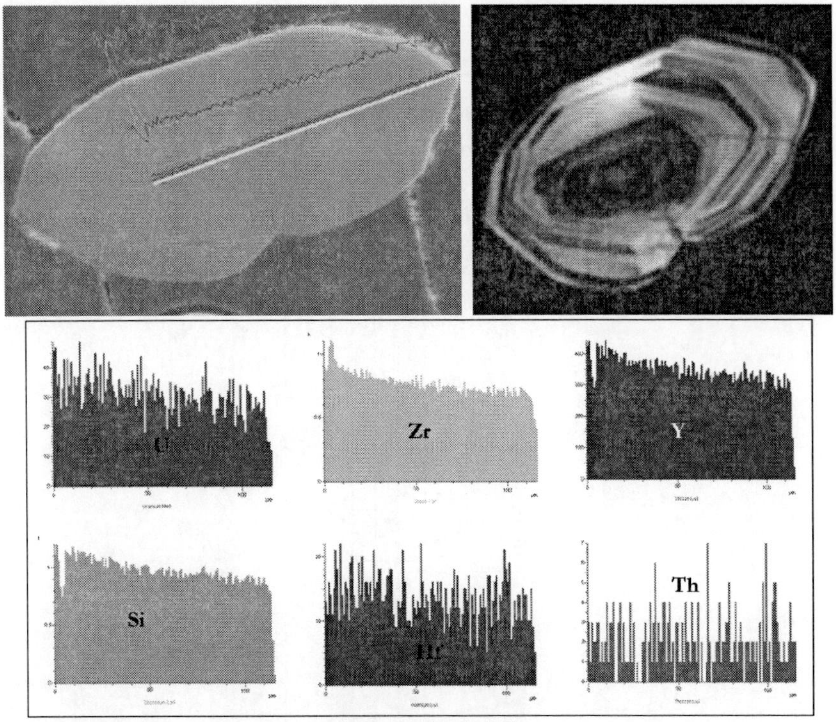

Figure 2.4.11. The curve line of spectral analysis in zircon Wh-015-2.

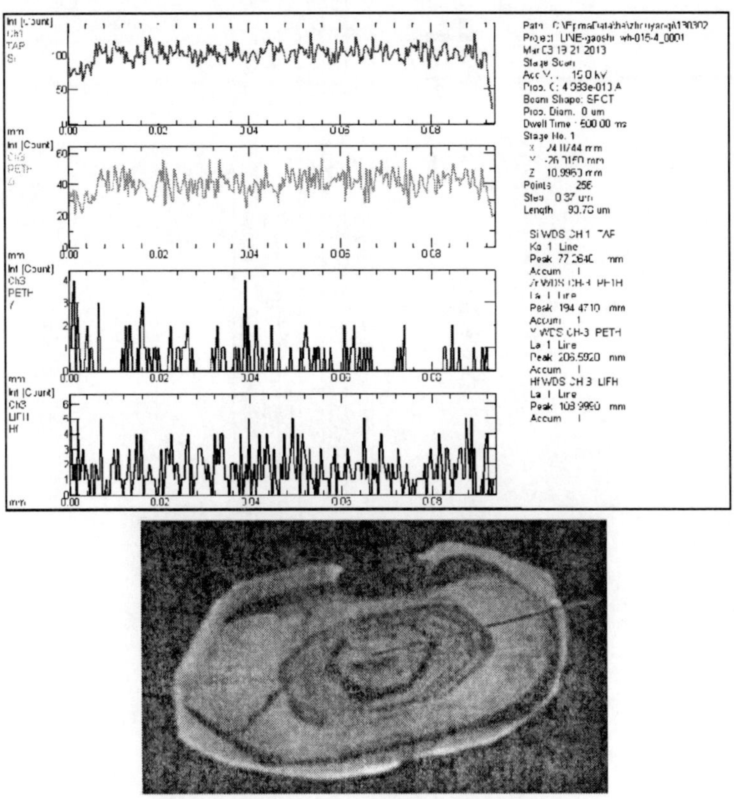

Figure 2.4.12. Linear scan of elements content in zircon Wh-015-4.

Table 2.4.5. The quantitative analysis results of major elements in zircon Wh-015-2

Element	1	2	3	4	5	6	7	8	9	10	11	12	Minimum	Maxmum	Average
SiO2	34.041	32.938	32.993	33.270	33.005	32.969	33.255	32.512	33.045	32.901	32.814	32.405	32.405	34.041	32.970
ZrO2	63.065	64.358	64.048	63.897	63.112	64.224	64.659	63.322	64.093	63.959	64.934	64.158	63.065	64.934	63.983
HfO2	2.058	2.012	2.130	2.172	1.800	1.309	1.143	0.831	1.321	1.777	1.382	1.451	0.831	2.172	1.638
Y2O3		0.139	0.026		0.232				0.015		-	-		0.232	0.06
Total	99.164	99.447	99.197	99.339	98.149	98.502	99.057	96.665	98.474	98.637	99.130	98.014	96.665	99.447	98.766

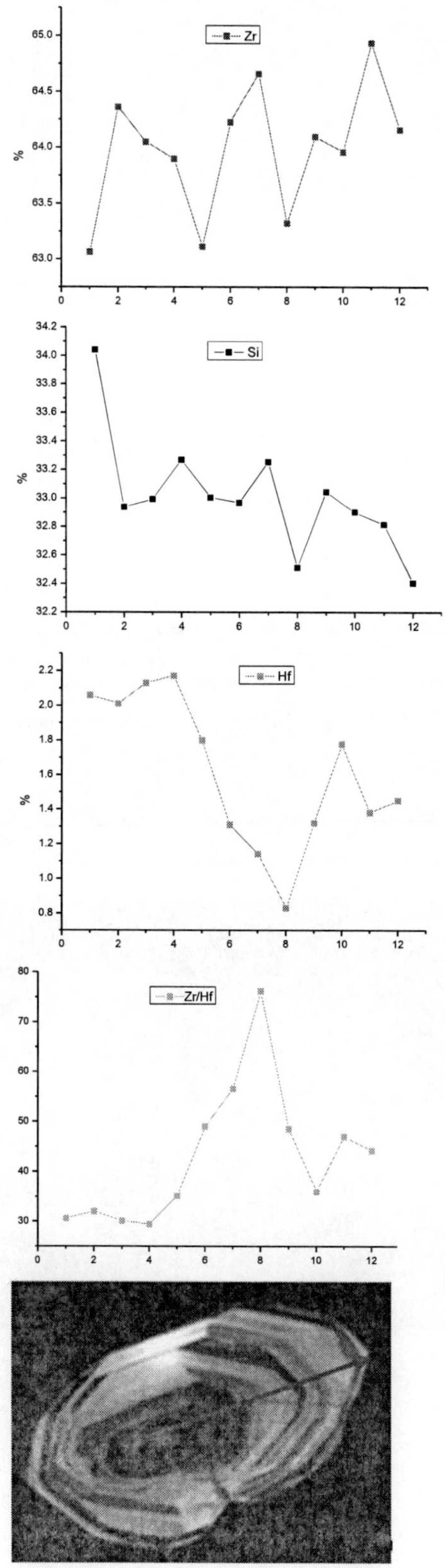

Figure 2.4.13. The changing trend of major elements in zircon Wh-015-2.

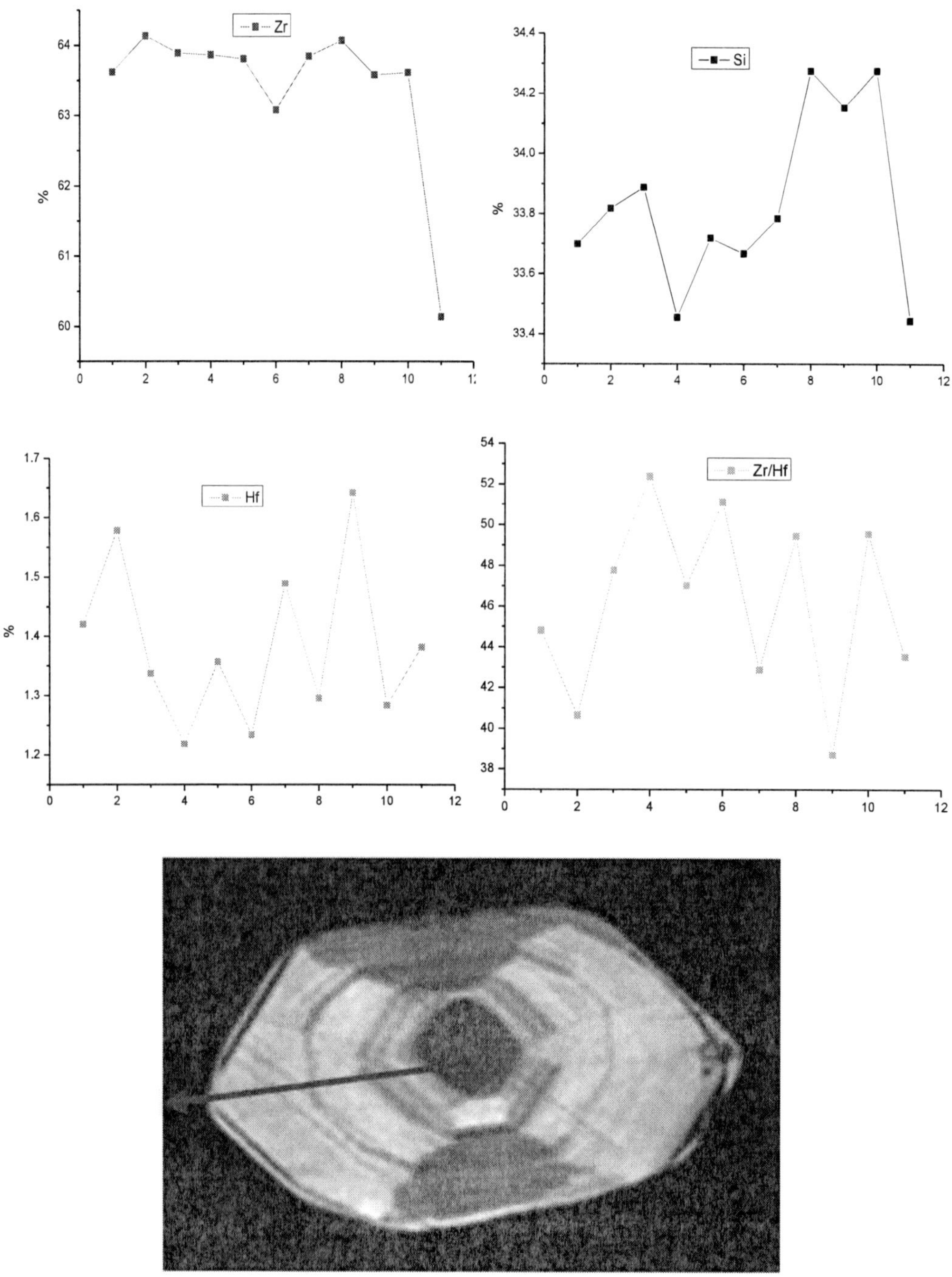

Figure 2.4.14. The changing trend of major elements in zircon Wh-015-5.

Table2.4.6. The quantitative analysis results of major elements in zircon Wh-015-5

Element	1	2	3	4	5	6	7	8	9	10	11	Minimum	Maximum	Average
SiO2	33.699	33.818	33.888	33.455	33.719	33.667	33.785	34.275	34.153	34.276	33.444	33.444	34.276	33.829
ZrO2	63.621	64.135	63.890	63.865	63.810	63.079	63.845	64.07	63.582	63.617	60.140	60.140	64.135	63.709
HfO2	1.420	1.578	1.337	1.219	1.357	1.234	1.489	1.296	1.642	1.284	1.382	1.219	1.642	1.375
Y2O3	-	0.026	-	-	-	-	-	-	-	-	-	-	0.026	-
Total	98.740	99.557	99.115	98.539	98.886	97.980	99.119	99.646	99.377	99.177	94.966	94.966	99.646	98.943

According to research of Bao Xuezhao, changing of the two elements, Zr and Hf in the zircon presents certain regularity, and the cause of this law has a certain instructive significance. Giving the magmatic zircon and metamorphic zircon electron probe composition analyzing respectively, finds that the content of HfO_2,UO_2+ThO_2 in the magmatic zircon, from the crystal central to the edge, from crystal band inside to the lateral, increases gradually, on the contrary, the content of ZrO_2, ZrO_2 / HfO_2 reduce gradually from inside to outside [40].

Making a comparation with the composition variation characteristics of magmatic zircon and composition change of magmatic evolution series, draws a conclusion that the composition of magma from the crystal nucleus parts to the edge, changing rule has the following features:

(1) the ZrO_2 / HfO_2 value of zircon in the basic - neutral - acid rocks drops gradually, consistent with the characteristics of ZrO_2 / HfO_2 value in magmatic zircon from the crystal center to the edge;
(2) the UO_2+ThO_2 content in zircon increases gradually in the ultrabasic - basic - neutral - acid rocks , coincident with the changing trend of UO_2+ThO_2 content in magmatic zircon from the crystal center to the edge;

We can learn from the two points above that magma evolution trend can control the composition changing law, such as ZrO_2 / HfO_2 et al, and ion itself properties can also be able to control the ingredient of zircon: Zr^{4+} (0.098 nm) radius is small. Compared with U, Th, large radius of U^{4+} (0.114 nm), Th^{4+} (0.118 nm) give priority to be rich in zircon (according to V.M. gold Schmidt rule); the electronegativity of Zr^{4+} (Zr) 1.21 is less than that of Hf^{4+} (Hf) 1.38,so Zr^{4+} gives priority to be rich in zircon (according to Lin wood added rule). In this case, Zr^{4+} content is high in the inside band of zircon or inside nuclear, and content of Hf^{4+}, U^{4+}, Th^{4+} is higher in the edge or the lateral band. In the evolution process of magma, the change trend is stable, and crystal chemical properties of the containing element of the zircon is relatively stable. In such conditions, crystallized zircon in magma will have relatively fixed composition change trend. So the composition, such as HfO_2, UO_2 + ThO_2 and ZrO_2 / HfO_2 value in magmatic zircon, will have important application value on determining the zircon genetic types.

Zircon crystals forming under the metamorphism, from the crystal central to the lateral edges, the change trend of ZrO_2, HfO_2, UO_2 + ThO_2,ZrO_2 / HfO_2 and so on, is the opposite to magmatic zircon, which is caused by different crystallization environments between them. Metamorphism process are related with temperature, and in most cases, the higher temperature is, the more stronger metamorphism is, resulting in the original rocks, metamorphic hydrothermal, metamorphic recrystallization occurred, etc.; If temperature continues to rise, the component of metamorphic zircon will remelt selectivity, producing similar-magmatic melting bodies. As to the magma evolution process, it is a dynamic process of transformation from liquid to solid rock in magma crystallization, accompanying with a cooling effect at the same time. According to the analysis above, we can find that the metamorphism process actually represents two complementary process, a temperature increasing and protolith evoluting to magma from solid to liquid. These differences will eventually lead to the reverse differences of composition change trend in magmatism and metamorphism process of zircon crystal.

The percentage content of the most granite zircon Hf ranges from 1.20% to 1.32%, Zr/Hf is about 35. In granite, Hf concentration, gathers in the form of full class quality with its geochemical behavior of magma crystallization of physical and chemical environment of an indirect response. Zr and Hf of low-temperature shell source transformation of granite zircon showed a negative correlation. The distribution coefficient of Zr is higher than that of Hf, magma for practical preparation of Zr/Hf value is inversely proportional to the degree of zircon crystal, zircon crystals, the better, zirconium, hafnium ratio is lower, so the magmatic zircon Zr/Hf rises from the edge to the center, gather Zr and Hf differences with the temperature decrease with increasing of the partition coefficient, reduce the displacement capacity of Hf. Granite middle-late Hf concentration in the residual magma, that the content of its rise. From the crust-mantle-sourced hybrid granite to crust-sourced granite, the degree of hafnium concentration is gradually declined (table 2.4.7). The Hf content in early zircons increases from mantle sourced granite to crust-mantle granite then crust sourced granite.

Table 2.4.7. Geochemical characteristics of Hf of zircon from three different types of granite

granite	crust sourced	mantle-crust	mantle sourced
hafnium content of early zircon (HfO_2)	1.41wt%	1.31wt%	1.04wt%
Enrichment of hafnium in zircon (%)	19	40	63

2.5. Laser Raman Features of Zircon

2.5.1. The Calculation of Raman Spectra of Single Zircon Crystal

This Raman experiment was completed in Key Laboratory of Tectonics and Lithosphere Dynamics in China University of Petroleum (East China) with the LabRam-010 Raman spectrometer. Its specific experimental conditions are the same as above.

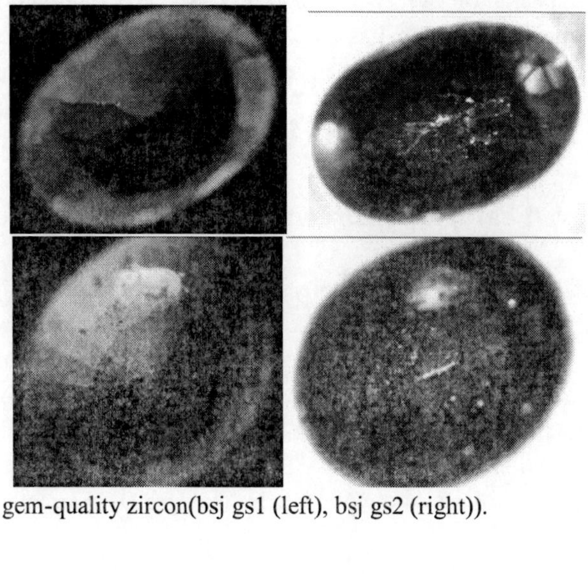

Figure 2.5.1. The two gem-quality zircon(bsj gs1 (left), bsj gs2 (right)).

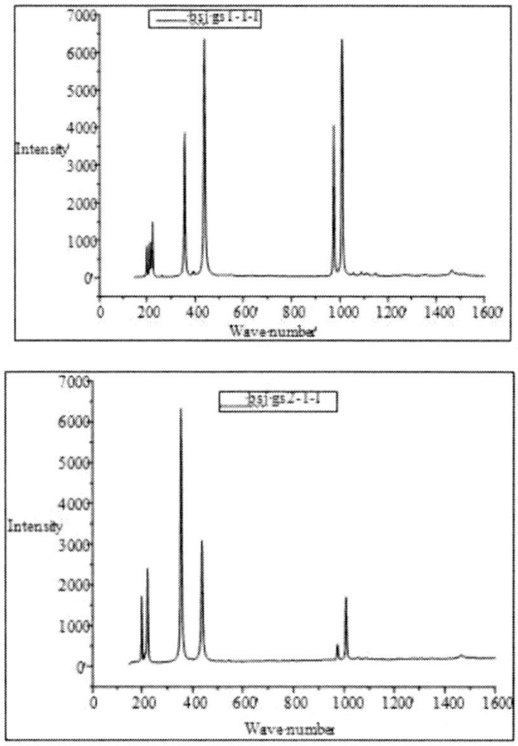

Figure 2.5.2. The Raman curves of two gem zircon.

We tested the two pure quality zircon crystals (figure 2.5.1)firstly with laser Raman irradiation, and obtained their Raman spectra as shown in figure 2.5.2. The laser Raman spectra of the gem-quality zircon posses obvious sharp characteristic peaks which are strong and slim, and distribute at 202-225 cm^{-1}, 357 cm^{-1}, 439 cm^{-1}, 974 cm^{-1} and 1008cm^{-1}. The peak at 800cm^{-1} is not develop which may be affected by the crystal orientation. Besides, the characteristic peak of No. 2 zircon at 974cm^{-1} is much weaker relative to other peaks. The Raman characteristic peaks of the two zircon crystals are relatively standard and their intensity is stabile, especially the No.1 is easier to measure than No.2 zircon. Therefore, we took the Raman spectra of No.1 zircon as an approximate standard zircon Raman spectra for comparison with experimental samples.

2.5.2. Raman Curve Features of Zircon in Tuff of Yakou, Mountain Laoshan

17 test points of LSYK-3 zircon sample in the direction showed in Figure 2.5.3 were exposed to Raman laser and Raman curves were obtained. Because intensity of the characteristic peak at 439cm^{-1} is strong and the signal of it is excellent and stable, parameters of characteristic peak at 439cm^{-1} are selected as the objects of study. Lorentzian method was adopted to fit the characteristic peak at 439 cm^{-1} using the origin software curve 439 cm^{-1}, and four characteristic quantities were obtained: y0 (intensity of characteristic peak at baseline), x0 (characteristic peak position), w (FWHM), A (integrated characteristic peak area). W

(FWHW), relative intensity H (intensity difference between the highest of curve and baseline) and characteristic peak position of 17 test points were calculated and listed in table 2.5.1.

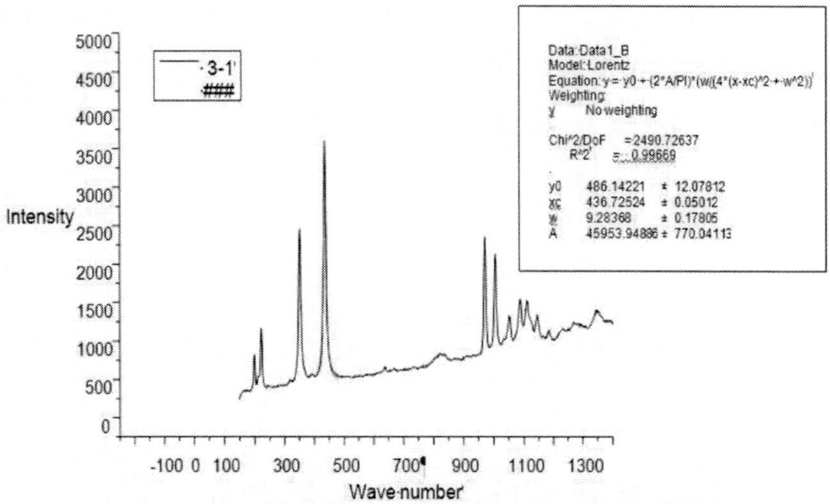

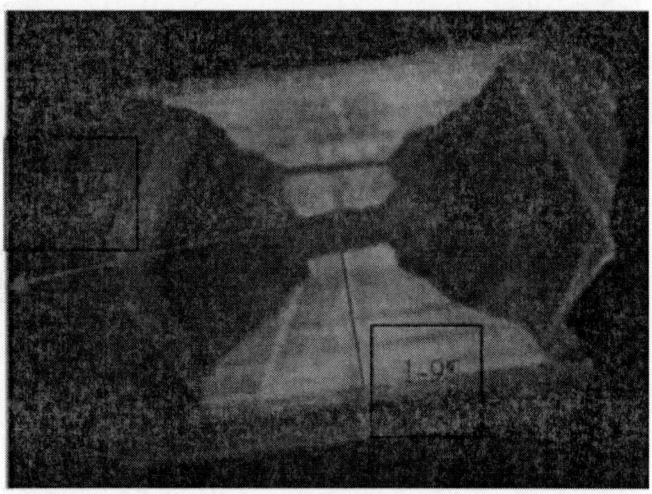

Figure 2.5.3. Test point direction of LSYK-3 zircon in Raman experiments (lower) and Raman curve of LSYK-3 zircon (upper).

Test points were divided into two groups (figure 2.5.4 and figure 2.5.5). 3-1 to 3-9 are test points in the vertical direction of zircon grain, Laser Raman results show that from the crystal core to rim in the vertical direction, relative intensity of characteristic peaks increase faintly before decrease faintly; FWHW remains unchanged; 439cm^{-1} peak position shows cyclicity change; the change of H / W ratio consistent with trend of relative intensity H, increase faintly before decrease faintly, but the change range is smaller.

Table 2.5.1. Calculated H, W, H / W and peak position of characteristic peak at 439cm^{-1} in LSYK-3 zircon after Raman curve is fitted

No. Test point	y0	y c	H	W	Peak position	A	H/W
1	486.14221	3596.90084	3110.75863	9.28368	436.7252	45953.9489	335.0782
2	381.75444	4497.75471	4116.00027	8.91782	436.5455	57687.687	461.5478
3	365.69672	4291.31995	3925.62323	8.92037	436.3542	55107.9595	440.074
4	338.77359	4382.74467	4043.97108	8.9641	436.5685	56946.9476	451.1296
5	350.29797	4559.44559	4209.14762	8.99844	436.8173	59566.8679	467.7641
6	335.63907	4440.30377	4104.6647	9.07494	436.4285	58604.242	452.3076
7	319.66196	4326.82573	4007.16377	8.96654	436.6029	56825.1656	446.9019
8	342.77373	4037.75473	3694.981	9.15873	436.2129	53268.3823	403.4381
9	447.08212	3616.5173	3169.43518	10.26232	436.4852	51554.2062	308.842
10	302.68136	4716.82092	4414.13956	8.71331	436.5696	60418.7056	506.5973
11	313.60757	4591.93964	4278.33207	8.84143	436.4884	59520.3432	483.8959
12	325.58138	4330.63631	4005.05493	9.11485	436.4454	57563.809	439.3989
13	373.01533	3816.13338	3443.11805	10.00397	436.4738	54726.2125	344.1752
14	392.55104	4140.93659	3748.38555	9.48911	436.748	56465.599	395.0197
15	649.31456	4570.03497	3920.72041	9.46665	436.3556	58911.8788	414.1613
16	709.53276	4477.69245	3768.15969	9.61441	435.9424	56916.4491	391.9283
17	743.18549	3287.9289	2544.74341	11.03862	435.5128	44552.7951	230.5309

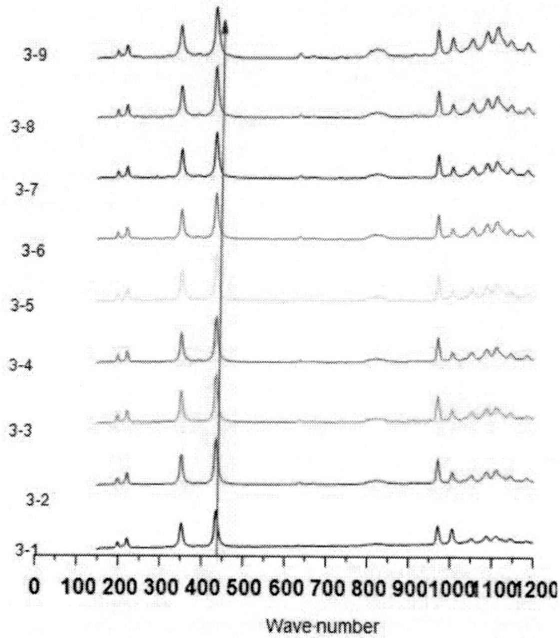

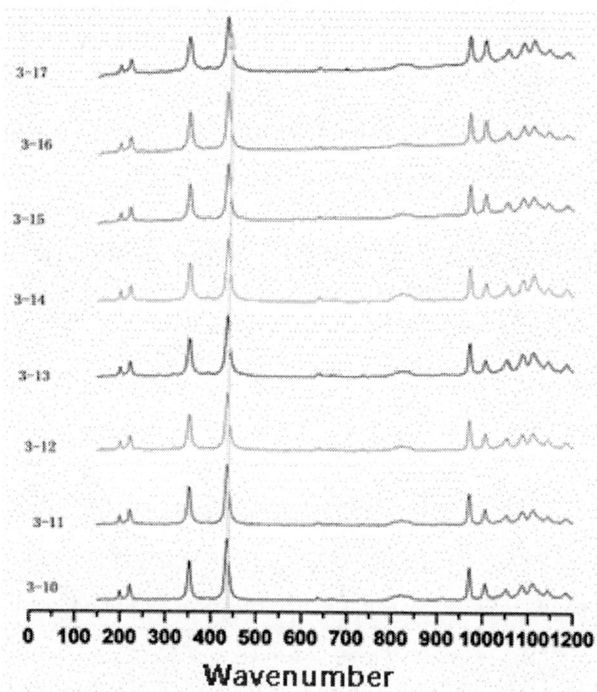

Figure 2.5.4. LSYK-3 the comparison chart of vertical and horizontal Raman curves in LSYK-3 apatite (3-1 to 3-9 :test points in vertical direction, 3 -10 to 3-17: test points in transverse direction).

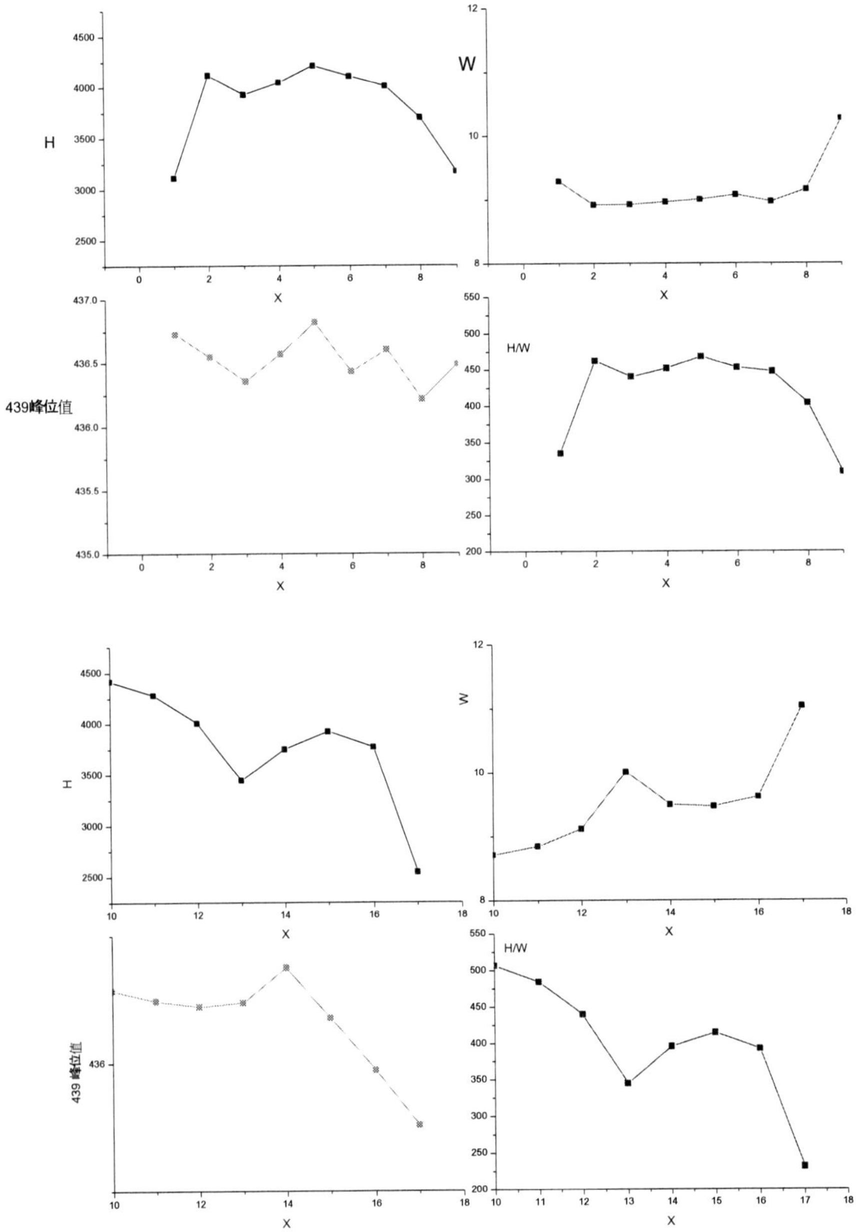

Figure 2.5.5. Variation of H, W, H / W and characteristic peak position in vertical (top) and transverse (lower) direction in LSYK-3 Zircon.

Characteristic quantities of 3-10 to 3-17 test points can be divided into two segments, relative intensity of characterisitc peaks decrease from core to rim, and in the rim area intensity increases; H/W ratio show the same trend; characteristic peak position shifts to lower wave numbers, in the rim area, the offset is much more obvious; FWHW increases from core to rim.

CL images of this zircon grain shows that 3-1 to 3-9 test points locate in the bright area of sector within zircon crystal, the H/W ratio generally ranges from 300 to 470; while 3-10 to 3-

17 test points locate in the dark area of fan section within zircon crystal, H/W ratio can be divided into 2 segments, the H/W ratio of 3-10 to 3-14 decreases from 500 to 300, 3-15 to 3-17 increase rapidly to around 400. In most cases, sector accompanies with oscillatory zoning, which is an important indication of magmatic zircons. According to Raman curve characteristic of zircon, it can be concluded this zircon is typical magmatic zircon.

2.5.3. Raman Curve Features of Zircon Volcanic Tuffs at Jiaonan Lingshan Island Area

By analyzing Raman experimental curves and characteristic quantities of LHZ-002-2(figure 2.5.6, figure 2.5.7 and table2.5.2), zonings of zircons are well developed and they can be divided into three phases. From the inside to the outside of the core crystal, characteristic peaks intensity increase. While the intensity of margin crystal and strength / FWHM ratios decrease, showing that from inside to outside, the metamorphism of core crystal increases and that of margin crystal demonstrates a opposite trending. According to the statistics of H, W and peak values, the H/W values at core vary roughly between 500-750, while the values at edge are more concentrated, most of the data falls between 300-500. Peak value at 439cm^{-1} is about 436 which is fewer than that of standard zircon, indicating the presence of weak metamorphism.

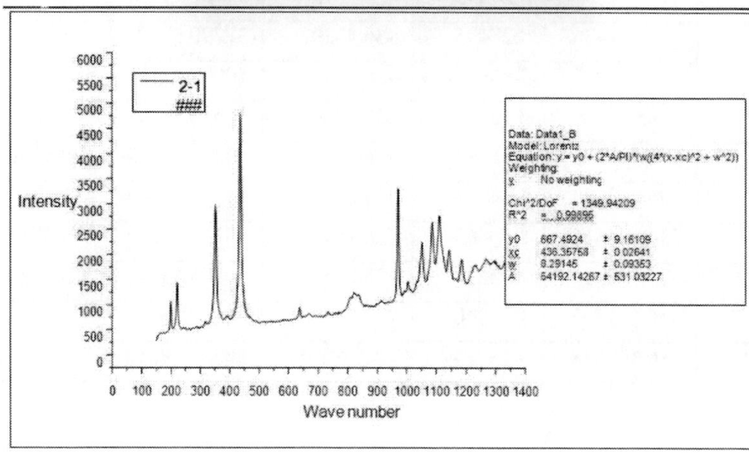

Figure 2.5.6. LHZ-002-2 zircon Raman experiments selected point direction (left) and Raman curve diagram (right).

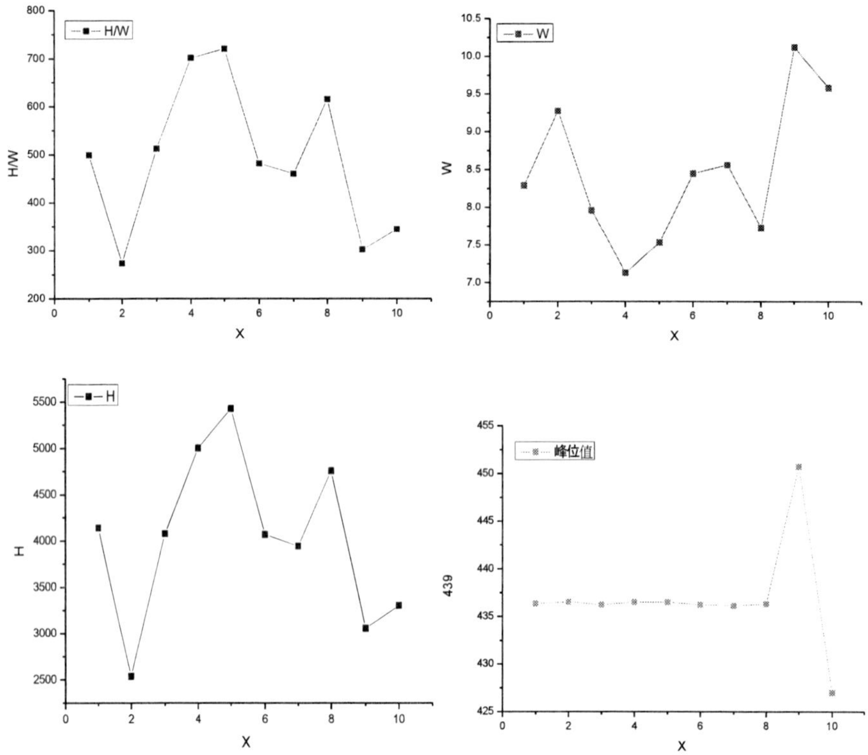

Figure 2.5.7. H, W, H/W and characteristic peak value variation of LHZ-002-2 zircon from core to rim.

Table 2.5.2. LHZ-002-2 zircon Raman curves' statistical table of characteristic peak value as well as H, W, H/W after fitting peaks at 439cm^{-1}

Number	H	W	439cm^{-1}	A	H/W
1	4139.07253	8.29145	436.35758	54192.14267	499.1976711
2	2537.41436	9.27435	436.53297	37299.07925	273.594846
3	4078.17037	7.96094	436.23471	50999.45174	512.2724666
4	5003.04759	7.13111	436.50846	56316.5819	701.5804819
5	5427.59878	7.53393	436.49679	64497.44644	720.4206543
6	4068.32383	8.44942	436.23698	54184.80694	481.4914905
7	3942.84518	8.56063	436.15784	53177.53853	460.5788569
8	4756.62324	7.73218	436.29532	57805.83648	615.1723369
9	3057.41557	10.12409	450.71366	48867.36714	301.9941121
10	3303.05228	9.58894	426.98934	50284.39058	344.464798

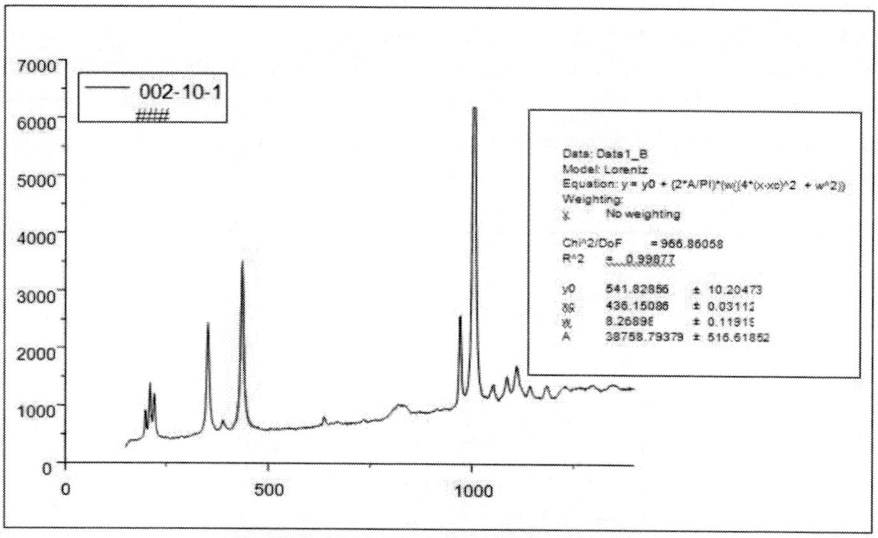

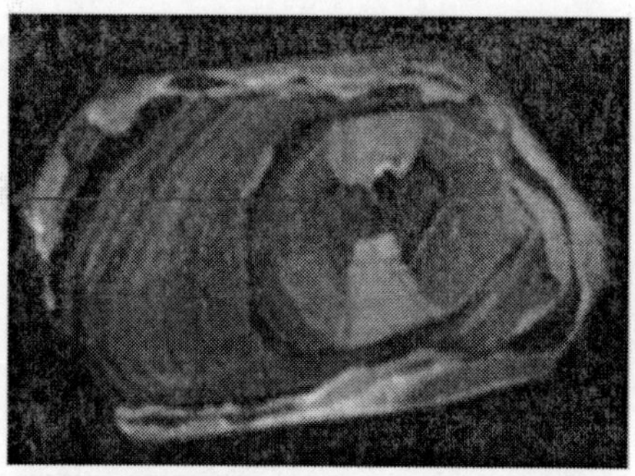

Figure 2.5.8. Testing direction and Raman characteristic curve of lhz-002-10 zircon.

Table 2.5.3. LHZ-002-10 zircon Raman curves' statistical table of characteristic peak value as well as H, W, H / W after fitting peaks at 439cm^{-1}

	Area	Center	Width	Offset	Hight	H/W
002-10-1	38759	436.15	8.269	541.83	2984	360.8658846
002-10-2	42304	436.32	8.1231	396.14	3315.4	408.1446738
002-10-3	46059	436.21	7.9585	387.35	3684.4	462.9515612
002-10-4	47136	436.23	7.553	292.67	3972.9	526.0029127
002-10-5	45495	436.19	8.0071	312.21	3617.1	451.7365838
002-10-6	51047	436.14	7.4836	266.64	4342.5	580.2688546
002-10-7	36239	435.95	8.088	301.28	2852.4	352.6706231
002-10-8	49700	436.23	7.6371	314.04	4142.9	542.4703094
002-10-9	42169	435.87	7.8793	671.16	3407.1	432.4115086
002-10-10	49010	436.09	7.5293	332.26	4143.9	550.3698883
002-10-11	46926	436.11	7.7565	394.87	3851.5	496.5512796
002-10-12	44177	435.98	7.6443	336.93	3679	481.2736287
002-10-13	53803	436.41	7.2873	445.73	4700.2	644.9851111
002-10-14	73339	436.08	8.3928	779.27	5563	662.8300448

By the comparative analysis of LHZ-002-10 zircon sample(figure 2.5.8, figure 2.5.9 and table 2.5.3), 3-4 magmatic zonings are found from core to margin. FWHM have a slightly decrease, peak value at 439cm^{-1} also decreases, but the relative intensity increases. Overall, H / W value increases. H / W value at core is relatively small and about 400. The value from intermediate to outer part increases to 650. Due to magmatism, metamorphism at margin is very slight

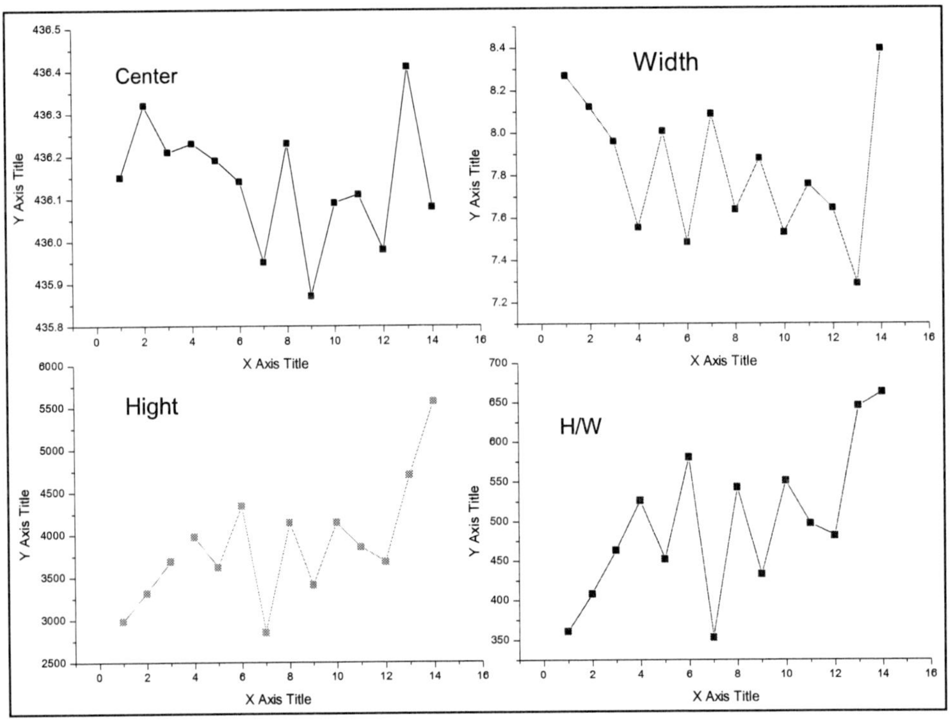

Figure 2.5.9. H, W, H/W and characteristic peak value variation of LHZ-002-10 zircon from core to margin.

2.5.4. Raman Curve Features of Zircons in Gneiss at Rongcheng, Weihan Area

Raman irradiations are conducted at the 20 selected points as the arrow shown in the figure2.5.10 and the Raman curve is obtained. As is shown in the CL images (figure 2.5.11, figure 2.5.12 and table 2.5.4), the zircon is divided into two parts: internal dark band and external bright band. According to the observation of Raman spectra, the intensity of 439cm^{-1} characteristic peaks decreases from the inner line 1 to outer line 10, however, that intensity of line 11-20 at the outer bright band.

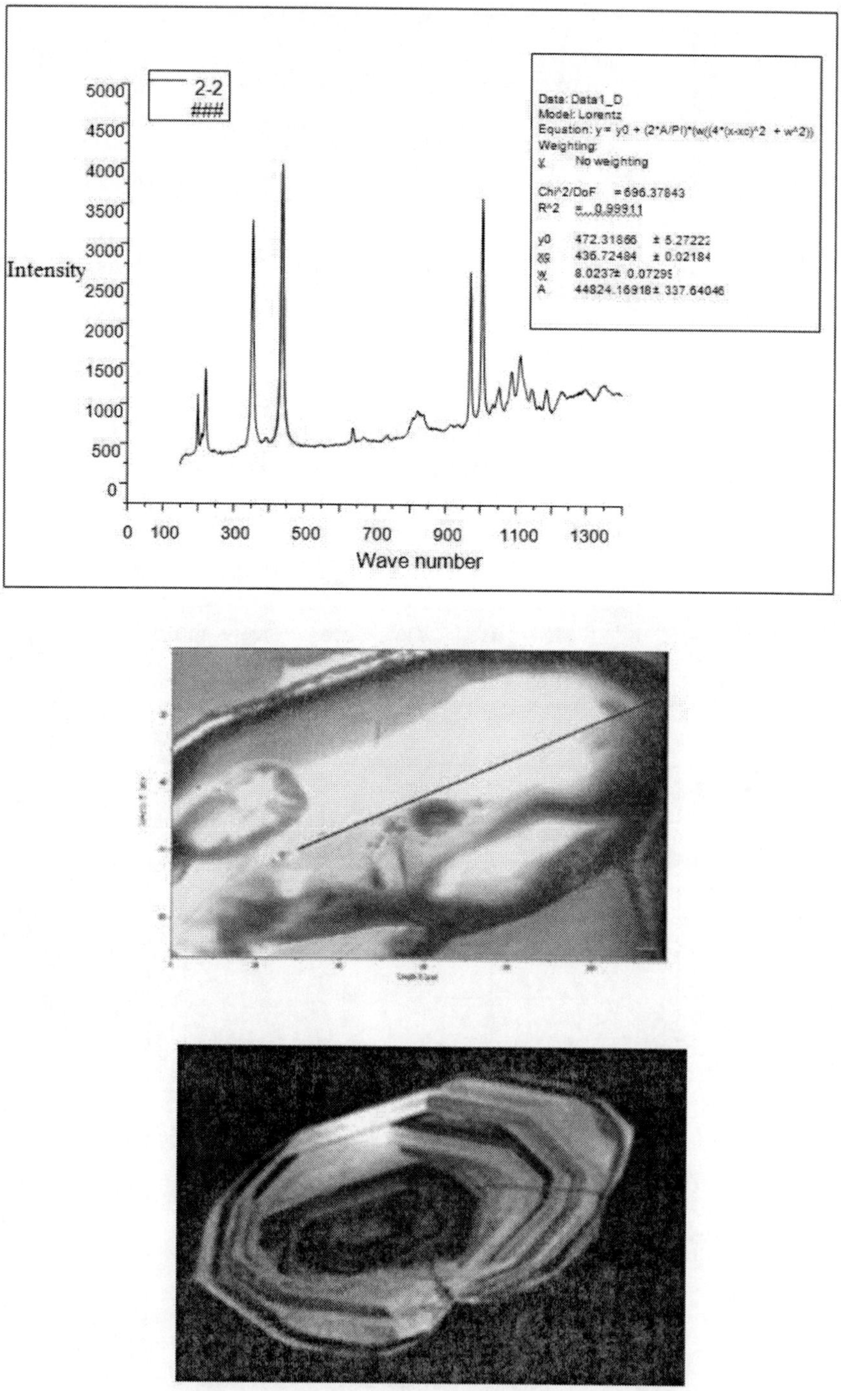

Figure 2.5.10. wh-15-2 test-point select direction in zircon Raman experiment(left) and schematic diagram of Raman curve (right).

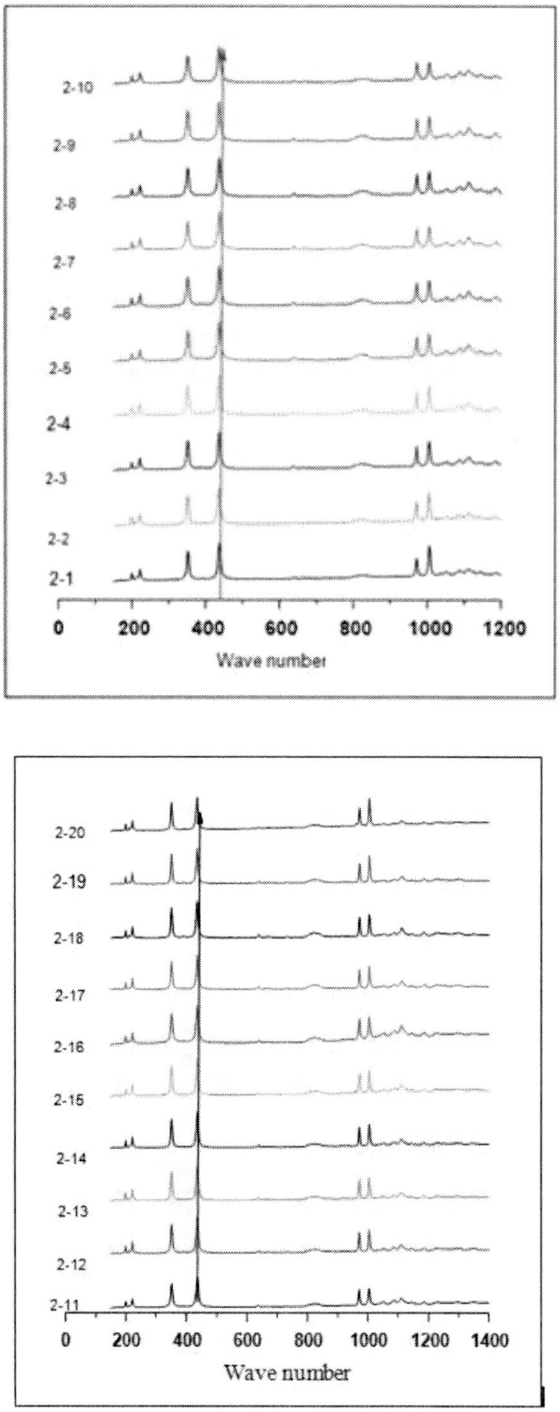

Figure 2.5.11. Raman curve comparison chart of 5-10 wh-15-2 zircon from core to margin.

Table 2.5.4. wh-015-2 zircon Raman curves' statistical table of characteristic peak value as well as H, W, H / W after fitting peaks at 439cm^{-1}

Number	W	y 0	y m	H	439cm^{-1}	A	H/W
1	7.93585	508.59645	4311.71702	3803.12057	437.04579	47992.19864	479.2329
2	8.0237	472.31866	3960.80564	3488.48698	436.72484	44824.16918	434.7729
3	7.92059	490.21922	4083.45	3593.23078	436.75655	44833.06671	453.657
4	7.93885	399.75819	3886.67427	3486.91608	436.70299	43961.6595	439.2218
5	8.56291	510.09218	3422.83545	2912.74327	436.70284	39278.33077	340.1581
6	8.2977	489.03496	3660.16038	3171.12542	436.71336	41810.25577	382.1692
7	8.25075	366.69413	3652.23911	3285.54498	436.72047	42746.67987	398.2117
8	8.42227	411.07058	3522.15132	3111.08074	436.63029	41178.19409	369.3874
9	8.4361	414.61311	3789.18931	3374.5762	436.63847	44775.73564	400.0161
10	9.29354	391.27177	3086.71849	2695.44672	436.84952	39420.93694	290.0344
11	8.07411	416.1688	3686.50765	3270.33885	436.85638	41652.32884	405.0402
12	7.03099	339.06852	4527.22189	4188.15337	436.89888	46842.22658	595.6705
13	7.12884	299.47394	4449.82283	4150.34889	436.79975	46585.97962	582.1913
14	7.03311	315.96623	4434.22671	4118.26048	436.88714	45944.40833	585.5533
15	7.17244	382.60659	4045.62372	3663.01713	436.89461	41804.15012	510.7073
16	7.04942	275.09658	4336.7203	4061.62372	436.75086	45006.77033	576.1642
17	7.09627	272.0352	4081.47203	3809.43683	436.59164	42556.6369	536.8224
18	7.06453	278.69971	4037.83499	3759.13528	436.59136	42330.31656	532.114
19	6.69674	262.49354	4501.13848	4238.64494	436.75037	45127.22597	632.9415
20	6.87832	455.08593	4343.51457	3888.42864	436.36253	42122.09626	565.3166

In addition, from the zircon's core to margin, there is trend that values at 439cm^{-1} Raman characteristic peak shifted to high wavenumber area. All of these features demonstrate that zoning structures are well developed in zircon, the core inherited residual zircon and a significant metamorphism occurred at the margin.

Bao Xuezhao et al. (1998) [41] defined Δ 356 = H / W (the relative intensity to FWHM ratio at 356cm^{-1} characteristic peak) and used it to judge zircon's metamict degree in studying the origin discrimination of zircon's micro-Raman spectroscopy.Δ 439 = H / W is similarly defined and proportional to zircon's metamict degree.

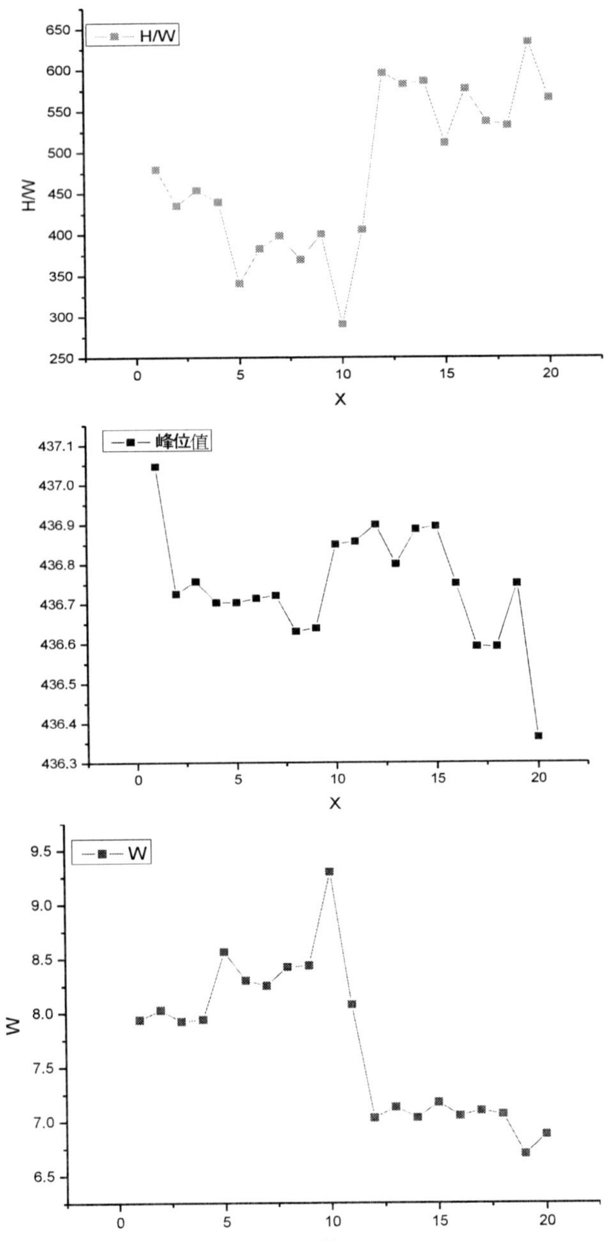

Figure 2.5.12. (Continued).

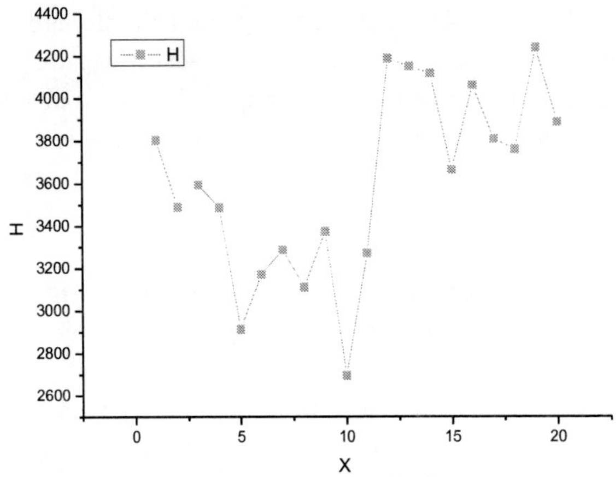

Figure 2.5.12. H, W, H / W and characteristic peak value variation of wh-015-2 zircon from core to margin.

Quantitative analysis of Raman characteristic quantities in Table 5.4, zircon with obvious core and rim, the variation of H, W and 439cm^{-1} peak position can be divided into two segments, from inside to outside within the core section, Δ439=H/W decreases gradually(reduce from 479 to 290); at the junction of core and rim, Δ439=H/W jumps(290-405-595); in the rim area, Δ439=H/W remains stable between 500 to 650, special point may be different, for example,Δ439=H/W of No.15 test point in non-luminous zone of bright area decreases sharply; the change of W consistent with variation trend of Δ439=H/W.

We can draw a conclusion that this zircon is of metamorphic origin, the metamorphism have two stages, the core area have gone through a stronger metamorphism compared to the rim part. In the core area, the metamorphism stronger from inside to outside; in the rim area, the intensity of metamorphism changes over the variation of oscillatory zoning, non-luminous zone represents stronger metamorphism compared with luminous zone.

2.6. Relationship between Raman Features of Zircon and Regional Tectonic Thermal History

2.6.1. Analysis of the Regional Tectonic Thermal History of the Research Area

Rongcheng in Weihai, Yakou in Qingdao, Lingshan Island in Jiaonan and other areas geographically belong to the eastern Shandong Province, structurally divided to Jiaonan-Weihai orogenic belt in Jiaodong of Sulu orogenic belt.

Sulu orogenic belt is in the eastern part of Qinling-Dabie orogenic belt. It is the product of Triassic subduction and collision between the Yangtze Plate and North China Plate. A lot of high pressure-ultrahigh pressure metamorphic rocks occur intermittently cut off by Meso-Cenozoic basins and granite [42-45].

Sulu HP - UHP metamorphic belt mainly consist of Neoproterozoic granite (gneiss) rocks with developed eclogite (Tang Jun et al, 2005) [46], and Sulu eclogite and rocks may have lived through the HP-UHP events [45]. Summarizing the isotopic age data, it is found that eclogite ages focused on three stages: 720-900 Ma, 418-495 Ma and 207-313 Ma. It can be inferred that this three historical periods played an important role in the evolution of eclogite, the following were analysis of three periods of tectonic evolution Sulu terrane.

(1) Neoproterozoic Jinningian

Metamorphic granitic gneiss accounted for more than 80 percent of the total content of Sulu HP - UHP metamorphic belt. Li Shuguang et al think they are intrusive rocks of Neoproterozoic Jinningian. Such metamorphic granitic gneiss has various types, uniform structure and many intrusive contact relationship outcrops, It constitutes regional rocks of eclogite. The original rocks was a product of magmatic activity in the tectonic setting of the early subduction - interim collision - late stretch, they are gneiss combination of granitic diorite - adamellite - alkaline granite, with a more complete evolution features of collision orogenic granite series of S-I-A type.

This collision during that time occurred within 130Ma ±intervals. Relativing to the long Proterozoic, the structure - metamorphic - magmatism with unexpected features is the mainly formed period of HP - UHP metamorphic belt. But the fast pyrolysis and separation of North China plate and Yangtze plate, was due to the strong extension in late orogenic and a large number of "A"-type granites were formed which stretched and rised, breaking the continent, and the southern cleavage was more strong.

(2) Paleozoic - Early Mesozoic Caledonian and Indosinian

After Neoproterozoic rifting, orogenic accepted Aurora - Cambrian sedimentary which had been eroded after a large-scale of orogenic uplift. No reliable Caledonian intrusive rocks have been found.

The presence of intrusive rocks in Indosinian has been confirmed by multi-information, the rock types are mainly diorite, monzonite granite and syenite series. SHRIMP study found that granitic gneiss in HP - UHP metamorphic belt has the features of multiple era evolution. The pinnacle of zircon ages are divied into two section: 600-859 Ma and 220-250 Ma. Granitic gneiss were mainly formed in Jinningian and Indosinian. After Caledonian, tectonic and magmatic activity became intense, which increased from north to south, this law performaned evidently in the Triassic magmatism [47].

During Indo movement, magmatic events were not developed and in the upper crust mainly existed land-land compressive stress. There were no subduction-related volcanic rocks and granitic magma intrusion events were developed in small-scale[47]. Dive and collage was slower and collision is not strong enough to cause strong magmatic activity. It is more similar to the mechanism of crust and mantle detachment and intracontinental subduction.

(3) Intracontinental orogeny in the Cenozoic

Tectonic - thermal event in Late Jurassic, Jiaonan – Weihai mainly occurred in the 150-160Ma. The rock's gneissosity feature were developed. According to the existing structural analysis of magmatic zircons, oscillatory zoning at zircon edge was developed, with typical magmatic characteristics. The presence of inherited zircon cores indicates that it experienced

UHPM. The magmatism was related to squeeze and deformation of continental margin caused by ancient Pacific plate's subduction and convergence to the mainland. It was the result of the crustal contraction and deformation in initial convergence.

In early Cretaceous, tectonic - thermal event lasted 135-125Ma, occurred in the tectonic setting of continental rifting. Magmatic activity during this period was in smaller-scale, mostly in the form of rock plants, and had bimodal features. The original rock came from partial melting of the North substrate [40].

In the middle and late of early Cretaceous, magmatism were and intense, occurred in the 125-105Ma. More of them developed in the eastern side of the Tan-Lu fault. Granite era was mainly in the 100-120Ma, granite porphyry era was mainly in 120-125Ma, and the most intense volcanic eruption occurred in the 110-115Ma. In this case it happened to be the peak of a large-scale lithospheric thinning and continental rifting in eastern China. The subduction of Ancient Pacific plate caused the asthenosphere's heat upwelling, metasomatism, erosion, melting of the materia at mantle bottom, and magma rising, underplating the crust, causing partial melting of crustal rocks at the bottom, forming a bimodal early Cretaceous igneous rocks [45, 48] .

Magmatic activity rarely showed during 102-92Ma. Late Cretaceous magmatic activity began in 92Ma or 85Ma, mainly developed in the eastern Shandong Jiaolai basin basalts (73Ma). Paleocene basalt (stratigraphic age is 65-45Ma) developing along Tan-Lu fault indicates that during late Cretaceous- early Tertiary, asthenosphere upwelled constantly, the crust tensile strength increased, and continued thinning [45, 47, 49].

2.6.2. Geological Significance of Raman Features of Zircon Micro Area

The samples from three areas have many differences in lithology, formation, age and many other aspects. According to the studies on geological age, blended granite-gneiss of Rongcheng area in Weihai ages 543-800Ma, which belongs to neoproterozoic erathem nanhua period and sinian period. A few samples age in 513-543Ma for the zircon edges may have experienced the early Cambrian metamorphism. Most Tuff zircons in Tiger Mouth, Lingshan Island, Jiaonan age in 96-137Ma (Early Cretaceous), while a few belongs to Middle-late Jurassic and some reaches 550-600Ma. Speculating that they are the residual pits of zircon in simian period, there may be zircon experienced Paleozoic complex metamorphism deformation between residual pit and new edge.

The newest zircon era of col in Mountain Laoshan is about 120Ma, which belongs to early cretaceous. While the older part is about 700Ma, which is speculated as neoproterozoic erathem residual pits. And the oldest part is 2000+Ma, which is speculated as palaeoproterozoic metamorphosed basement residual pits by excluding other disturbance.

According to the dating data above, combined with the Laser Raman features of zircon micro area, we can conclude that different parts of zircon micro area have different genesis geological features, which are determined by zircon formation geological conditions, in other words, by the tectonics background at that time. Outward manifestation of micro area structure and Laser Raman curve features are connected with mineral genesis background.

Table 2.6.1. statistical table of 439cm-1 Raman peak Δ 439 of zircon from different regions

	H/W	〈300	300-400	400-500	500-600	600-700	〉700	Age
Laoshan-Yakou	lsyk-3bright band		22.20%	77.80%				118.9 Ma
	lsyk-3dark band	12.50%	37.50%	37.50%	12.50%			
	lsyk-7	8.30%	16.70%	25.00%	41.70%	8.30%		
	lsyk-8		37.50%	50.00%	12.50%			
Jiaonan-Lingshandao	lhz-002-1 lengthways	66.70%	33.30%					119.2 Ma
	lhz-002-1 crossise	75.00%	25.00%					
	lhz-002-2	10%	20%	30%	10%	10%	20%	
	lhz-002-6	90.90%	9.00%					
	lhz-002-5	42.90%		57%				
	lhz-002-10		14.30%	42.90%	28.60%	14.30%		
Weihai-Rongcheng	wh-015-1crossise			13.30%	46.70%	40.00%		682-787Ma
	wh-015-1 lengthways			10.00%	30.00%	60.00%		
	wh-015-2	5%	20%	30%	40%	5%		
	wh-015-3	10.50%	21.10%	26.30%	21.10%	21.10%		
	wh-015-4	21.40%	21.40%		21.40%	28.60%	7.10%	
	wh-015-5 lengthways				36.40%	63.60%		
	wh-015-5 crossise				40%	40%	20%	

Therefore we can speculate that forms of zircon micro area structure and Laser Raman features have a relationship with historical conditions like tectonic backgrounds. That is the final aim of the thesis.

As shown in table 2.6.1, accounting the characteristic quantity Δ439 of zircon grains micro area Raman curve in Mountain Laoshan Yakou, Lingshan Island and Rongcheng area, we have found that characteristic quantity of zircon grains from different eras and areas has a certain range. The Δ439 value of metamorphic rocks of Rongcheng in Weihai is over large on the whole, it is larger than 500 in zircon grains developed fan-shaped zone, sometimes reaches above 700, just a little part between 400 and 500. 500 is a boundary for the common zircon grains without fan-shaped zone, two parts are in equal proportion.

The Δ439 values in Raman curve of zircon grains in Lingshan Island are lower than 500 mostly.

The zircon grains with good crystal morphology and clear oscillatory zoning have a Δ439 lower than 400, while those zircon grains with complicated inner structure and having experienced multiperiod magmatism have a scattered Δ439 value.

The Δ439 values in Raman curve of zircon of tuff in Mountain Laoshan focus between 300 and 600, indicating an intensive magmatism. Based on the above analysis, we conclude that the stratigraphic age grow younger from Rongcheng area to Mountain Laoshan and Lingshan Island. And the Δ439 value in Raman curve of zircon grains has a decreasing trend.

Combine with the tectonic evolution history of sulu orogenic belt, we conclude the reason of the rules above: zircons in Rongcheng area formed early, having experienced many tectono-magmatic events, such as the metamorphic recrystallization, hyperplasia of metamorphism, metamorphic freshman and other geological processes since proterozoic era, resulting in a complicated internal partition and a higher Δ439. However, zircons of tuff in mountain Laoshan Yakou and Lingshan Island formed late, mainly influenced by magmatism, resulting in a regular appearance, an easier internal structure and a lower Δ439 relatively.

2.7. Conclusion

Studying the systematic mineralogical characteristics, petrography, internal structure, major elements, trace elements, and laser Raman spectral features of zircon by analyzing the rock samples from Wei Hai Rongcheng, Laoshan Yakou and Jiao Nan, together with the background tectonic evolution investigation, a detailed study of the internal structure of the sample, geochemical and laser Raman features analysis, and a combination analysis of the regional tectonic settings in eastern Sulu orogen Jiaonan - Weihai orogenic, we concluded the following conclusions:

(1) Different geological formation process of zircon leads to its different internal structure. In cathodoluminescence images, most magmatic zircons crystals are euhedral - subhedral, in which the magmatic oscillatory zoning is characteristically formed. Metamorphic zircon crystals vary in shape, from anhedral to euhedral. Its

internal structure varies and its shape and internal structure changes with changing metamorphic conditions;

(2) With the electronic probe, major elements and trace elements in different causative zircon have different variation trend. In the magmatic zircon, from the core to edge in the crystals (the band is not obvious in the crystal) and in each band of the crystal, the content of the HfO2 increases from the inner side to the outside.the value of ZrO2 and the rate of ZrO2 / HfO2 decrease; In the metamorphic zircon, from the center to the edge crystal the content of ZrO2 increases;the level of HfO2 drops, and with the characteristics showing ZrO2 / HfO2 value increasing. Thus this is opposite to the magmatic zircon.

(3) Laser Raman microprobe is a good way to detect the geological cause of zircon formation. From the core to the edge of magmatic zircon, the characteristic peak of Raman curve gradually declines, and the ratio between relative intensity and half-width height increases with the increasing metamorphic intensity. From the core to the edge of metamorphic zircon, the characteristic peak of Raman curve intensity increases or basically remain unchanged. The stronger the metamorphism is, the higher the ratio between relative intensity and half-width height would be, and the half high width will increase generally.

(4) The eastern Shandong region in Sulu orogenic belt has experienced complex tectonic and magmatic activities in the geological history, causing complex internal structure characteristics of zircon in geological body. Using laser Raman curve, we can better understand the internal structure of zircon and its origion. In the direction of Rongcheng-Yakou-Lingshandao, the stratigraphic age gradually becomes newer, the value of delta 439 in zircon Raman characteristic peak has a declining trend, the metamorphic intensity changes from strong to weak.

SECTION 3. LASER RAMAN CHARACTERISTIC OF ZONING WITHIN APATITE

As a common accessory mineral in various types of rocks, apatite is commonly used in studies of isotope tracer and low temperature geochronology due to its stability over a wide range of temperature and pressure and abundance of trace and rare-earth elements. Changes in composition, temperature and pressure of surrounding melt result in the chemical zoning within apatite.

In specific temperature and pressure condition, chemical equilibrium exist between outer zone of zoning apatite and matrix, but it does not exist between inner zone and outer zone of zoning apatite or between inner zone and matrix, so the variation of growth environment can be recorded by the regular change in content of trace elements from core to edge of apatite.

In this experiment, laser Raman spectroscopy was applied to analyze zoning within apatite separated from sedimentary rocks and metamorphic rocks, according to characteristic peak's variation in laser Raman spectra of different zones within apatite and the known thermal history of apatite, the features and rules of different zones within apatite were inferred, which could help explore new methods applied on the low temperature geochronology of apatite.

3.1. Sample Selecting

There are differences between the apatite have gone through middle to high temperature metamorphism and apatite from clastic rocks just have gone through low to middle temperature. 4 metamorphic rock samples of Sulu orogenic belt, eclogite 10-9, granitic gneiss WH-015, gneissic migmatitic granite WH-014,brick-red migmatitic granite WH-006, and 5 sandstone samples of Songliao basin, YC-2, SL-1, SL-2, SL-3, SL-4, were selected as representatives respectively. Samples were grinded and apatite were selected. The size and content of apatite in those samples are different, apatite volume percentages in different samples vary from less than 1% to 5%.

3.2. Test Method

Though microscopy, the structure of apatite and characteristic of adjacent mineral were observed.

The zoning and morphological characteristics and crystal orientation of apatite can be obtained through Cathodoluminescence (CL) characteristics. Electronic probe (EPMA) can be used to detect trace element content of apatite, which can help research the correlation and coupling alternation of elements within apatite, combined with zoning structure showed in CL images, rare element distribution characteristic in the core, mantle and rim or the luminescence band and non- luminescence band of apatite can be analyzed.

Depending on the zoning and growth characteristic of apatite, test point can be designed. Using Laser Raman spectrometer, some valuable spectrogram can be got.

The variation in shifts, intensity, FWHM of characteristic peaks can be compared and anlysed, the envelope curve can be fitted and stripped, the site and intensity of the original peaks can be analysed rationally, then, the reasons for the variation can be got with the chemical crystal characteristics and crystal structure characteristics of apatite.

Combined the research on zoning within apatite and trace element distribution in apatite with the basic knowledge of distribution and diffusion of trace element and crystal chemical characteristics of apatite, the relationship between temperature and pressure changes in mineral growth with the zoning within apatite and trace element distribution in apatite can be got. Then, the corresponding variation in laser Raman spectra caused by the change of the growth environment can be discussed if the thermal history of the apatite have been known.

3.3. Structure Characteristic of Zoning within Apatite

After making chosed single grain apatite into sample targets, the cathodoluminescence technology was used to observe the microstructure of apatite and the luminescence characteristics of apatite were got, which can help us identify the apatite with bands or zoning characteristic.

The luminescence characteristics of apatite separated from different samples were different, this may be the combined effect of the stimulation and inhibition of partial ions. Bands or zoning were observed in some apatite, which may be caused by the inhomogeneous distribution of trace element in apatite.

3.4. Experimental Process of Cathodoluminescence

Cathodoluminescence characteristic of apatite were tested in the laboratory of China University of Petroleum (East China), the ELM-3RX cathodoluminescence system and Nikon DS-Ril microscope were used in this experiment. Test conditions: operating voltage is 10KV, electric current is 0.3mA, time of exposure ranges from 60ms to 600ms depending on the condition of sample.

The magnification was adjusted to the size of apatite grain, the acceleration voltage and electron beam of incident beam which used to stimulate the apatite were regulated to 15KV and 220μA, respectively.

3.4.1. Structure Characteristic of Apatite Separated from Sandstone Samples of Songliao Basin

In CL (cathodoluminescence) images of apatite separated from sample SL-1, the apatite were dark green, bright yellow green, green, with dark green being the main color. 3 apatite single grains could be observed the zoning structure which can not be identified in the general microscope (figure 3.4.1) and we numbered them 1-1, 1-2, 1-3.

In CL (cathodoluminescence) images of apatite separated from sample SL-2, the apatite were dark yellow green, yellow, with dark yellow green being the main color. 1 apatite single grain could be observed the zoning structure obviously (figure 3.4.2), and we numbered it 2.

In CL images of apatite separated from sample SL-3, the apatite were dark green or non-luminance. 2 apatite single grains could be observed the zoning structure, and we numbered them 3-1,3-2 (figure 3.4.3).

In CL images of apatite separated from sample SL-4, the apatite were yellow, yellow green, green, with yellow green being the main color. One apatite single grain could be observed the zoning structure obviously (figure 3.4.4) and we numbered it 4.

Figure 3.4.1. CL image of sample SL-1 ,a:No.1-1 ; b:No.1-2 ; c:No.1-3.

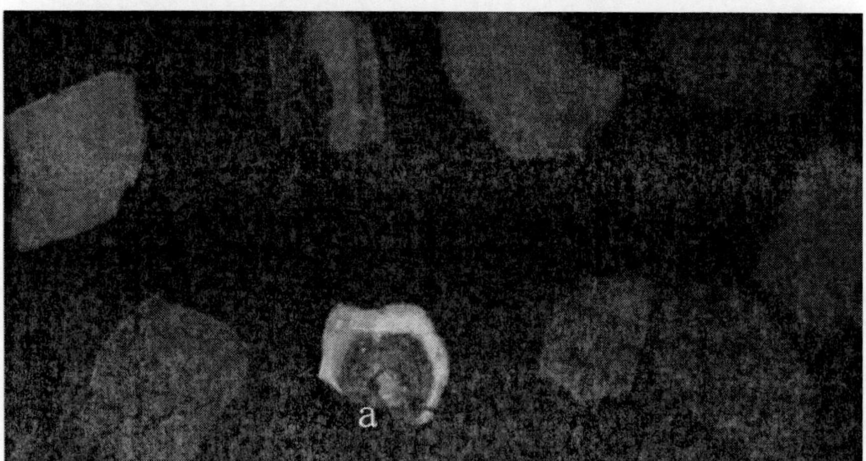

Figure 3.4.2. CL image of sample SL-2 ,a:No.2.

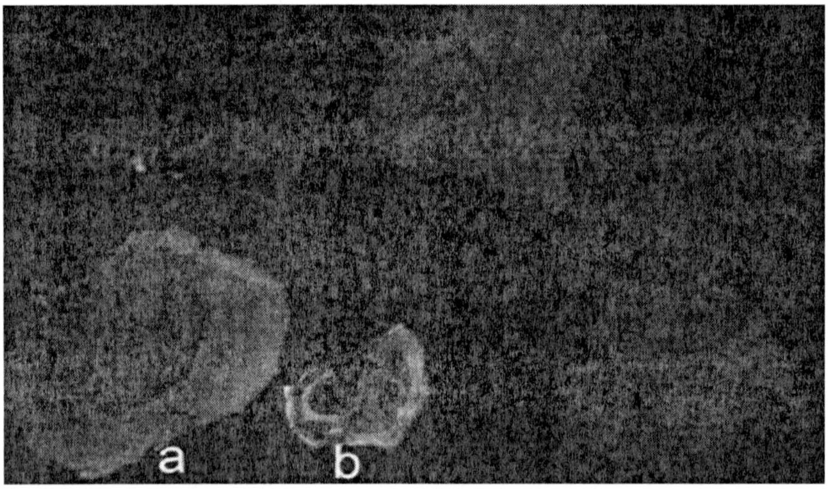

Figure 3.4.3. CL image of sample SL-3, a: No.3-1,b: No.3-2.

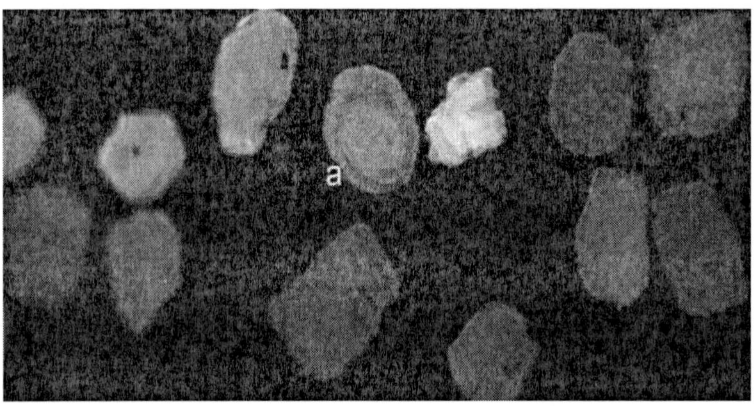

Figure 3.4.4. CL image of sample SL-4, a: No.4.

In CL images of apatite separated from sample YC-2, no apatite single grain with zoning structure were observed.

In conclusion, the band and zoning structure observed in apatite separated from Songliao Basin sandstone are center-based and the rim of those apatite is luminescence zone. In the core of those apatite always lack fine scales vibration band, but high frequency vibration band can be observed in the mantle part.

Those indicate that the apatite grow slowly in core-forming stage and grow rapidly in mantle-forming stage. The luminescence zones have many different types, asymmetric growth of zones may be controlled by apatite's structural position in sample and may related to the interaction with the surrounding paragenetic mineral.

3.4.2. Structure Characteristic of Apatite Separated from Eclogite Samples of Sulu Orogenic Belt

In CL images of apatite separated from eclogite sample 10-9, the apatite were bright yellow green, green, 60% of those apatite grains' color are bright and uniform.

Most of apatite grains with inhomogenous color have zoning structure from core to rim, most of the boundaries between zones are not regular and they are not parallel to the crystal growth face. 7 apatite single grains could be observed the zoning structure, and we numbered them 5-1, 5-2, 5-3, 5-4, 5-5, 5-6, 5-7, respectively (figure 3.4.5).

Figure 3.4.5. CL image of eclogite sample 10-9,a,b,c,d,e,f,g: No.5-1, No.5-2, No.5-3, No.5-4, No.5-5, No.5-6, No.5-7.

In CL images of apatite separated from granitic gneiss WH-015, gneissic migmatitic granite WH-014,brick-red migmatitic granite WH-006, the apatite were uniform yellow, yellow green, no apatite single grains could be observed the zoning structure obviously.

In CL images of apatite separated from Sulu orogenic belt, apatite were yellow, yellow green and brown, those apatite which can be observed zoning structure do not have vibration band and the core-mantle boundaries of them are not straight, the outlines of boundaries are always smooth oval-shaped or embayment, which are normal in the early formed zones, this may be contributed to the partial dissolution may have happened in phase interval during crystal early growth stage. Some other scholars, Hodson and Finch considered this may be contributed to significantly different environment that the apatite go through in the early and late stage and the growth of apatite is related to episodic hydrothermal activity [50].

3.4.3. Conclusion

Because the crystal orientation of apatite is random, different section direction may influence the color and shape of the zoning in CL image. Cathodoluminescence microscopic imaging technology can image the internal structure and variation of natural apatite, which can offer information about rock formation and mineralization, especially the evolution of the fluid.

Apatite in certain condition exist weak zoning structure, CL imaging can help identity zoning. Through research on man-made and natural apatite, some scholars discover characteristic emission band of activator ions such as Mn^{+}、Eu^{2+}、Ce^{3+} and location of the emission peak of rare earth elements ion [51-58]. U. Kempe research on Kovdor, Mushugai and Svensken, conclude that cathode emission spectrum is activated by rare earth elements [59]. Blue cathode emission spectrum (about 410-430 nm) is related to Eu^{2+} [60], cathodoluminescence activity Ce^{3+} result in lavender (about 360nm, near ultraviolet), red and orange cathode emission spectra is related to rare earth ion, the other cathode emission spectra may be caused by Dy^{3+} and Nd3+ (Figure 3.4.6) .If AS^{5+}、Fe^{2+} and other ions are included in apatite structure, this can inhibit luminescent intensity [61, 62].

In this paper, the zoning structure of apatite is related to the variation of chemical component and element content in rim, mantle and core of apatite.

Zoning within apatite can be divided into three types: (1) Oscillatory zoning exist in the core, mantle and rim of apatite. Boundary between core and mantle and boundary between mantle and rim are distinguished by overall light and shade degree of oscillatory zoning. Zoning is center based, the boundaries between different zones are always parallel to the crystal growth face (figure 3.4-1a, figure 3.4-1c, figure 3.4-2a), the outlines of zoning boundary are always straight, the crystalline form of apatite develops well.(2) The core is an uniform ensemble, oscillatory zoning always develops in the mantle and rim. Boundary between core and mantle is obvious, while there is a gradual transition from mantle to rim and do not exist obvious boundary.

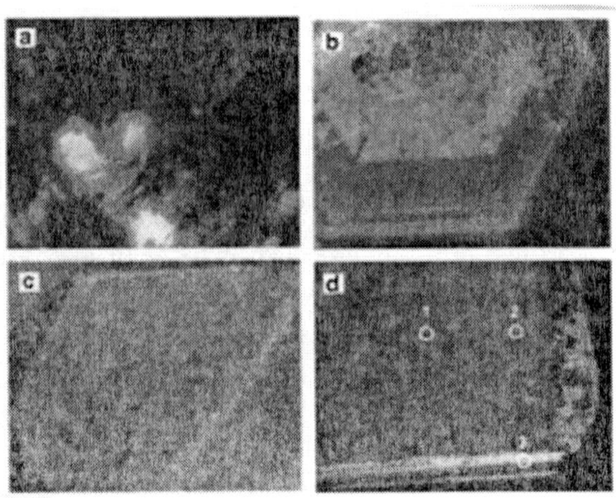

Figure 3.4.6. Different color showed in CL images of apatite.

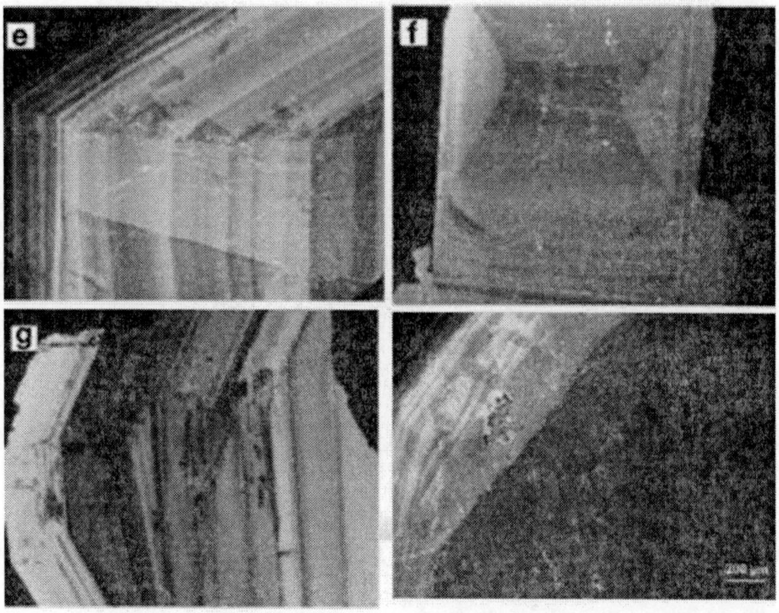

Figure 3.4.7. Zoning structure showed in CL images of apatite.

The core generally has a regular outline, oscillatory zoning is parallel to the crystal growth face and the shape of it is regular (figure 3.4-6d, figure 3.4.1b, figure 3.4-2a). Apatite with those characteristics may crystallize in early stage and continue growing in later period, apatite may suspend in the fluid during growth and excitation elements which control the luminescent effect of apatite can only diffuse limitedly [63]. (3) There is almost no oscillatory zoning existing in the core, mantle and rim, those parts can be distinguished by different luminescent colors, boundaries are obvious but not regular, sometimes, they can be embayment (figure 3.4.5), dissolution may have happened during crystal growth.

3.5. Geochemistry Characteristic of Apatite

Electron probe can test elements composition of apatite, through this technology, composition zoning structure (content of trace elements vary from the core to the edge) in the apatite separated from Sulu orogenic belt is observed clearly.

3.5.1. Experimental Procedure of Electron Microprobe

Instrument JEOL, JXA-8230 was used for this electron microprobe experiment.

Accelerating voltage was set to 15KV and electric current was 10mA. Before test, JEOL, JEE-420 plated carbon analyzer was used to plated carbon to the fixed target of apatite and thickness of the carbon film was controlled to about 20nm.

Then calibrated the instrument with the international standard sample, in order to ensure the accuracy of the experimental test results, we used the Si comes from chlorite of West Chester Co., PA and USA, the Ca, P, F, S come from apatite of Wilberforce, Ontario, Canada, the Cl comes from tugtupite of Greenland, the La comes from monazite of Greenland, the Mn comes from zinc ore of Franklin Furnace, NJ, USA, and the Na comes from chrome diopside of Batbjerg, E.Greenland to calibrate the elements to be test.

Then placed the target apatite with carbon film to the sample compartment, adjusted the electron gun and the working distance, and chose the site to test. Keep the temperature at a constant condition of 20°C in the laboratory, during the experiment, calibrate accurately the working distance of each point to ensure the validity of the results.

3.5.2. The Elements Distribution of Apatite in Sandstone in Songliao Basin

We conducted a line scan as the arrow showed in figure 3.5.1 to the No.1-1 apatite with electron probe. As can be seen in the figure of elements distribution, the area of F content decreases corresponds to the area of Cl content increases, and the area of Si content increases corresponds to where the P content decreases. Such shift phenomenon may occur due to the Si-O tetrahedral occupying the position of P-O tetrahedral and Si substituting the P element.

As can be seen in the figure of elements distribution of No.1-2 apatite, the area of F content decreases corresponds to the area of Cl content increases, and the area of Si content increases corresponds to where the P content decreases. The Mn and La content change obviously in a small range, which curve are jagged; the next are Cl and F with relatively small fluctuation, which curve are jagged; the fluctuation of Si and P content curve are the smallest (figure 3.5.2).

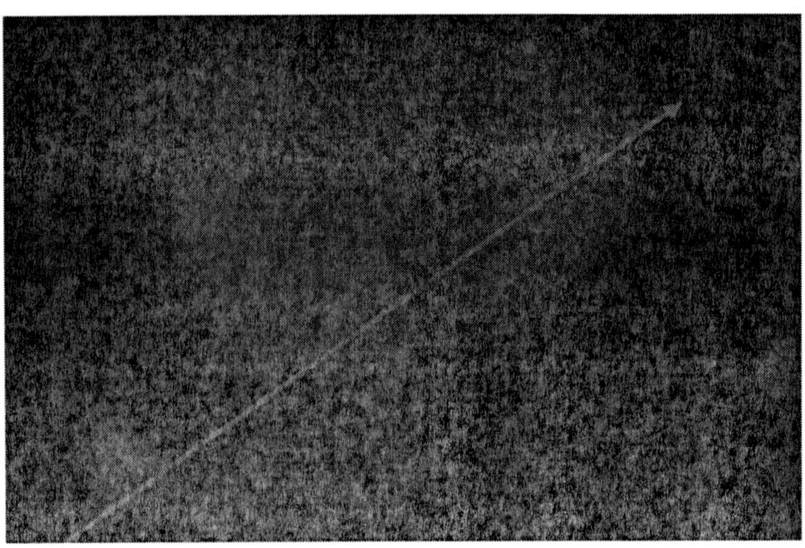

Figure 3.5.1. (Continued).

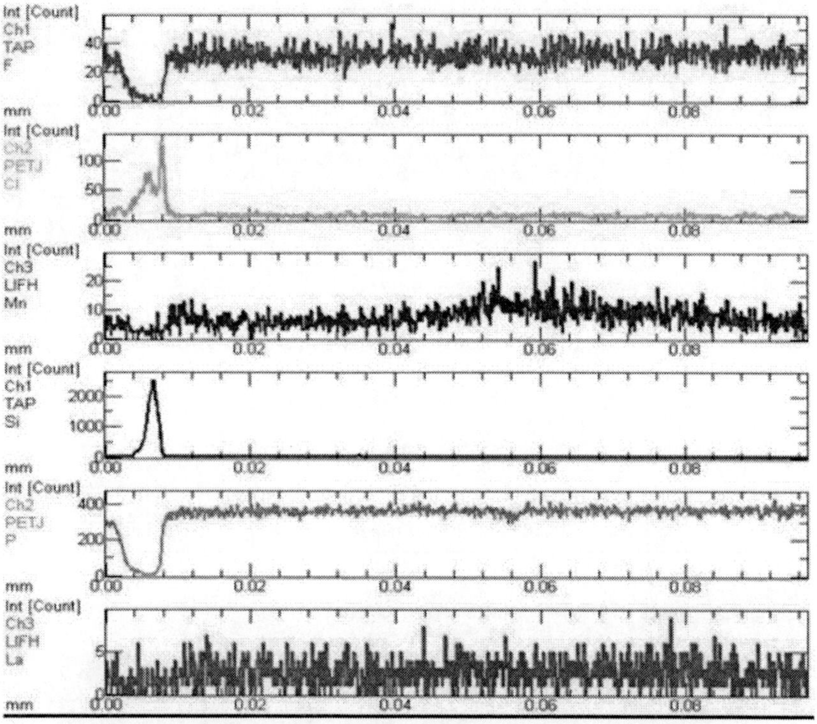

Figure 3.5.1. Elements chart of No.1-1 apatite cross section.

The Si content of No.1-3 apatite decreases obviously in the range from 0.00mm to 0.01mm. The F, Cl, Mn, P and La content change slightly, but the fluctuation of F, Cl, Mn and La content curve are obvious, which are jagged. While the fluctuation range of Si and P content is relatively slight (figure 3.5.3).

Figure 3.5.2. (Continued).

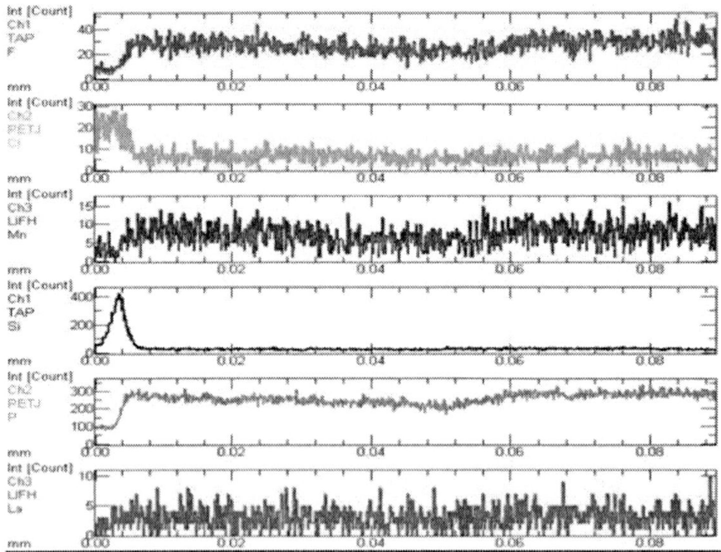

Figure 3.5.2. Elements chart of No.1-2 apatite cross section.

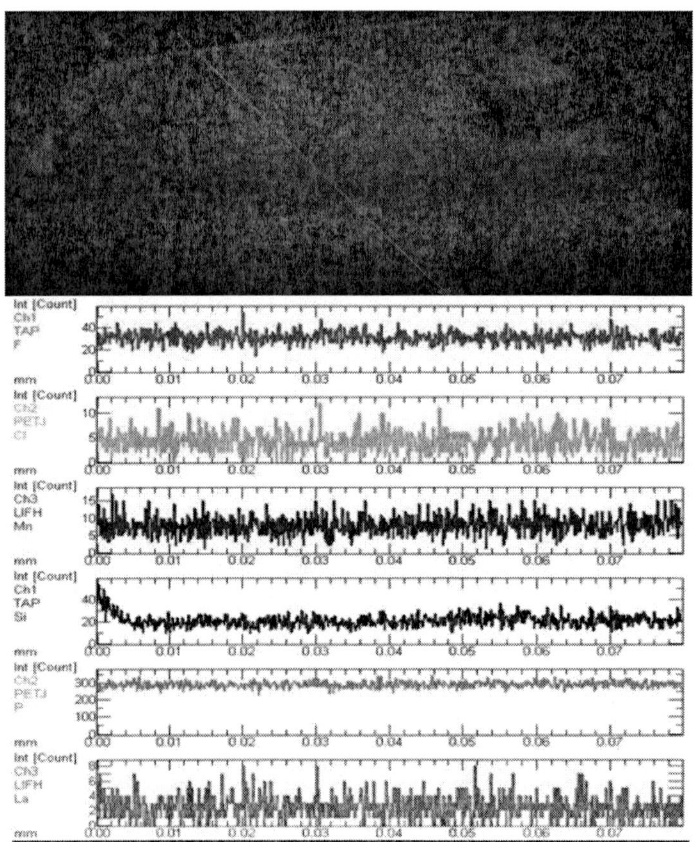

Figure 3.5.3. Elements chart of No.1-3 apatite cross section.

We tested the No.2 apatite as showed in the figure 3.5.4, Overall, F, Cl, Mn and La elements content charts'jagged fluctuation are relatively obvious while the curve of Si and P are on the contrary. This similar complementary characteristic of F and Cl elements may due to the different elements selection of channel-ions in apatite crystals during different growth stages. It may occurred that Si-O substituted P-O inside the crystals in where the zoning of Si content increases significantly. Mn and La content are also different in different zoning.

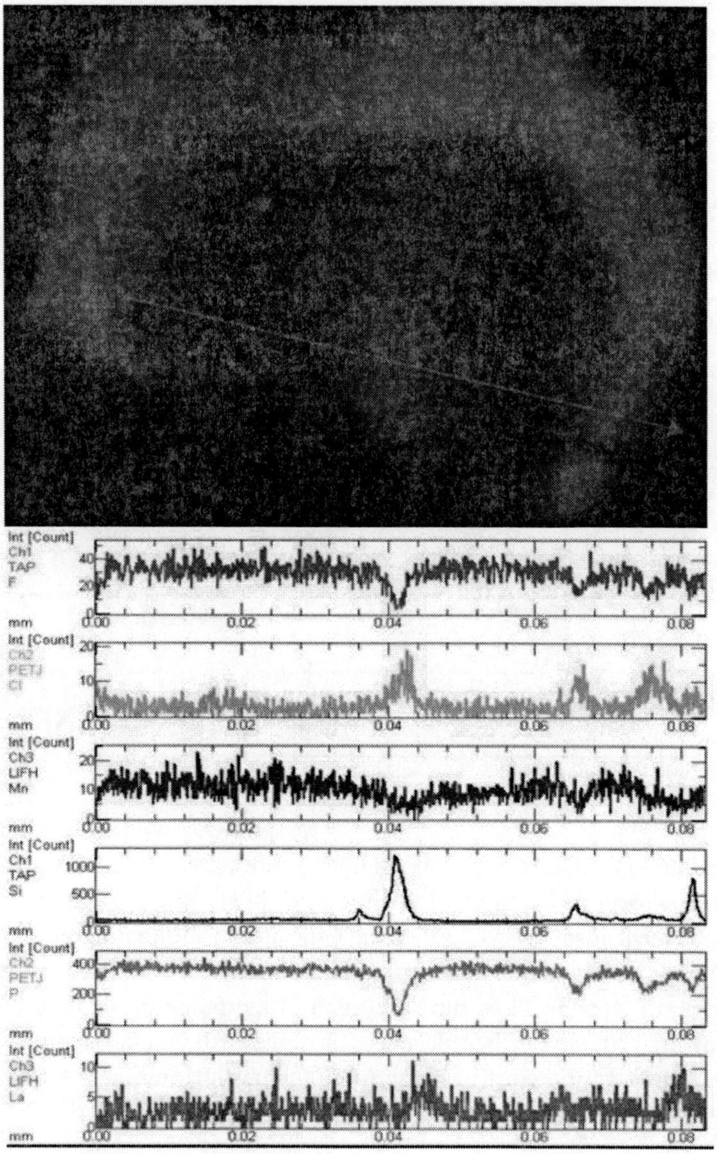

Figure 3.5.4. Elements chart of No.2 apatite cross section.

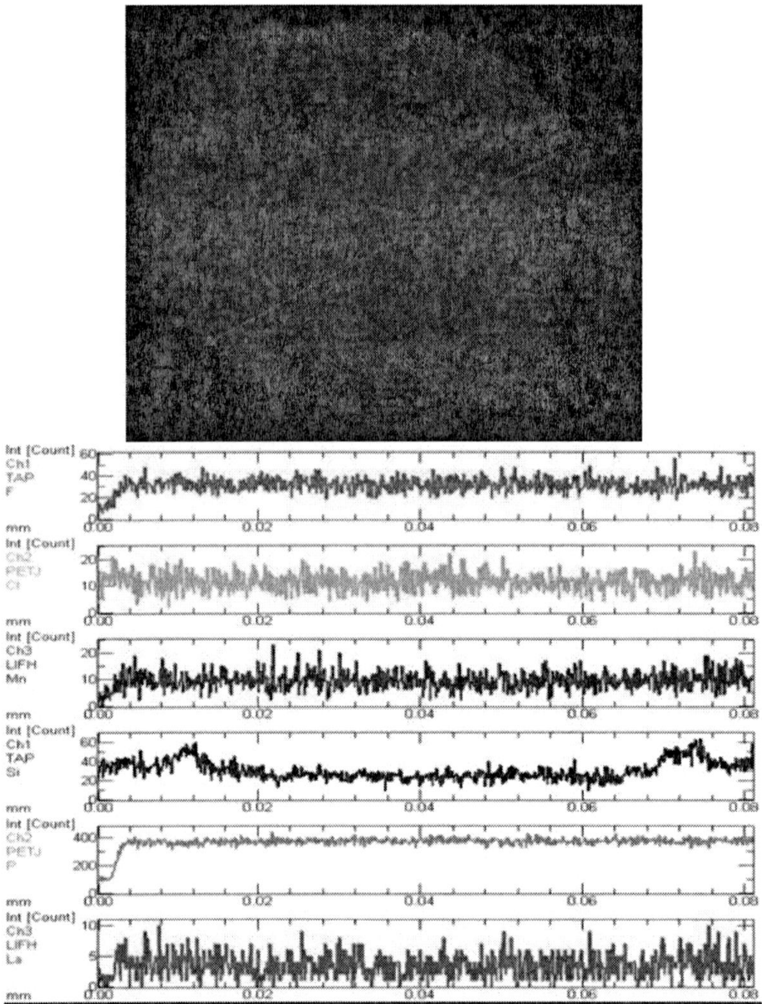

Figure 3.5.5. Elements chart of No.4 apatite cross section.

Corresponding to the CL image of apatite, the yellow-green light zone may represent that a relatively high content of Cl, Si and La while a relatively low content of F, Mn and P. Mn might not the element inspire a yellow-green light due to the inhibition of some rare earth elements in apatite for the activation of light-emitting of Mn. On the cross section of No.4 apatite, F, Mn and P content show increase trend in the range of 0.00mm-0.004mm, the Cl and La content firstly increase then decrease in range of 0.00mm-0.05mm. The Si content curve appears respectively a crest in range of 0.005mm-0.015mm and 0.064-0.078mm. The curve of F, Cl, Mn and La content are jagged which have relatively large fluctuation (Figure 3.5.5).

3.5.3. The Elements Distribution of Apatite in Sulu Eclogite

From the Figure 3.5.6 we can see, the fluctuation of the Mn and La content curve of No.5-3 apatite are relatively large, which look jagged.

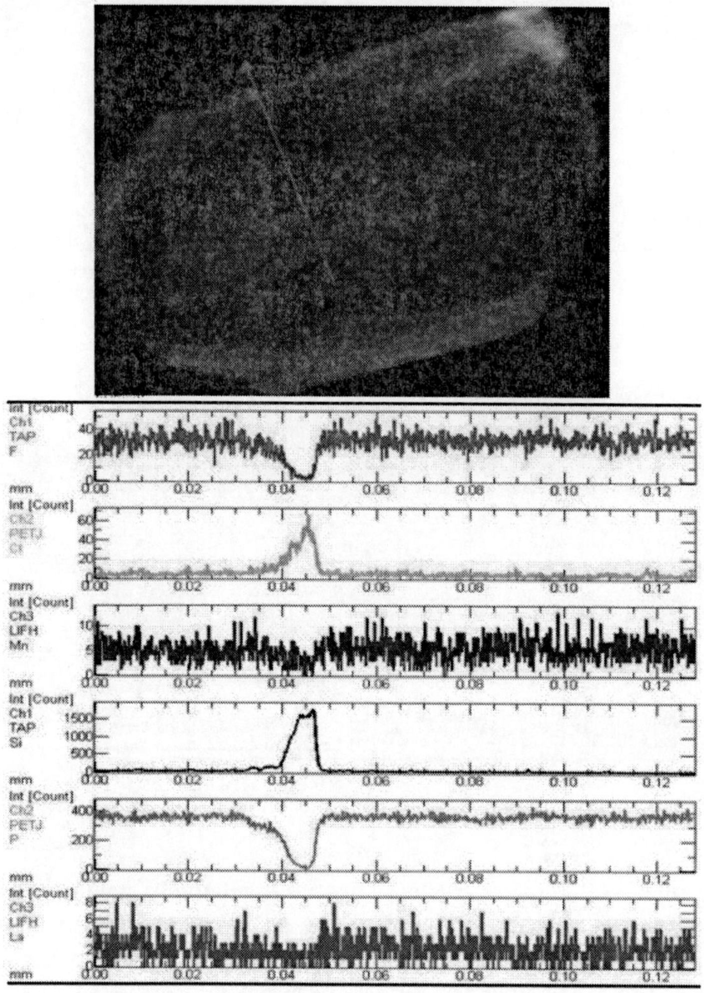

Figure 3.5.6. Elements chart of No.5-3 apatite cross section.

The crests and troughs of F and Cl elements, Si and P elements content charts also showing correspondence. The distribution of P content of No.5-4 apatite is relatively stable over the whole cross section, while the F, Mn and La content distribute like jagged with relatively large fluctuation. The Si content is low, but its curve appears a crest in range of 0.16mm-0.20mm, the maximum is 140 ppm. The Cl content is also low, and appears several crests in range of 0.08mm-0.10mm, 0.15mm-0.17mm, the content increase to about 20ppm then decrease (Figure 3.5.7).

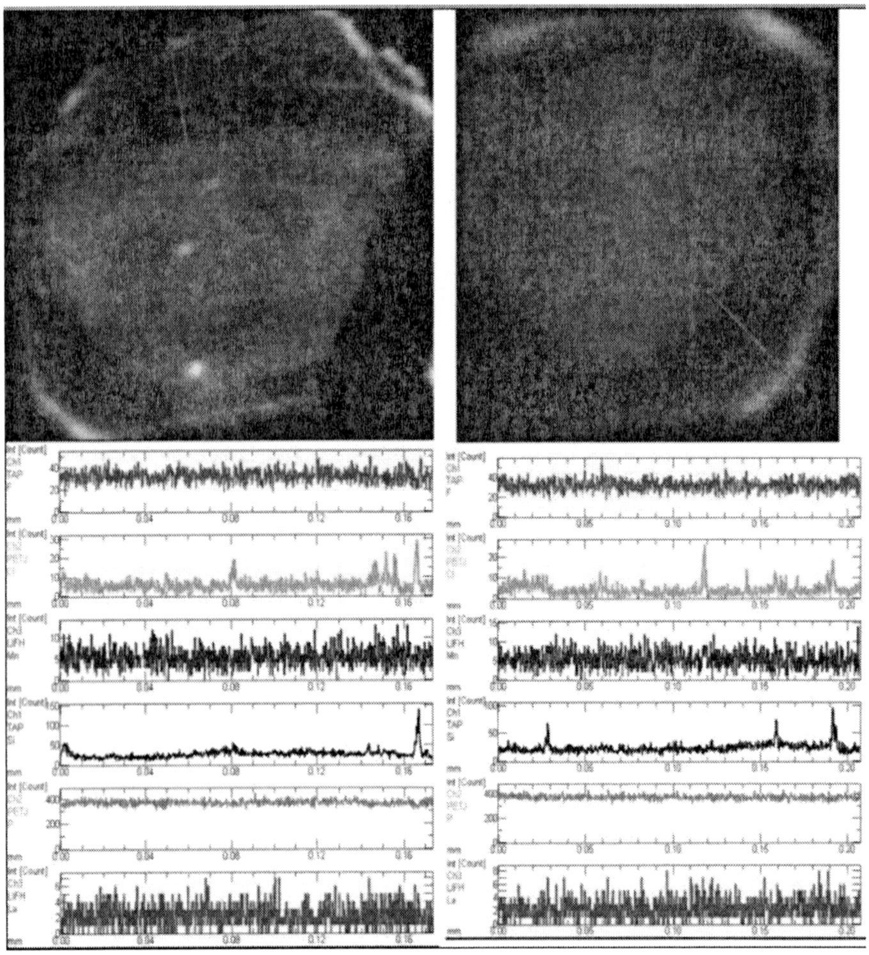

Figure 3.5.7. Elements chart of apatite cross section. The left is No.5-4 apatite, the right is No.5-5 apatite.

The distribution of P content of No.5-5 apatite is relatively stable over the whole cross section, while the Mn and La content distribute like jagged with relatively large fluctuation.

The F content distributes as jagged over the whole cross section, but its fluctuation is weaker than that of La and Mn. The curve of Cl and Si content appear several crests of narrow width (figure 3.5.7).

In the cross section of No.5-6 apatite, only the curve of Si content appears four crests, while the curves of the others are relatively stable beside P element and the fluctuation of F, Cl, Mn and La content charts are relatively large (figure 3.5.8).

In the cross section of No.5-7 apatite, the Cl content showing ups and downs in the core of apatite.

The Mn content is constantly in a volatile state without obvious law, the Si content curve appears three crests at the end of the test location, and the F and La content are always in a volatile state (figure 3.5.8).

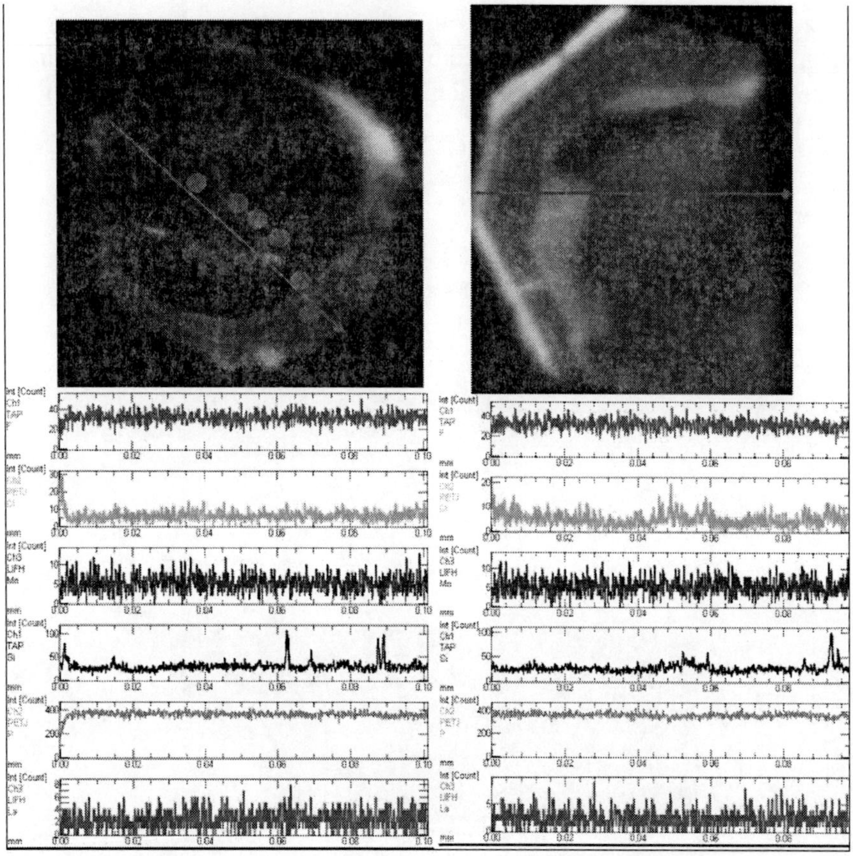

Figure 3.5.8. Elements chart of apatite cross section. The left is No.5-6 apatite, the right is No.5-7 apatite.

3.5.4. Conclusion

Overall, it is happened only sometimes, not universal, that wave crest or trough in element zoning profiles of apatite coincides with the light and shade zones within apatite. Generally, the cathodoluminescence characteristic is consistent with the composition characteristic of trace element of apatite. The results tested by U.Kempe et al. showed that the Mn and Fe content change along the section which vertical to the zoning, and the low Mn/Fe value will inhibit the intensity of CL while the area of high Mn/Fe value will appear bright zone (Figure 3.5.9) [69].

But because of the uncertainty of CL excitation mechanism, we can only give a qualitative description but not necessarily be able to identify all of the color that ions are inspired, that is, the simple and effective quantitative relationship between trace elements content and CL intensity is not exist now.

Besides, REE and trace elements used in the experimental are not complete, so this relationship can not be proved. T.J. Dempster [63] and other studies suggest that fine oscillatory zoning of CL images indicates that the diffusion of elements may be limited. Accordingly, the elements distribution of the cross section in apatite in Songliao Basin is relatively stable may be due to the diffusion process is relatively weak during the crystal growth.

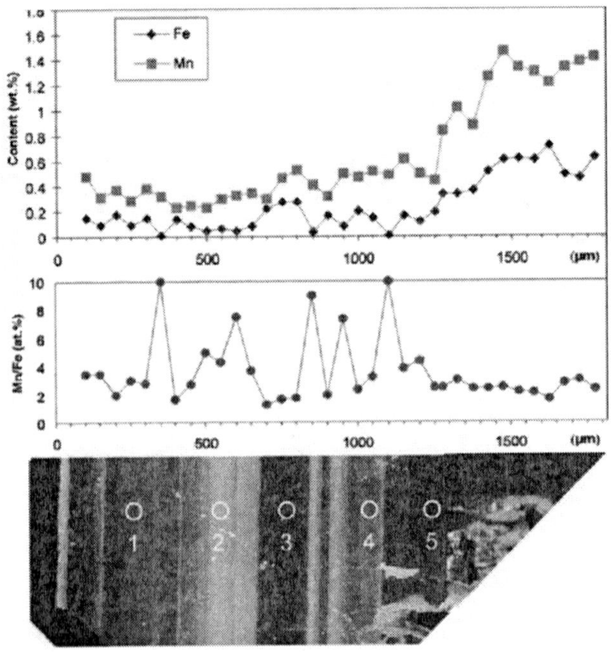

Figure 3.5.9. The relationship between CL intensity and Mn, Fe content, CL intensity and Mn/Fe along the cross section vertical to the zoning in apatite crystal. (modify according to U.Kempe, 2002) [59].

3.6. Laser Raman Spectroal Features of Apatite

This experiment was completed in Lithosphere and Dynamics Laboratory of China University of Petroleum (East China) with Micro-Raman spectrometer. The range of Raman spectroa is 150cm^{-1}-3000cm^{-1}, the laser power is controlled by a series density filter, minimum lateral and depth resolution is about 2μm (×100 eyepiece). Slit width is 400μm, scan interval is 1cm^{-1}, integration time is 5s. The generation and calculation process of various laser Raman spectra of apatite was accomplished by geological software origin 7.5.

3.6.1. Analysis of Raman Spectrogram

Analysis on apatite using laser Raman spectroscopy is still in exploration. Compared with infrared spectrum, laser Raman spectrum is more sensitive to crystal structure and give more narrow band, the changes in microstructure and bonding features of crystal can be obtained by analyze the variation on laser Raman spectra. In laser Raman spectrum of apatite, nine characteristic peaks can be identified in 200-1600cm^{-1} region: peak at 1127cm^{-1} (weak),1084cm^{-1} (weak), 1059cm^{-1} (strong), 969cm^{-1} (strongest), 756cm^{-1} (strong), 612cm^{-1} (weak), 595cm^{-1} (strong), 455cm^{-1} (weak) and 435cm^{-1} (strong), sometimes, peaks can be detected near 288cm^{-1} and 205cm^{-1}. Anionic complexes in apatite, $[PO_4]^{3-}$, have four shift peaks: peak v1 at 962cm^{-1}-965cm^{-1} is caused by asymmetric stretching vibrations; peak v2 (relative strong peak) at 419cm^{-1}-431cm^{-1} results from bending vibration; peak v3 at 1040cm^{-1}-1049cm^{-1} is caused by asymmetrical stretching vibration; peak v4(weak absorption band) at 575cm^{-1}-593cm^{-1} result from asymmetric bending vibration. Strong absorption band at 1061cm^{-1}-1088cm^{-1} can be identified in laser Raman spectra of apatite in which $[CO_3]^{2-}$ replace partial $[PO_4]^{3-}$, this band indicates $[CO_3]^{2-}$ occupy lattice positions in apatite, its appearance cannot be attributed to simple mixture of carbonate into apatite. Stretching vibration band of $[OH]^-$ in apatite is around 3568 cm^{-1}, which is lower than that of $[OH]^-$ in typical hydroxide (3630 cm^{-1}-3700cm^{-1}). The lower stretching frequencies of $[OH]^-$ in apatite may be attribute to bond length between $[OH]^-$ and other ions and bond length of $[OH]^-$ itself. Vibration of $[F]^-$ is at 4136cm^{-1}-4142cm^{-1}, belonging to stable strong band. Yu Liu et al. researched on the vibration mode outside the lattice(variation vibration outside the tetrahedron) of the manmade fluorapatite single crystal and found that frequency of external vibration mode was low($<$400cm^{-1}) and easy to be covered by Rayleigh line [64].

Xianyong Liu et al. have researched on apatite of different genetic types [65], variation on laser Raman peak of ionic group in different subspecies of apatite separated from phosphorite have been analyzed [64,66,67]. The v4 vibration band of $[PO_4]^{3-}$ in different subspecies shifts largely, the other three bands show minor shift; vibration band of $[OH]^-$ shifts largely, which attribute to the content of $[OH]^-$; vibration band of $[F]^-$ is relatively stable, minor shifts exist in different subspecies.

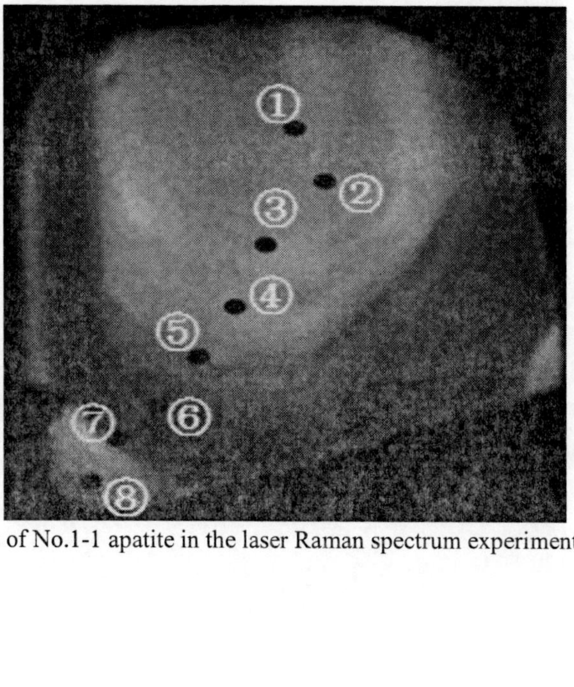

Figure 3.6.1. Test point of No.1-1 apatite in the laser Raman spectrum experiment.

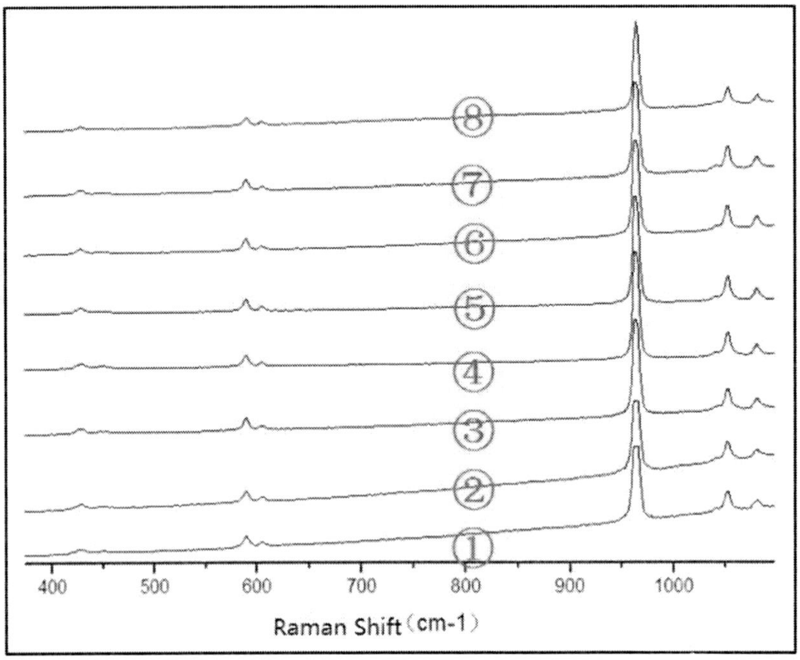

Figure 3.6.2. Laser Raman spectrogram of No.1-1 apatite.

Table 3.6.1. The experiment data of Laser Raman spectrum of No.1-1 apatite

Test point	Characteristic peak/cm^{-1}	Relative intensity	FWHM /cm^{-1}
8	964	3858	7.5
7	963	3749	7.6
6	963	4837	6.2
5	964	5360	5.7
4	963	5350	5.6
3	963	4867	6.1
2	963	4819	5.8
1	964	4518	4.9

According to designed test point of apatite zoning (Figure 3.6.1), we got the laser Raman spectrogram of No.1 apatite (Figure 3.6.2). From the spectrogram we can see that there is a strong peak at 964cm^{-1}, and its offset is not obvious along the direction of the arrow on the apatite particles.

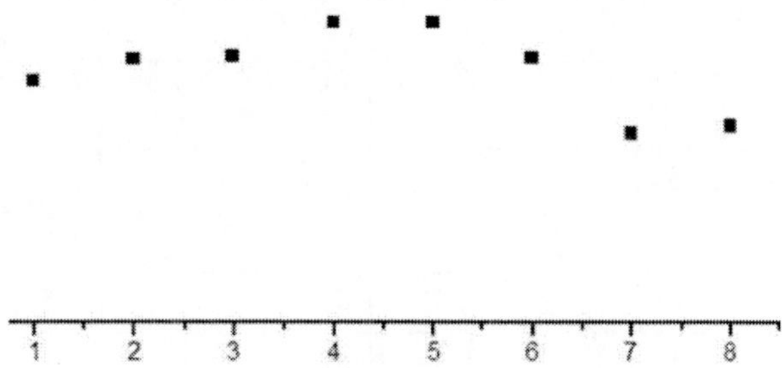

Figure 3.6.3. The value of relative intensity at 964cm^{-1} characteristic peak.

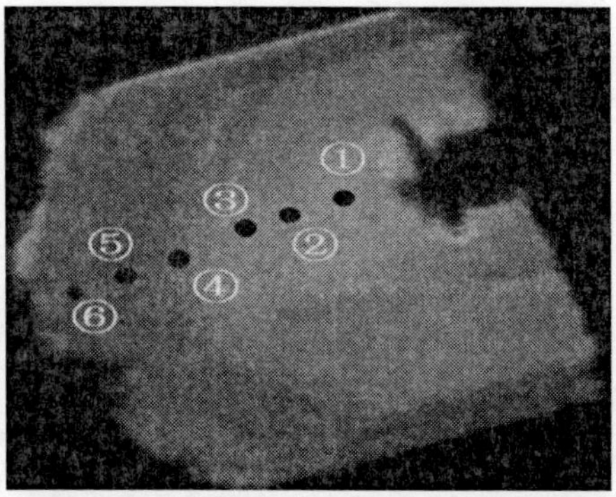

Figure 3.6.4. Test point of No.1-2 apatite in the laser Raman spectrum experiment.

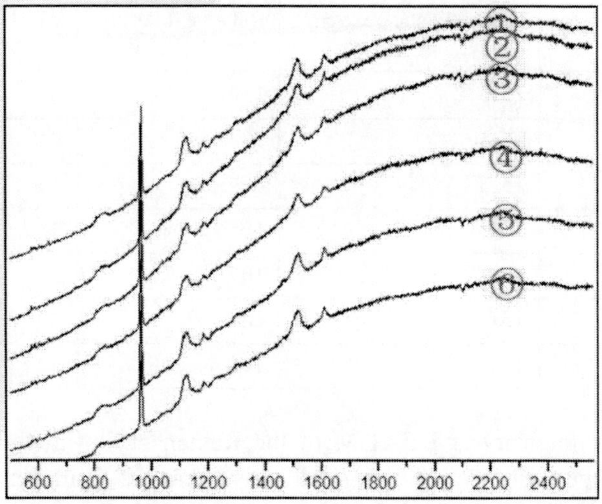

Figure 3.6.5. Laser Raman spectrogram of No.1-2 apatite.

Table 3.6.2. The experiment data of Laser Raman spectrum of No.1-2 apatite

Test point	Characteristic peak/cm^{-1}	Relative intensity	FWHM /cm^{-1}
1	964	384	5.9
2	964	532	5.2
3	964	534	5.5
4	964	521	5.1
5	964	575	5.6
6	964	400	6.2
7	964	220	5.8
Test point	Characteristic peak/cm^{-1}	Relative intensity	FWHM /cm^{-1}
1	1122	173	26
2	1121	114	26
3	1121	32	26
4	1121	78	20
5	1121	96	20
6	1121	139	25
7	1121	264	31
Test point	Characteristic peak/cm^{-1}	Relative intensity	FWHM /cm^{-1}
1	1515	130	24
2	1515	93	24
3	1515	67	30
4	1515	60	28
5	1515	107	24
6	1515	185	30
7	1515	164	39

But the relative intensity and FWHM of the Raman peak value changed (table3.6.1), main characteristic peak showing a rise - fall - rise changes from the core to the edge of particle (Figure 3.6.3). We also found that the trough position of curve of the relative intensity and FWHM ratio corresponds to the dark areas of the CL image.

For the No.1-2 apatite particle, we tested the apatite particle in turn from the core to the edge as shown in figure3.6.4, and appeared Raman peaks at 964 cm^{-1}, 1121 cm^{-1} and 1515cm^{-1} (figure 3.6.5). From some parameters of the table3.6.2, we can see that the intensity of Raman peaks at 964 cm^{-1} firstly becomes strong then weaken from the core to the edge of the particle, and the weakest intensity is the 7th test point, while the FWHM is floating between 5-6. The intensity of Raman peaks at 1121 cm^{-1} firstly becomes weak then increases gradually, and the strongest intensity is the 7th test point, while the FWHM is about floating between 25-30. The intensity of Raman peaks at 1515cm^{-1} has the same change trend as that at 1121 cm^{-1}, but the peak value of the 7th slightly below that of 6th test point.

For the No.1-3 sample, we tested the apatite particle in turn from the core to the edge as showed in figure 3.6.6, the Raman peaks show in 429 cm^{-1}, 446 cm^{-1}, 579 cm^{-1}, 589 cm^{-1}, 604 cm^{-1}, 638 cm^{-1}, 964 cm^{-1}, 1114 cm^{-1}, 1126 cm^{-1}, 1185 cm^{-1}, 1518 cm^{-1}, 1609 cm^{-1} and 3068cm^{-1}, but the intensity of the Raman peaks at 429 cm^{-1}, 446 cm^{-1}, 579 cm^{-1}, 589 cm^{-1}, 604 cm^{-1}, 638 cm^{-1} are very weak. For the peaks at 1114 cm^{-1}, 1126 cm^{-1}, 1185 cm^{-1}, 1518 cm^{-1}, 1609 cm^{-1}, we can see the Raman shift values almost did not change much from test point 1st to 7th, but the intensity of Raman peaks show a gradually increasing trend. The ratio of intensity and FWHM of characteristic peak at 964cm^{-1} also have some changes (Figure 3.6.7). We observed the oscillatory zoning on the mantle and rim part of the apatite particles from the CL image, may be due to the boundaries of the core and the edge portion is not particularly clear, the intensity and FWHM ratio is not change obviously.

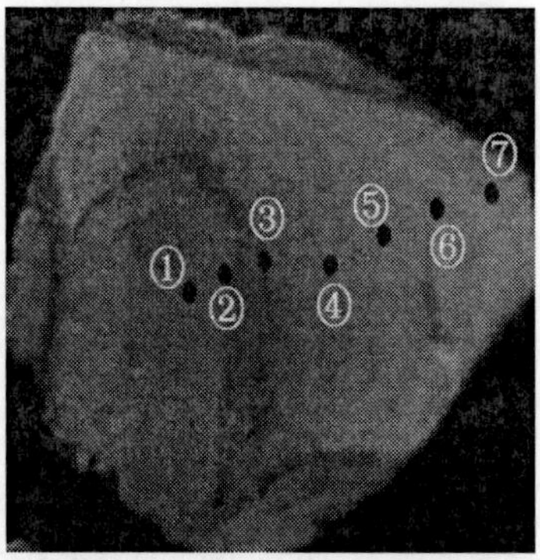

Figure 3.6.6. Test point of No.1-3 apatite in the laser Raman spectrum experiment.

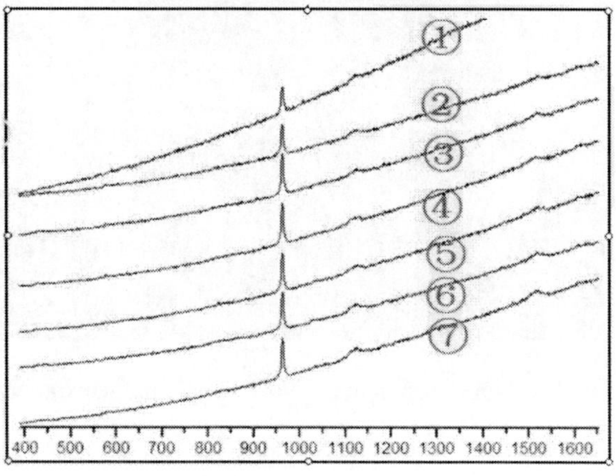

Figure 3.6.7. Laser Raman spectrogram of No.1-3 apatite.

Table 3.6.3. The relative intensity and FWHM data of the characteristic peak of No.2 apatite

Test point	Characteristic peak/cm^{-1}	Relative intensity	FWHM /cm^{-1}
1	963	1486	4.2
2	963	1788	4.4
3	963	1795	4.6
4	963	2071	5.1
5	963	1984	5.1
6	963	1861	5.0
7	963	1605	5.1

The Raman peak at 3068cm^{-1} indicated that the apatite particle has a certain amount of hydroxyl radical.

Through the experimental results of No.2 apatite (Table3.6.3), we found that the Raman shift of each test point at 963cm^{-1} has no obvious change, but the intensity of Raman peaks and the ratio of intensity and FWHM are decrease firstly and then increase. Corresponding to the CL image of apatite, the yellow-green zone may represent high peak intensity and ratio of intensity and FWHM, while the dark zone shows low peak intensity and ratio of intensity and FWHM.

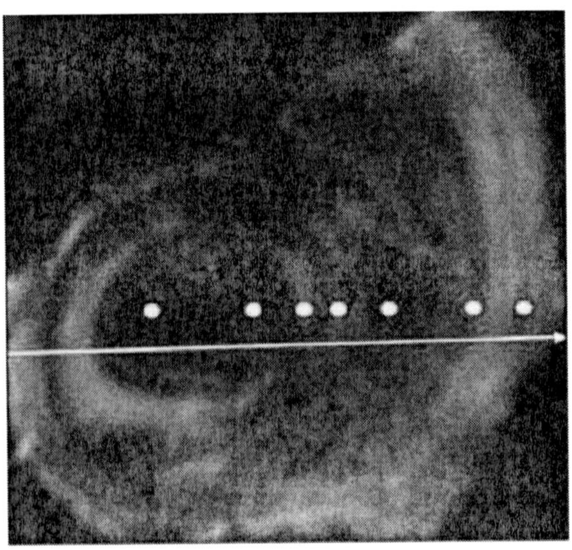

Figure 3.6.8. Location and direction of test point of No.3 sample in the experiment.

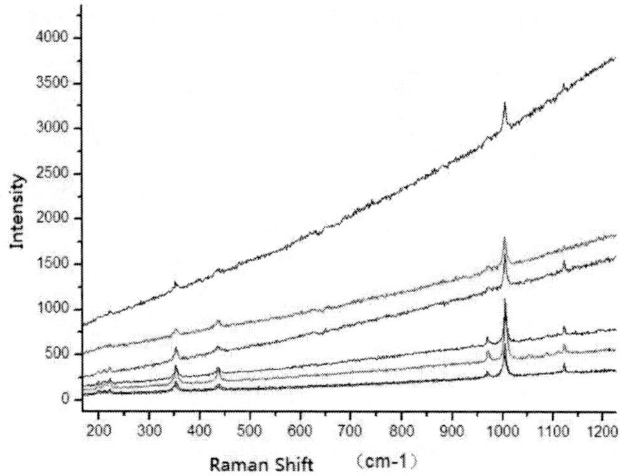

Figure 3.6.9. Laser Raman spectrogram of No.3 apatite.

For the No.3 sample, we tested the apatite particle in turn from the core to the edge as the arrow showed in figure 3.6.8, the Raman peaks show in 198cm^{-1}, 211cm^{-1}, 223cm^{-1}, 352cm^{-1}、437cm^{-1}, 972cm^{-1}, 1005cm^{-1}, 1123cm^{-1} (figure 3.6.9 and table 3.6.4). These intensity of Raman peaks are very low, the intensity of Raman peak at 1005cm^{-1} is strongest, and there are no peaks appeared at 198cm^{-1}, 211cm^{-1}, 223cm^{-1} and 1123cm^{-1} of the 3rd and 4th test point.

The peaks at 1005cm^{-1}, 1123cm^{-1} are the vibration band of SO_4, and the peak at 972 cm^{-1} is the vibration band of PO_4, indicates that SO_4 may substitute the PO_4 in the apatite crystals. According the study of Khorari et al. (1994) which found that as the SO_4 substitute PO_4 gradually, the vibrational bands of PO_4 Raman peaks will gradually widen, and the relative intensity will gradually decrease, so the rim portion occurred stronger substitution effect than that in the core of this apatite [68].

For the No.4 sample, we tested the apatite particle in turn from the core to the edge portion, the Raman peaks appears at 964 cm^{-1}, 1123 cm^{-1} and 1047cm^{-1} (figure 3.6.10), and some parameters of the Raman peaks at 964 cm^{-1} and 1123 cm^{-1} are shown in table3.6.5. This apatite has vibrational band of SO_4 at peak of 1123 cm^{-1}, which also represents that there are some SO_4 substitute PO_4 in the crystals. We can also found that the intensity of vibrational bands of PO4 in the apatite crystals core portion is higher than that in the edge portion.

We tested the No.5-1 apatite as the point showed in figure 3.6.11, and found that the Raman shift peaks appears at 430cm^{-1}, 590cm^{-1}, 610cm^{-1}, 964cm^{-1}, 1053cm^{-1}, 1083cm^{-1}, 1608cm^{-1} and 3541 cm^{-1} (Figure 3.6.12), the relative intensity value at 964cm^{-1} is the largest. From the Raman spectrogram we can see that the Raman shift have no obvious change at 430cm^{-1} , 590cm^{-1} and 610cm^{-1}, but the intensity of peaks gradually decrease from the core to the apatite edge, while it increase gradually from the core to the edge at 1608cm^{-1}. The peak at 3541cm^{-1} is the characteristic peak of OH, indicated that this apatite is hydroxyapatite.

Table 3.6.4. The experiment data of Laser Raman spectrum of No.3 apatite

Test point	Characteristic peak/cm^{-1}	Relative intensity	FWHM /cm^{-1}
1	1003	230	8.5
2	1004	365	6.5
3	1004	309	6.6
4	1005	290	5.2
5	1005	362	5.5
6	1005	401	6.1
7	1006	688	5.8
Test point	Characteristic peak/cm^{-1}	Relative intensity	FWHM /cm^{-1}
1	1122	100	2.7
2	1123	88	2.0
3	1123	43	2.2
4	1123	79	2.1
5	1123	110	3.1
6	1123	100	3.6
7	1123	106	4.8
Test point	Characteristic peak/cm^{-1}	Relative intensity	FWHM /cm^{-1}
1	351	70	8.9
2	353	98	8.3
3	353	75	9.5
4	352	95	4.1
5	353	134	6.5
6	353	88	7.3
7	353	223	7.8

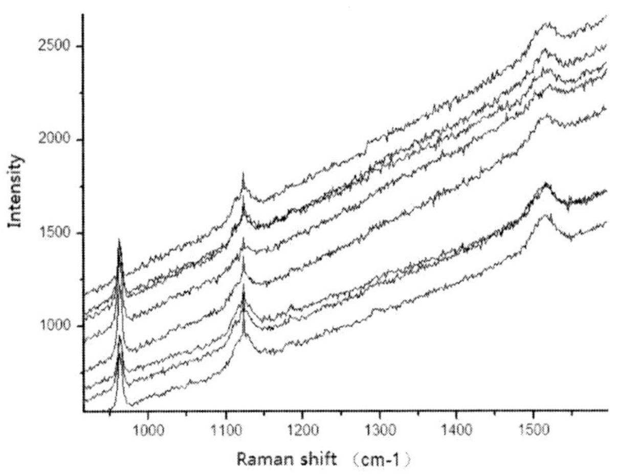

Figure 3.6.10. Laser Raman spectrogram of No.4 apatite.

Table 3.6.5. The experimental data of laser Raman spectra of No.4 sample

Test point	Characteristic peak/cm^{-1}	Relative intensity	FWHM /cm^{-1}	Characteristic peak/cm^{-1}	Relative intensity	FWHM /cm^{-1}
1	964	87	6.6	1123	204	26
2	964	210	7.3	1123	229	18
3	963	313	7.0	1123	202	18
4	963	337	7.0	1123	263	18
5	963	425	6.8	1123	184	10
6	963	370	7.9	1123	209	10
7	964	274	7.1	1123	142	16
8	963	399	6.9	1123	127	18

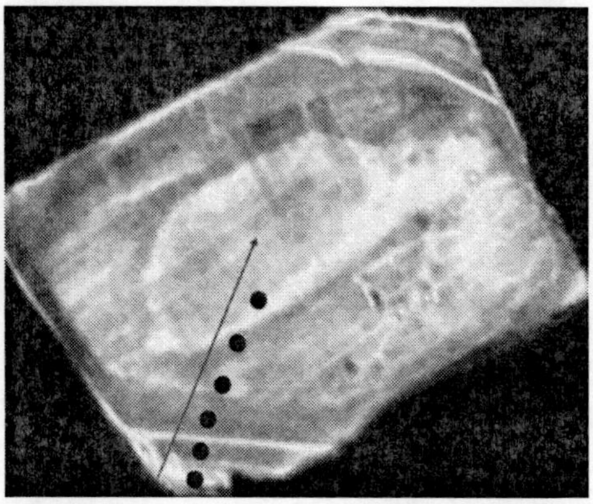

Figure 3.6.11. Location and direction of test point of No.5-1 apatite in the experiment.

For the No.5-2 apatite, we tested it as the arrow showed in figure 3.6.13, and found that the strong peak of PO_4 only appears at 966 cm^{-1}, while other positions with no obvious peaks. The Raman shift of characteristic peak at 966 cm^{-1} with no evident shift, but the intensity of the peaks changes obviously (Figure 3.6.14). For the No.5-3 apatite, we tested it as the arrow showed in figure 3.6.15, and found that the strong peak of PO_4 only appears at 966 cm^{-1}, while other positions with no obvious peaks. The Raman shift of characteristic peak at 966 cm^{-1} with no evident shift, but the intensity of the peaks changes obviously. The intensity of Raman peaks of 1st, 2nd, 3rd, 4th test point gradually decreases in turn; the 5th test point hits the junction of without emitting band and yellow-green luminescence band, the intensity of 966cm^{-1} peak is very low; the 6th test point hits the yellow-green luminescence band, and the intensity of 966cm^{-1} peak lower than intensity of apatite non-luminous band, but higher than that of Raman peaks of junction (Figure 3.6.16).

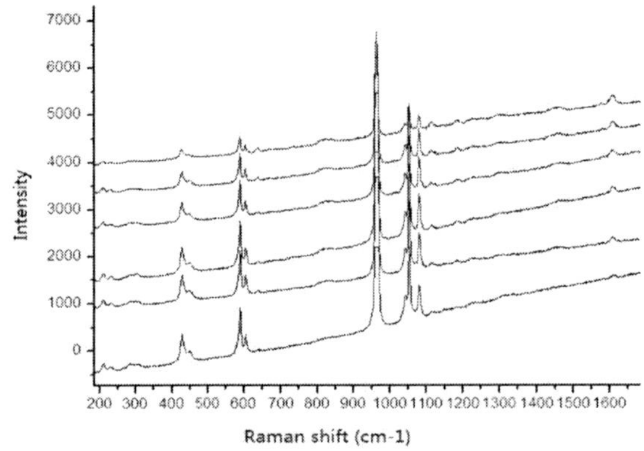

Figure 3.6.12. Laser Raman spectrogram of No.5-1 apatite.

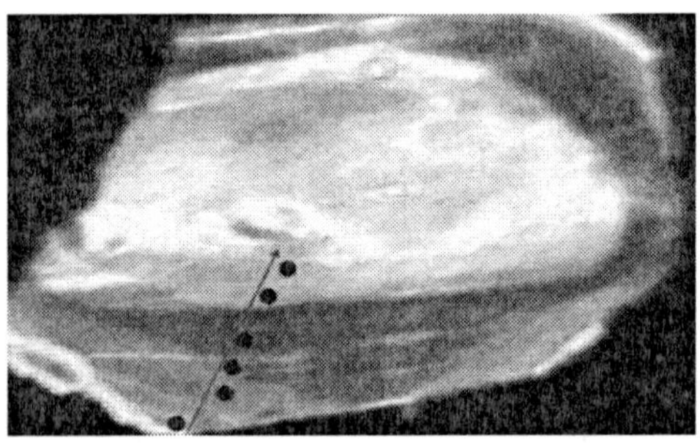

Figure 3.6.13. Location and direction of test point of No.5-2 apatite in the experiment.

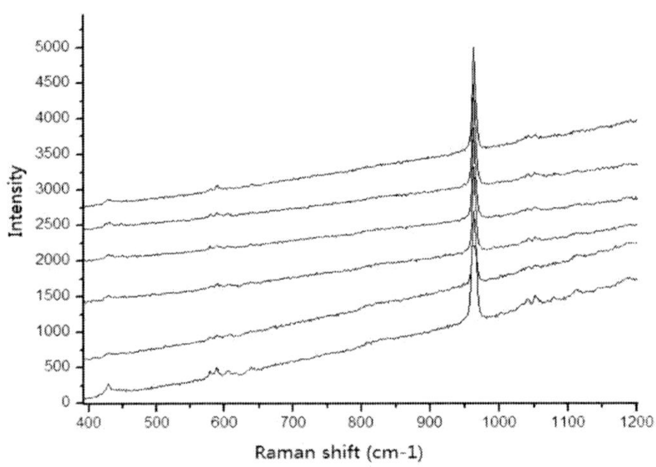

Figure 3.6.14. Laser Raman spectrogram of No.5-2 apatite.

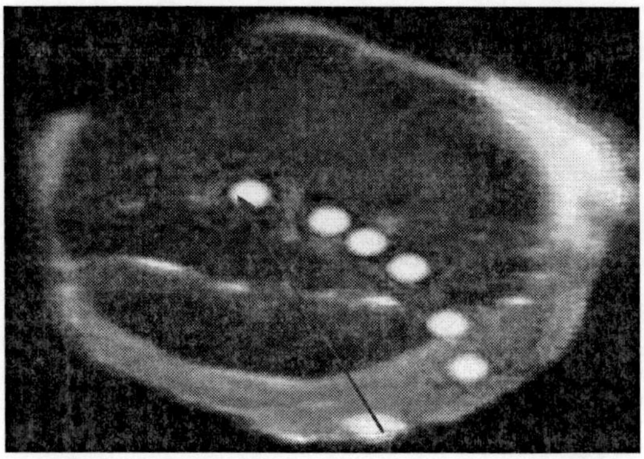

Figure 3.6.15. Location and direction of test point of No.5-3 apatite in the experiment.

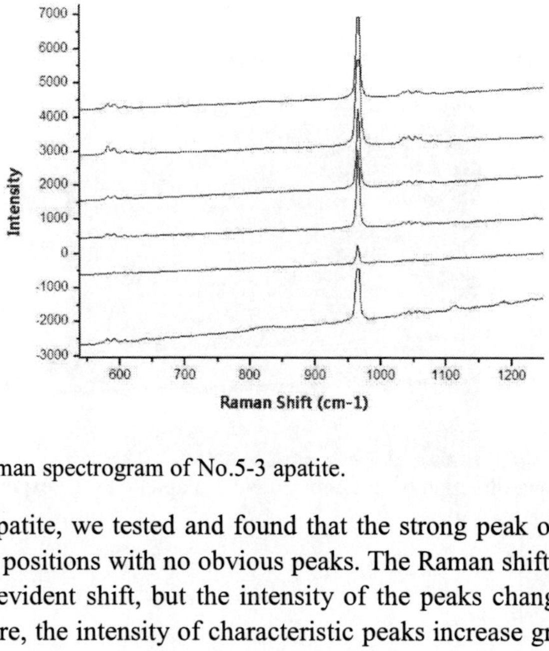

Figure 3.6.16. Laser Raman spectrogram of No.5-3 apatite.

For the No.5-4 apatite, we tested and found that the strong peak of PO_4 only appears at 966 cm^{-1}, while other positions with no obvious peaks. The Raman shift of characteristic peak at 966 cm^{-1} with no evident shift, but the intensity of the peaks changes slightly. From the apatite edge to the core, the intensity of characteristic peaks increase gradually then decrease gradually, which just correspond to the CL image, the intensity of characteristic peak is weak in where the area with yellow-green light while the intensity of characteristic peak is strong in where the area without yellow-green light (Figure 3.6.17).

For the No.5-5 apatite, we tested it as showed in figure 3.6.18, found that the strong peak of PO_4 appears at 966 cm^{-1}, and weak peaks appear at 430cm^{-1}, 591cm^{-1}, 1042cm^{-1}, 1053cm^{-1}, 1080cm^{-1}. The Raman shift of characteristic peak at 966 cm^{-1} with no evident shift, but the intensity of the peaks changes obviously, the intensity of characteristic peak decrease gradually from the apatite core to the rim portion (figure 3.6.19).

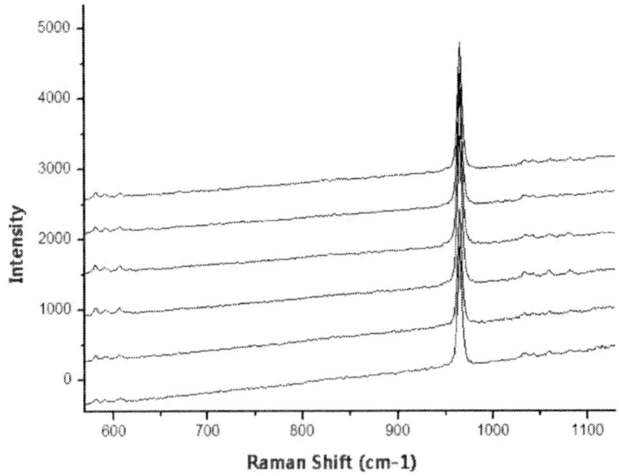

Figure 3.6.17. Laser Raman spectrogram of No.5-4 apatite.

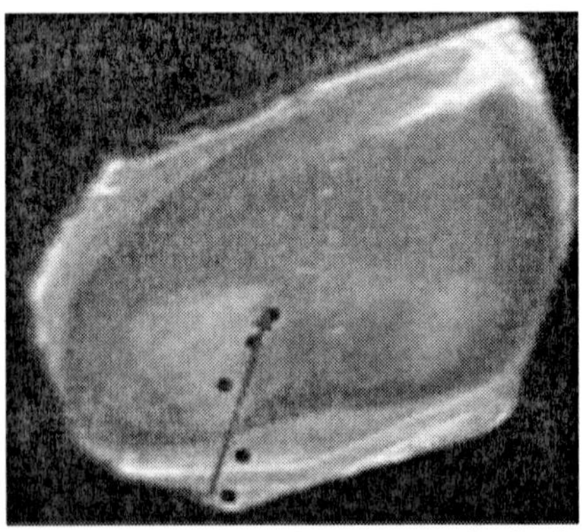

Figure 3.6.18. Location and direction of test point of No.5-5 apatite in the experiment.

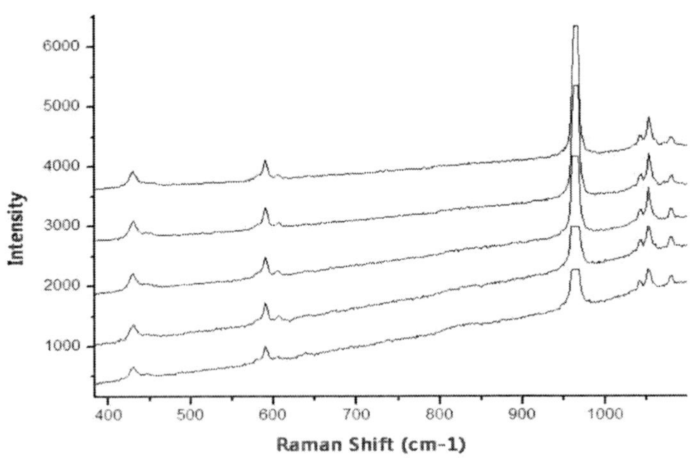

Figure 3.6.19. Laser Raman spectrogram of No.5-5 apatite.

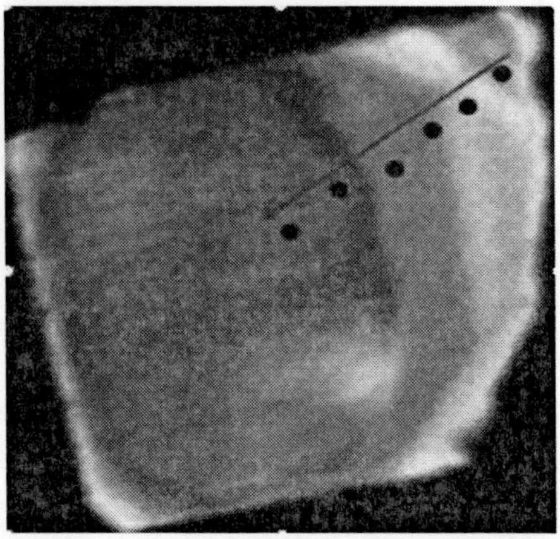

Figure 3.6.20. Location and direction of test point of No.5-6 apatite in the experiment.

For the No.5-6 apatite, we tested it as showed in figure 3.6.20, and got the laser Raman spectrogram (figure 3.6.21) and table 3-6-7 which consists of the relative intensity and FWHM of characteristic peak at 966cm^{-1}. Overall, the apatite has a simple Raman spectrogram, and only appears a relative strong vibration peak of PO_4. As can be seen from the data in the table, within the error range, the Raman shift and intensity of peaks do not shift and change obviously.

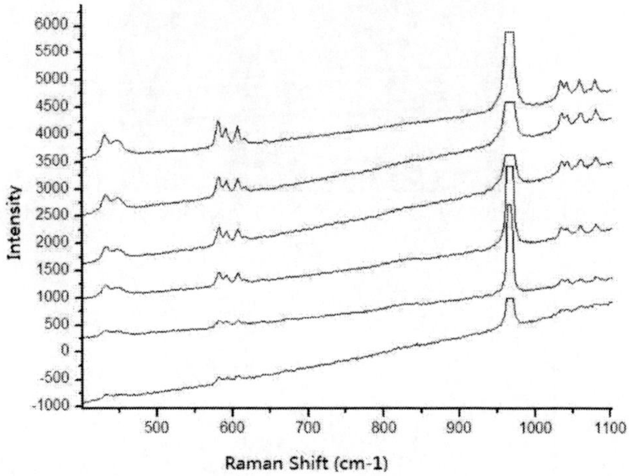

Figure 3.6.21. Laser Raman spectrogram of No.5-6 apatite.

For the No.5-7 apatite, we tested it as showed in figure 3.6.22, and got the laser Raman spectrogram (figure 3.6.23) and table 3.6.7 which consists of the relative intensity and FWHM of characteristic peak at 966cm^{-1}. Overall, the apatite has a simple Raman spectrogram, and only appears a relative strong vibration peak of PO_4.

As can be seen from the data in the table, the Raman shift does not change, but the intensity of peaks and FWHM change slightly which decrease gradually from the apatite rim to the core, the reason may be the crystallinity degree of the rim portion is better than that of the core.

Table 3.6.6. The experimental data of laser Raman spectra of No.5-6 apatite

Test point	Characteristic peak/cm^{-1}	Relative intensity	FWHM /cm^{-1}
1	966	799	4.5
2	966	790	4.5
3	966	825	4.3
4	966	908	4.3
5	966	970	4.4
6	966	941	4.5
7	966	931	4.4

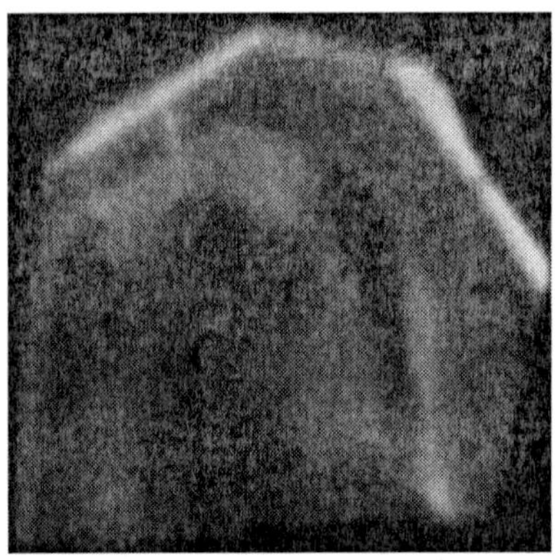

Figure 3.6.22. Location and direction of test point of No.5-7 apatite in the experiment.

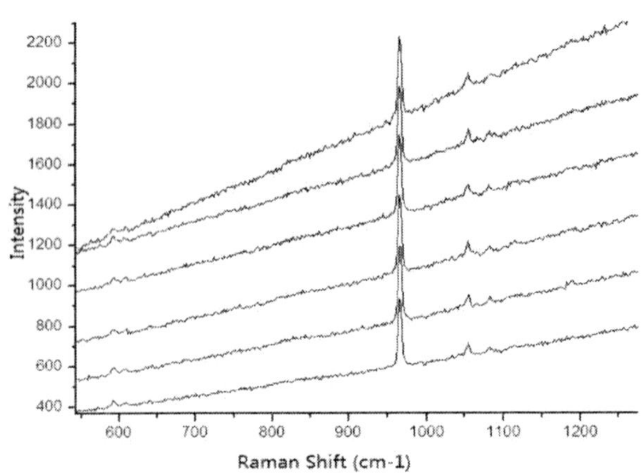

Figure 3.6.23. Laser Raman spectrogram of No.5-7 apatite.

Table 3.6.7. The experimental data of laser Raman spectra of No.5-7 apatite

Test point	Characteristic peak/cm^{-1}	Relative intensity	FWHM /cm^{-1}
1	966	412	4.8
2	966	399	4.8
3	966	394	4.7
4	966	393	4.7
5	966	388	4.5
6	966	356	4.5

Through observation and analysis, generally, the Raman spectrogram of apatite in eclogite are more complex than that of apatite in sandstone, and intensity of main characteristic peak at 964cm^{-1} of apatite in eclogite is higher than that of apatite in sandstone.

3.7. Relationship between Laser Raman Spectra and Thermal History

From CL image of apatite from eclogite, obvious boundaries between core and rim in apatite can be observed. Through electronic probe, element distribution of apatite can be got, which indicates apatite's growth environment have changed during crystallization or apatite compete with mineral association for element that leads to element's uneven distribution in apatite.

Watson(1979,1980) quantitative research crystallization conditions of apatite from common magma, he point out P_2O_5 content of magma that needed for apatite crystallization from magma is mainly determined by the temperature and Si content of magma, pressure factor take second place [69, 70].

The preference temperature for apatite crystallization is 900-950°C [71]. From the eclogite's P-T-t path of Sulu orogenic belt (figure 3.7.1) we can see, in the fastigium of last phase during prograde metamorphism, the temperature reaches 800-1000°C, which is enough for apatite crystallization, the zoning structure of apatite is original apatite's hyperplasia of edge in metamorphism. The boundary between hyperplasia of edge and original apatite have gone through several million, it may be homogenization because of element diffusion in apatite, so the zoning structure may disappear. So, the rapid exhumation of eclogite can help preserve the obvious boundaries of core and rim in apatite. We can infer the rim of apatite may form in the stage from prograde metamorphism to fastigium of metamorphism.

Chopin (1981) ② Newton (1986) ③ Holland (1980) ④ Massone (1984) ⑤ Green 和 Ringwood (1972).

Laser Raman data of apatite shows that from apatite's core to rim, the shift of characteristic peak does not change obviously but the intensity of peak decrease. So, the decrease of characteristic peak's intensity may illustrate apatite have gone through high temperature and high pressure during crystallization.

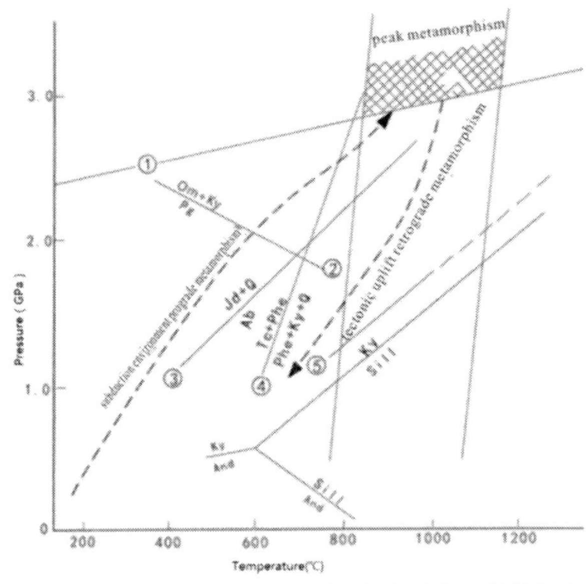

Figure 3.7.1. PTt trace of Sulu eclogites (modified after Han et al. 1992 [72].

Observe the zoning of No.1-1, No.1-2, No.1-3, No.2, No.3 and No.4 apatite particles, we found that their shapes are relatively flat, and the particles were rounded without unrounded core, so that we can deduce that the apatite in the sandstone were formed after the formation of the source region and did not growth again in sedimentary basins. Otherwise, it should have a line of rounding, and the overall shape of apatite should be the rule. Combined with the thermal history information of these basins, we know that it is impossible for the apatite to reach the condition of recrystallization, and we can get the thermal history of the apatite in source region rather than that in the basins if we use the Raman spectrogram to study the thermal history.

3.8. CONCLUSION

(1) Zoning within apatite can be divided into three types: 1) Oscillatory zoning exist in the core, mantle and rim of apatite. Boundary between core and mantle and boundary between mantle and rim are distinguished by overall light and shade degree of oscillatory zoning. Zoning is center based, the boundaries between different zones are always parallel to the crystal growth face, the outlines of zoning boundary are always straight, and the crystalline form of apatite develops well. 2) The core is a uniform ensemble; oscillatory zoning always develops in the mantle and rim. Boundary between core and mantle is obvious, while there is a gradual transition from mantle to rim and do not exist obvious boundary. The core generally has a regular outline, oscillatory zoning is parallel to the crystal growth face and the shape of it is regular. Apatite with those characteristics may crystallize in early stage and

continue growing in later period, apatite may suspend in the fluid during growth and excitation elements which control the luminescent effect of apatite can only diffuse limitedly. 3) There is almost no oscillatory zoning existing in the core, mantle and rim, those parts can be distinguished by different luminescent colors, boundaries are obvious but not regular, sometimes, they can be embayment, dissolution may have happened during crystal growth.

(2) Element content in different location of apatite varies. It can be found that only sometimes, not universal, that wave crest or trough in element zoning profiles apatite coincides with light and shade zones within apatite. Generally, cathodoluminescence characteristics coincides with trace element characteristics, but because there are still uncertainties existing in the excitation scheme of cathodoluminescence, the cathodoluminescence characteristics cannot identify all luminescence stimulated by ions, but they can give us partial qualitative inference. For the time being, the simple and effective quantitative relation between trace element content and cathodoluminescence intensity cannot be obtained. In this experiment, only partial rare earth elements and trace elements were obtained, so the correspondence cannot be amply demonstrated.

(3) Cathodoluminescence technique can help us detect zoning within apatite that cannot be seen under general microscope. Zoning is not directly correlated to distribution of F, Cl, P, Si, La in apatite, which illustrates the zoning shown in CL image is stimulated by some other rare earth elements and trace elements besides those elements.

(4) Laser Raman characteristics of apatite separated from Sulu eclogites are much more complex than that of apatite separated from Songliao basin sandstone. The former one just show the strong peak in $964cm^{-1}$ of PO_4, while the characteristic peak of SO, CO can be detected in the later one. This illustrates the two have big difference in crystal structure and elementary composition and some other aspects.

(5) It can be qualitative inferred that the variation of characteristic peak intensity can reflect the thermal history which apatite have gone through.

REFERENCES

[1] Ribbe P H, Hofmeister A M. Feldspar mineralogy[M]. Mineralogical Society of Amer, 1983.

[2] Pan Zhaolu. Crystallography and mineralogy[M]. Beijing: Geological Publishing House, 1998 (in Chinese).

[3] Wright T. L, Stewart D. B. X-ray and optical study of alkali feldspar. I Determination of composition and structural state from refined unit-cell parameters and 2V [J]. *American Mineralogist,* 1968, 53(1-2):112-134.

[4] Ye Danian, Jin Chengwei. X-ray powder method and its application in petrology [M].Beijing: *Science Press,* 1984,223-265 (in Chinese).

[5] T. P. Mernagh. Use of the Laser Raman Microprobe for Discrimination Amongst Feldspar Minerals [J]. *Journal of Raman Spectroscopy,* 1991, 22:453-457.

[6] Xie Jun, Yu Xuehui, Zhang Jian, etc. Raman Spectra Study on Feldspar's Ordering Degrees [J]. *Bulletin of Mineralogy, Petrology and Geochemistry,* 2007, 26: 227-229 (in Chinese).

[7] Xie Jun. A Raman spectroscopy study of hyperfine structure of aluminosilicate and feldspar [D]. Beijing. Department of Gem, China University of Geosciences, 2008 (in Chinese).

[8] Tian Guhui, Chen Yajie, Feng Qingmao. Development and application of Raman technology [J]. *Chemical Engineer,* 2008, 148(1):34-36 (in Chinese).

[9] Long D A, Long D A. Raman spectroscopy [M]. New York: McGraw-Hill, 1977.

[10] Zhang Meizhen, Shi Weijun, Zhang Zhirong. Laser Raman Microscopy and its Application in Geology [J]. Petroleum Geology & Experiment, 2008,30(3):37-41(in Chinese).

[11] Chen Shiyue. Mineralogy Petrology [M]. Dongying: China University of Petroleum Press, 2006 (in Chinese).

[12] Sun Yijian. The problems on the compositions and grade of order in feldspar of volcanic rooks [J].Bulletin of the Chinese Academy of Geological Sciences. Nanjing Institute of Geology and Mineral Resources.1989,(1). (in Chinese).

[13] Chen Xiaoming, Liu Changshi. Methods for determining the Al-Si distribution in alkali feldspar [J]. Geological Review, 1997,43(6):601-605 (in Chinese).

[14] Smith J. V. Crystal structure and Physical Properties Feldspar Mineralogy [J]. 1974, 1.

[15] Ragland P C. Composition and structural state of the potassic phase in perthites as related to petrogenesis of a granitic plution [J]. Lithos, 1970, 3(2): 167-189.

[16] Bowen N L, Tuttle O F. The system $NaAlSi_3O_8$-$KAlSi_3O_8$-H_2O [J]. Jour, Geol, 1950, 58, 489.

[17] Orville, P M. Unit-cell parameters of the microcline-low albite and the sanidine-high albite solid solution series [J]. *Amer. Min.,* 1967, 52, 55-86.

[18] Jones, J B, Nesbitt R W, Slade, P G. The determination of the orthoclase content of homogenized alkali feldspars using the ($\bar{2}$ 01) X-ray method [J]. *Min. Mag.,* 1969, 37, 487-496.

[19] Swanson S E. Relation of nucleation and crystal-growth rate to the development of granitic textures [J]. *American Mineralogist,* 1977, 62:966-978.

[20] Longhi J, Fram M S, Auwera J V, et al. Pressure effects, kinetics, and rheology of anorthositic and related magmas [J].*American Mineral-ogist,* 1993, 78:1016-1030.

[21] Sisson T W, Grove T L. Experimental investigations of the role of H_2O in calc-alkaline differentiation and subduction zone magmatic arcs from basalt at Galunggung, Indonesia [J]. *Nature,* 1993, 391:883-886.

[22] Wang Chen, Yu Xuehui, Mo Xuanxue et al. Raman spectra for structure of Tremolite [J].*Journal of the Chinese Ceramic Society,* 2006,34(12): 421-427 (in Chinese).

[23] Sun Qiang, Zheng Haifei, Xie Hongsen et al. The advance in silicate structure research and its significance [J]. *Acta Petrologica Sinica,*17(2): 332-336(in Chinese)

[24] Miao Chunsheng, Gao Xinguo. The determination of order degree of feldspar and its application on geology [M].Beijing: *Geological Publishing House,* 1983,9-52,119-133(in Chinese).

[25] Li Shanpo, Xu Hong, Gao Di. Structure and forming temperature of K-feldspar and their metallogenic significance in the Jinchang Gold deposit, Heilongjiang [J]. *Bulletin of Mineralogy, Petrology and Geochemistry,* 2012,31(1)：31-37(in Chinese).

[26] Turrell G, Corset J. Raman Microscopy. Developments and Applications [M], New York, Academic Press, 1996.

[27] Zhang Bingliang, Zhao Bo, Yu Hongmei et al. Structure characteristic of feldspar from Holocene volcanic eruptive rock on the Tianfeng profile on Tianchi and their geological implications [J]. *Seismology and Geology,* 2009,31(4):628-638 (in Chinese).

[28] Ma Hongwen. Composition of K-feldspar ordering by X-ray powder diffraction and infrared spectroscopy and its relations to Al occupancy and equilibrium temperature [J]. *Acta Petrologica Sinica,* 1988, 8(2):143-150(in Chinese).

[29] Wu Pingxiao, Wu Jingping, Xiao Wending, Li Caiwei. Genetic Mechanism of Plagioclase Zoning[J]. *Geology-Geochemistry,* 1997, 4:40-49(in Chinese).

[30] Homma F. Über das Ergebnis von Messungen an zonaren Plagioklasen aus Andesiten mit Hilfe des Universaldrehrisches [J]. *Schweiz. Min. Prtr. Mitt., 1932*, 12:345-351.

[31] Rubatto D, Gebauer D. Use of cathodoluminescence for U-Pb zircon dating by ion microprobe: some examples from the Western Alps [M]//Cathodoluminescence in geosciences. *Springer Berlin Heidelberg,* 2000: 373-400.

[32] Vavra G, Schmid R, Gebauer D. Internal morphology, habit and U-Th-Pb microanalysis of amphibolite-to-granulite facies zircons: geochronology of the Ivrea Zone (Southern Alps)[J]. *Contributions to Mineralogy and Petrology,* 1999, 134(4): 380-404.

[33] Belousova E, Griffin W L, O'reilly S Y, et al. Igneous zircon: trace element composition as an indicator of source rock type [J]. *Contributions to Mineralogy and Petrology,* 2002, 143(5): 602-622.

[34] Wang Yinxi, Gu Lianxing,Zhang Zunzhong et al.Sr-Nd-Pb isotope geochemistry of Rhyolite of the Late Carboniferous Dashitou group in eastern Tiansha [J], *Acta Petrologica Sinica,* 2007, 23(7): 1749-1755.(in Chinese).

[35] Song Mingchun, Wang Laiming,Wang Lanzhong, et al. Ludong Rongcheng gneiss suite and its origin [J]. *Geology of Shandong,* 1995, 11(2): 32-44. (in Chinese).

[36] FuYongtao, Yu Ziye. Metamorphosed marine clastic rocks in Qingdao: Tectonic attribute and implication [J]. *Chinese Journal of geology,* 2010, 45(1): 207-227.(in Chinese).

[37] Qiu Liangui, Ren Fenglou, Cao Zhongxiang, et al. Late Mesozoic magmatic activities and their constraints on geotectonics of Jiaodong region [J]. *Geotectonica Et Metallogenia,* 2008,32(1):117-123. (in Chinese).

[38] Luan Guangzhong, Li Anlong, Wangjian, et al. The geological origin division of the main sea island in Qingdao area and environment analysis [J]. *Periodical of Ocean University of China (Natural Sciences),* 2010 (8): 111-116. (in Chinese).

[39] Gao Z, Pan J. Factors affecting zircon metamictization in granites [J]. *Chinese Journal of Geochemistry,* 1986, 5(2): 180-188.

[40] Zheng Yongfei. Magmatism and global change in Neoproterozoic [J]. *Science Bulletin,* 2003,48(16):1705-1720.

[41] Bao Xuezhao, Li Huiming, Lu Songnian. A Raman spectroscopic study of zircons on micro-scale and its significance in explaining the origin of the zircons [J]. *Scientia Geologica Sinica,* 1998,4. (in Chinese).

[42] Zheng Yongfei, Chen Fukun, Gongbing, et al. Features of UHP metamorphic rock in Dabie-Sulu orogenic belt: Evidence of oxygen isotope and U-Pb aging on zircon[J]. *Science Bulletin,* 2003,48(2):110-119.(in Chinese).

[43] Xu Zhiqin, Qi Xuexiang, Yang Jingsui. Deep Subduction Erosion Model for Continent-Continent Collision of the Sulu HP-UHP Metamorphic Terrain [J]. *Earth Science,* 2006,31(4):427-436. (in Chinese).

[44] Song Mingchun, Han Jingmin, Gong Shulin. Diversified origin granitic gneiss in the Sulu ultra-high pressure metamorphic belt: evidence for neoproterozoic large scale magmatism [J]. *Mineral Petrol,* 2007,27(2):22-32. (in Chinese).

[45] Liu Fulai, Xu Zhiqin, Song Biao. Determination of UHP and Retrograde Metamorphic Ages of Sulu Terrane: Evidence from SHRIMP U-Pb on Zircons of Geneissic Rocks [J]. *Acta Geologica Sinica,* 2003,77(2):229-237(in Chinese).

[46] Tang Jun, et al. A Study of Zircon U-Pb Dating and Oxygen Isotopes in UHP Granitic Gneiss from the Weihai Region in Northeast Sulu [J]. *Earth Science,* 2005,30(6):692-706 (in Chinese).

[47] Qiu Liangui, Ren Fenglou, Cao Zhongxiang, et al. Late mesozoic magmatic activities and their constraints on geotectonics of Jiaodong region [J], *Geotectonica Et Metallogenia,* 2008,32(1):117-123.(in Chinese).

[48] Xu Wenliang, Wang Qinghai, Wang Dongyan,et al. Processes and mechanism of Mesozoic lithospheric thinning in eastern North China Craton: Evidence from Mesozoic igneous rocks and deep-sea tedxenoliths [J]. *Earth Science Frontiers,* 2004,3 (in Chinese).

[49] Tctsumi Y, et al. 1982. Geological and geonemical evolution of the Archaean continental crust. *Geology-geochemistry,* 1988, (3):24—31, 12: 53—59.

[50] Portnov A M, Gorobets B S. Luminescence of apatite from different rock types [J]. *Doklady Akademii Nauk SSSR, 1969*, 184: 110-13.

[51] Portnov A M, Gorobets B S. Luminescence of apatite from different rock types [J]. *Doklady Akademii Nauk SSSR,* 1969, 184: 110-13.

[52] Kuznetsov G V, Tarashchan A N. Luminescence centers in natural apatites [J]. *Constitution and Properties of Minerals,* 1975, 9: 120-124.

[53] Marfunin A S, Schiffer V V. Spectroscopy, luminescence and radiation centers in minerals [J]. 1979.

[54] Roeder P L, MacArthur D and Ma X-P, et al. Cathodoluminescence and microprobe study of rare-eath elements in apatite [J]. *American Mineralogist.* 1987, 72:801-811.

[55] Marshall D J, Mariano A N. Cathodoluminescence of geological materials [M]. Boston etc.: Unwin Hyman, 1988.

[56] Blanc P, Baumer A, Cesbron F, et al. Les activateurs de cathodoluminescence dans des chlorapatites préparées par synthèse hydrothermale[J]. Comptes rendus de l'Académie des sciences. Série 2. *Sciences de la terre et des planètes,* 1995, 321(12): 1119-1126.

[57] Rakovan J, Reeder R J. Intracrystalline rare earth element distributions in apatite: Surface structural influences on incorporation during growth [J]. *Geochimica et Cosmochimica Acta,* 1996, 60(22): 4435-4445.

[58] Blanc P, Baumer A, Cesbron F, et al. Systematic cathodoluminescence spectral analysis of synthetic doped minerals: anhydrite, apatite, calcite, fluorite, scheelite and zirco [M]//Cathodoluminescence in Geosciences. *Springer Berlin Heidelberg,* 2000: 127-160.

[59] Kempe U, Götze J. Cathodoluminescence (CL) behaviour and crystal chemistry of apatite from rare-metal deposits[J]. *Mineralogical Magazine,* 2002, 66(1): 151-172.

[60] Mariano A N. Some further geological applications of cathodoluminescence [J]. *Cathodoluminescence of geological materials,* 1988: 94-123.

[61] Filippelli G M, Delaney M L. The effects of manganese (II) and iron (II) on the cathodoluminescence signal in synthetic apatite [J]. *Journal of Sedimentary Research,* 1993, 63(1).

[62] Perseil E A, Blanc P, Ohnenstetter D. As-bearing fluorapatite in manganiferous deposits from St. Marcel–Praborna, Val d'Aosta, Italy [J]. *The Canadian Mineralogist,* 2000, 38(1): 101-117.

[63] Dempster T J, Jolivet M, Tubrett M N, et al. Magmatic zoning in apatite: a monitor of porosity and permeability change in granites [J]. *Contributions to Mineralogy and Petrology,* 2003, 145(5): 568-577.

[64] Liu Yu, Peng Mingsheng. Advances in the researches on structural substitution of apatite [J]. *Acta Petrologica Et Mineralogica,* 2003,22(4):413-415. (in Chinese).

[65] Liu Yongxian, Ge Dingyi, Zeng Yunfu et al. The enrichment characteristics of apatite from phosphorites in east Yunnan [J]. *Journal of Mineralogy and Petrology.* 1994, 14（4） : 17~36. (in Chinese).

[66] Liu Yu, Peng Mingsheng. An Experimental Investigation for the Apatite Lattice Vibration by Raman-Activation [J]. *Bulletin of Mineralogy, Petrology and Geochemistry,* 2005(1): 58-61. (in Chinese).

[67] Li Yu, Xu Huanyan.The effects of structural channel ions of apatite on their lattice parameters [J] .*Journal of Mineralogy and Petrology.* 2001， 21(1):1-4.(in Chinese).

[68] Khorari S, Cahay R, Rulmont A, et al. The coupled isomorphic substitution 2 $(PO_4)^{3-}$→$(SO_4)^{2-}$+$(SiO_4)^{4-}$ in synthetic apatite $Ca_{10}(PO_4)_6F_2$: a study by X-ray diffraction and vibrational spectroscopy[J]. *European journal of solid state and inorganic chemistry,* 1994, 31(10): 921-934.

[69] Watson E B. Apatite saturation in basic to intermediate magmas [J]. *Geophysical Research Letters,* 1979, 6(12): 937-940.

[70] Watson E B. Apatite and phosphorus in mantle source regions: an experimental study of apatite/melt equilibria at pressures to 25 kbar [J]. *Earth and Planetary Science Letters,* 1980, 51(2): 322-335.

[71] Ewart A, Bryan W B, Gill J B. Mineralogy and geochemistry of the younger volcanic islands of Tonga, SW Pacific [J]. *Journal of Petrology,* 1973, 14(3): 429-465.
[72] Han Zongzhu, Wang Laiming, Zhang Zihuan. Genesis and evolution of eclogites from Shandong and Jiangsu provinces [J]. *Contributions to Geology and Mineral Resources Research,* 1992,7(4):13-23.(in Chinese).

In: Horizons in Earth Science Research. Volume 11
Editors: Benjamin Veress and Jozsi Szigethy
ISBN: 978-1-63321-296-1

Chapter 7

STATISTICAL PREDICTION OF WINTER RAINFALL IN PATAGONIA (ARGENTINA)

***Marcela Hebe González*[1,2] *and Natalia Herrera*[3]**
[1]Departament of Atmospheric and Oceanic Science - FCEN-University of Buenos Aires
[2]Research Center of Ocean and Atmosphere (CIMA) – CONICET/UBA; UMI-IFAECI/CNRS
[3]National Meteorological Service of Argentina

ABSTRACT

This chapter describes rainfall evolution in southern Argentina (Patagonia) and faces with the possibility to predict winter rainfall using oceanic and atmospheric predictors. The Andes Mountain ranges along the west of Patagonia and a large plateau extends towards the east. A decrease of rainfall was observed in all seasons in central and east Patagonia meanwhile an increment is observed in southern part of Los Andes. The study area was regionalized using a principal component analysis and four regions could be defined. Almost all regions showed an annual cycle with maximum rainfall in autumn-winter. Therefore, several different statistical models: an autoregressive integrated moving average model (ARIMA), Holt Winter (HW), Climate Prediction Tool (CPT) and an ensemble of all, called multimodel were applied to predict MJJ (May to July) Patagonia rainfall in four regions defined in Patagonia. They were designed with MJJ rainfall for the period 1975-2007 in CPT model and for the period 1988-2007 in ARIMA and Holt Winter because this last two models use 1975-1987 values to generate the final scheme. All the models were proved for 2008-2012. The main result of this work is that the consideration of ARIMA and HW in the multimodel ensemble improves the efficiency obtained using CPT. This result was supported by the calculation of efficiency coefficients which represent the ability to detect over and sub normal rainfall.

INTRODUCTION

Argentinean Patagonia region is located in the southeastern continental extreme of America between 30° and 55°S, where westerly mean flow prevails all year around, being more intense during winter. As a consequence of the interaction between the mean flow and the orography, maximum precipitation occurs in the vicinity of the mountains; meanwhile it decreases eastward over Patagonian region especially in winter (Prohaska 1976, Russian et al 2010). In such dry region the availability of drinking water is scarce and generally the rivers which start in the Andes and flow towards the Atlantic Ocean, are the most important fresh water sources.

Rainfall variability in Patagonia has been studied by some authors. Castañeda and González (2008) observed positive annual rainfall trends in northern and southern Patagonia, and a decrease in western and central area using last century data. They applied a non-linear methodology which identify the years, called "breakpoints", when trends change their behavior. They showed that northeastern Patagonia experienced a change in the annual rainfall trend later 1960 decade but a vast region in northwestern Patagonia denotes a change in precipitation in 1970 decade. Minetti et al (2003) also performed an analysis of non-linear precipitation trends over Argentina and Chile using polynomial functions. Russian et al (2012) studied the relationship between rainfall interannual variability in Northern Patagonia and circulation, and they found some connections with El Niño- Southern Oscillation (ENSO) and the Southern annular mode. Recently, González (2013) investigated rainfall and circulation connection and showed that cyclonic anomalies over the surrounding Atlantic and Pacific Oceans and over the continent enhance precipitation all over Patagonia in all seasons. Aravena and Luckman (2008) also detected relation between rainfall in Patagonia and circulation modes like ENSO and Antarctic Oscillation. Gonzalez and Vera (2010), Gonzalez et al (2010) and Gonzalez and Cariaga (2011) studied the Comahue region, located in northwestern Patagonia, where there are located most of the energy resources of Argentina, and derived a scheme prediction for winter and spring rainfall. Pasquini and Depetris (2007) analyzed the Patagonia rivers discharge and detected some significant cycles associated with rainfall.

The physical basis of seasonal climate predictability lies in the fact that slow variations in earth's boundary conditions (i.e. sea surface temperature or soil wetness) can influence atmospheric circulation and thus precipitation. In order to better understand these processes, models have been developed in some different regions of the world. This subject has motivated a great deal of research in the last 30 years or so. As a consequence, enormous progress has been made in seasonal forecasting, which allows the estimation of climate anomalies for several months ahead. To address this issue dynamical and statistical models are derived. Therefore, the technology of numerical weather prediction is extended to longer scales and produces numerical climate predictions using state-of-the-art of global atmospheric models. Generally, atmospheric general circulation models are forced by predicted sea surface temperature in different part of the world, especially the tropics. However, there is still a great deal of uncertainty about the efficiency of dynamical models in seasonal time scales. The statistical forecasts are based on the detection of the relationship between rainfall and circulation patterns observed previously.

Progress in seasonal forecasting is far from saturated and further improvements may be expected. We are not sure if the failures of individual seasonal predictions reflect the inherent natural limit of seasonal climate predictability or if we can improve methodologies to contribute to decrease errors. Therefore, this chapter shows the performance of some statistical models in Patagonia (Argentina) and compares results obtained using one model called Climate Prediction Tool (CPT) with these ones using a multimodel which adds autoregressive and exponential smoothing techniques.

DATA AND METHODOLOGY

Monthly rainfall data at 18 stations in Patagonia for the period 1975-2012 are used in this study (Figure 1). The data are derived from different sources including the National Meteorological Service (SMN) of Argentina, the Secretariat of Hydrology of Argentina (SRH) and Territorial Authority of the Limay, Neuquen and Negro rivers basins (AIC). The area of study is Argentinean Patagonia, the southern part of South America. All the selected stations have less than 20% of missing monthly rainfall data and their quality has been carefully proved.

Monthly sea surface temperatures (SST), 500 HPa (G500), 1000 Hpa (G1000) and 200Hpa (G200) geopotential heights, zonal (U) and meridional (V) winds at 850 Hpa from National Center of Environmental Prediction (NCEP) reanalysis were also used (Kalnay et al, 1996). Monthly anomalies were determined removing the climatological monthly means from the original values.

Precipitation low frequency variability was analyzed using a linear trend method of minimum squares, and statistics significance was tested using a T-Student test. The leading patterns of rainfall variability on monthly time scales were obtained through a Principal Component Analysis (PCA) (Richman, 1986) in the S-mode (Green, 1978) and they were used to regionalize Patagonia in four sub-regions. This methodology allows a good grouping of meteorological stations with similar characteristics. Walsh and others (1982) suggest that the accuracy of seasonal rainfall forecasts should be higher if applied on homogeneous regions. Therefore, the regionalization emerges as a useful tool.

Several large scale oceanic and atmospheric forcing (sea surface temperature, low, middle and high levels geopotential height and low level winds) in April were used to evaluate their influence on winter (May to July, MJJ) Patagonia rainfall using Climate Predictability Tool technique (CPT). It was developed by The International Research Institute for Climate and Society (IRI, Columbia University, cpt@iri.columbia.edu) provides a computational package for constructing a seasonal climate forecast model, using a set of predictors which must be carefully selected. It calculates the correlation between the empirical orthogonal functions (EOF) of predictors in April and EOF of MJJ rainfall in an earlier period and then derives a linear regression model to predict precipitation in later years. Others models were also applied: an autoregressive integrated moving average model (ARIMA), Holt Winter technique (HW) (Box and Jenkins, 1976; Holt, 1957) and an ensemble of all (CPT, HW and ARIMA), called "multimodel". ARIMA model predicts future values of a time series by a linear combination of its past values performing a maximum likelihood fit of them. HW model is a technique which uses a set of exponential weights that decrease each time by a constant ratio giving more weight to the more recent data values and less weight to the data values from further in the past. All data values in a series contribute to the calculation of the prediction model. The advantage of the models in which a variable is explained using only its own past (ARIMA and HW) is the simplicity but the inclusion of a wider set of explanatory variables (CPT) improves forecast skill by considering the physical relationship between them and precipitation.

All the models were used to predict seasonal rainfall in winter (MJJ) in the four regions defined in Patagonia. They were designed with MJJ rainfall for the period 1975-2007 in CPT model and for the period 1988-2007 in ARIMA and HW because this last two models use the previous values to generate the final scheme. All the models were proved for 2008-2012 as verification period. The CPT model uses oceanic and atmospheric predictors and therefore, it contains the influence of oceanic and atmospheric mechanisms associated with rainfall, meanwhile ARIMA and HW uses only the values of rainfall series and thus they face with the memory of the data.

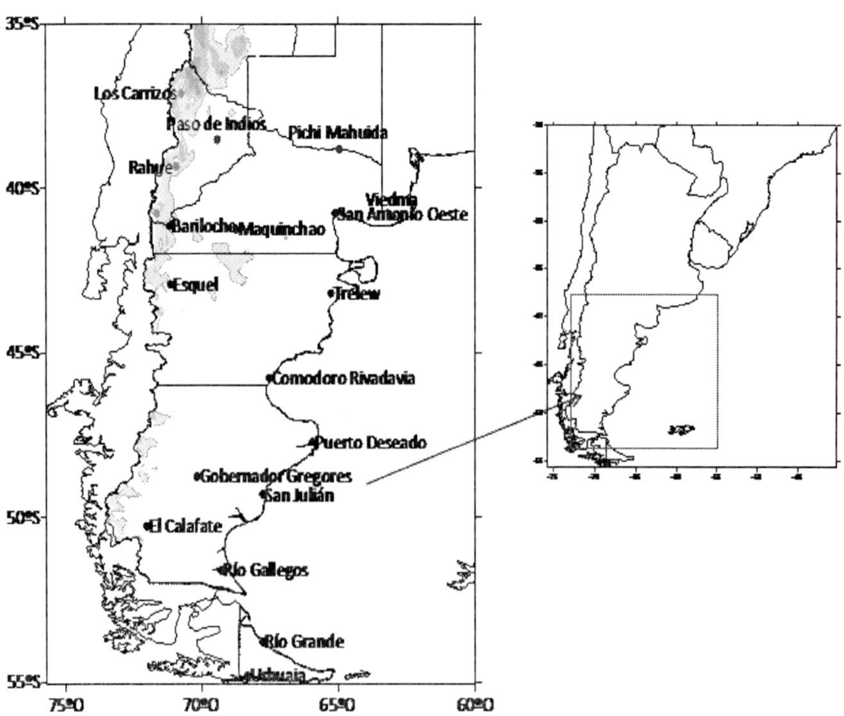

Figure 1. Stations used in the study.

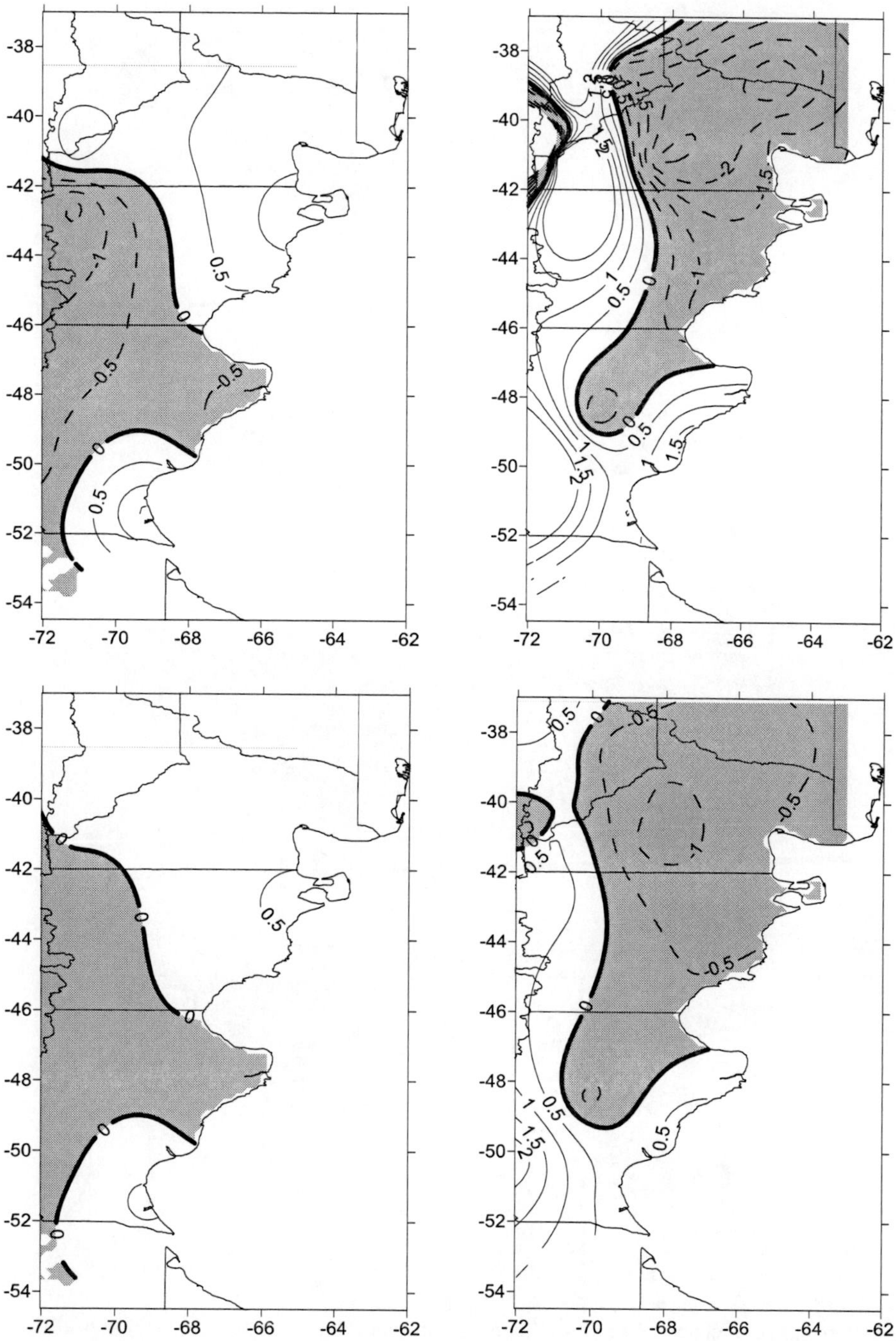

Figure 2. Annual rainfall trends for the periods (1961-2012) (top left panel) and for (1981-2012) (top right panel) in mm/year. Percentage of change respect to the mean annual rainfall for the periods (1961-2012) (lowleft panel) and for (1981-2012) (low right panel).

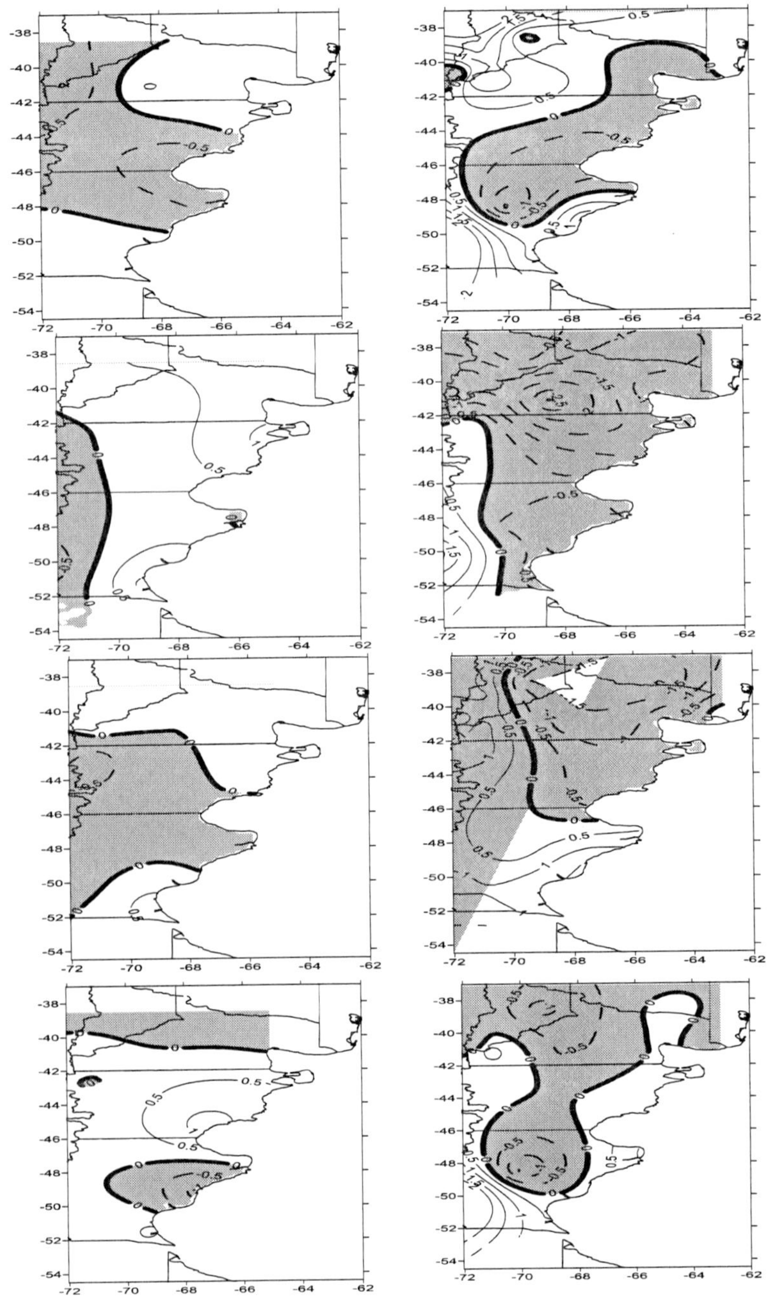

Figure 3. Percentage of change respect to the mean summer (December to February) rainfall for the periods (1961-2012) (top left panel) and for (1981-2012) (top right panel); autumn (March to May) (second panels), winter (June to August) (third panels) and spring (September to November) (bottom panels).

RESULTS

1. General Rainfall Patterns

To better understand the spatial distribution of annual rainfall evolution, annual trends (in mm/year) were calculated for the periods (1961-2012) in 13 stations (figure 2 top left panel) and for (1981-2012) in 18 stations (figure 2, top right panel) using a linear adjustment. The number of stations surveyed in each period depends on the record of the stations. An increase of 0,5 mm/year in 1961-2012 and a decrease of -2mm/year when only 1981-2012 period was considered, was observed in northeastern and central Patagonia. Meanwhile rainfall increase was present in Los Andes Mountain, especially in the southern part where trend was -1 mm/year in 1961-2012 and 2 mm/year in 1981-2012. These trends were significant at the 95% confidence level only in Rio Gallegos and Trelew in 1961-2012 and in El Calafate in 1981-2012. In fact, the percentage of the mean annual rainfall that these changes represent, was calculated (bottom panels) and it was less than 1%, therefore we can consider that the changes were not relevant. Figure 3 shows the percentage of change respect to the mean seasonal rainfall for both periods: summer (top panel), autumn (second panels), winter (third panels) and spring (bottom panels). The changes are less than 2,5% of the seasonal precipitation and trend behavior are similar in all seasons.

As all the stations show a mean rainfall annual cycle with maximum in late autumn and winter, a principal component analysis (S-mode) was applied to MJJ (May to July) accumulated rainfall anomalies for the period 1975-2012. The first four loadings explained 66% of the total variance and they let define four different regions (figure 4).

Therefore, the mean annual rainfall cycle of each region is shown in figure 5. Region 1 is called the Comahue Region. Some of the main Argentinean dams are in the western part of this region and most of the electrical resources derived from them. Electricity production is highly dependent on MJJ seasonal rainfall. Precipitation is lower in regions 2, 3 and 4 with maximums in March, May and April respectively. Region 2 tends to have a secondary maximum in spring. The mean rainfall in region 4 is slightly greater than in region 3.

2. Prediction Schemes for Winter Rainfall in Patagonia

Three different models were used to predict MJJ rainfall: ARIMA, HW and CPT. Then, the ensemble of them was proved (multimodel). Table 1 shows the April predictors used in each one of the regions to predict MJJ rainfall using CPT model. All they are carefully detailed in Gonzalez and Cariaga (2011) and in Gonzalez (2013). They reflect the influence of sea surface temperature in Pacific and Indian Oceans and the Rossby wave train from tropical Pacific Ocean towards western Argentina over winter precipitation.

Figure 6 shows the correlation between observed and forecast MJJ rainfall with each one of the three models and the multimodel ensemble, for the common period (1988-2007). Correlations greater than 0.46 are significant with the 95% confidence level using a T-Student test (Panofsky and Brier, 1958). ARIMA provides forecast values highly correlated (greater tan 0,7) with the observed ones. CPT has a good performance in regions 1 and 4; HW in regions 1 and 3 and the multimodel in the Comahue region (region 1). It is important to note that the inclusion of ARIMA and HW in the multimodel, improves the efficiency of the CPT forecasts.

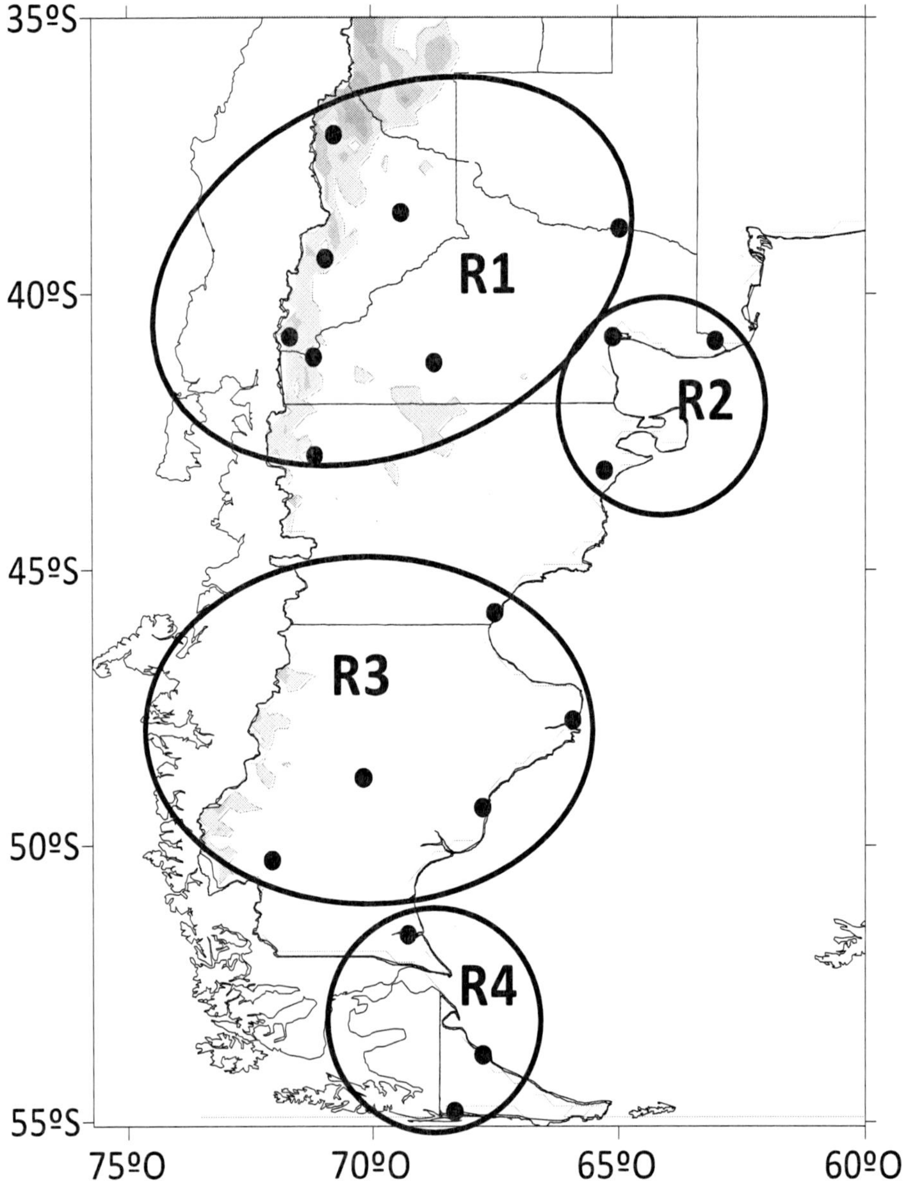

Figure 4. Regions derived from the principal component analysis of MJJ rainfall for the period 1975-2012.

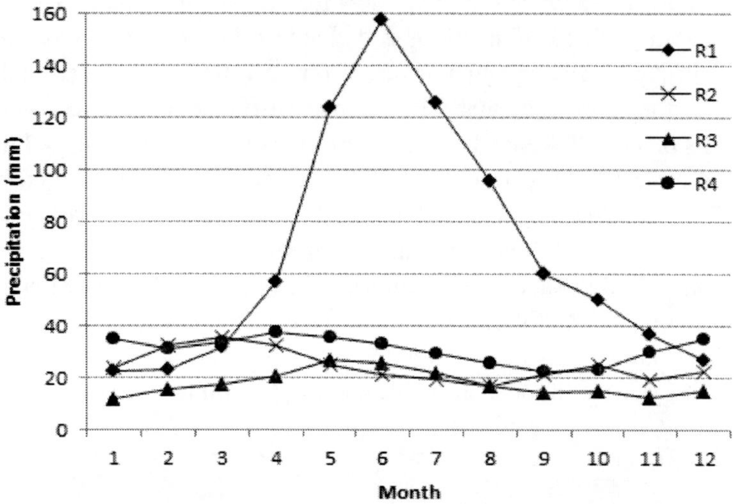

Figure 5. Annual rainfall cycle in the four regions defined in figure 4.

Table 1. April predictors associated with MJJ precipitation in the four regions from Gonzalez and Cariaga (2011) and Gonzalez (2013) (*)

Predictor	Variable	Pressure level	Area	Region
SST7	Sea surface temperature	Surface	[35ºS-50ºS;100ºE-130ºE]	R1
GEO8	Geopotential height	200 hPa	[40ºS-60ºS;110ºE-150ºE]	R1
Z1	Zonal wind	850 hPa	[35ºS-55ºS;170ºE-210ºE]	R1
M1	Meridional wind	850 hPa	[5ºS-15ºS;110ºW-90ºW]	R1
SST2	Sea surface temperature	Surface	[38ºS-47ºS;85ºE-110ºE]	R2
GEO12	Geopotential height	200 hPa	[15ºS-30ºS;50ºW-10ºW]	R2
SST5	Sea surface temperature	Surface	[5ºS-25ºS;40ºW-20ºW]	R3
GH5*	Geopotential height	500 hPa	[36ºS-55ºS;82ºW-72ºW]	R3
SST1	Sea surface temperature	Surface	[7ºS-20ºS;80ºE-120ºE]	R4
GEO6	Geopotential height	1000 hPa	[45ºS-60ºS;130ºE-150ºE]	R4

The skill of the forecast was measured in semi-quantitative terms as follows: the observed and forecast rainfall distribution was categorized in terciles and they were compared by constructing a contingency table (not shown). The "wet" interval refers to values greater than the second tercile; "dry" refers to values less than first tercile, and "normal" to values greater than first and lower than second terciles.

The events "normal", "dry" and "wet" were calculated collapsing the 3x3 contingency table into two 2x2 tables. Each one is constructed by considering the "forecast event" (normal, dry or wet) as opposed to the complementary "non forecast event" (non-normal, non-dry or non-wet). Some measures of accuracy (Wilks 1995) were calculated for normal, wet and dry cases in each station for both models. The hit rate (H) or the right proportion is the fraction of all the cases when the categorical forecast correctly anticipated the subsequent event. The probability of detection (POD) is defined as the fraction of those occasions when the forecasted event occurred on which it was also forecast. The false alarm relation (FAR) is the proportion of forecast events that fail to happen. Efficiency is perfect when POD and H are 1 and FAR is 0. The difference between these statistics values calculated for CPT and for multimodel ensemble for the period 1988-2007 was shown in figure 7. Therefore negative (positive) difference values of H and POD (FAR) indicate that the multimodel ensemble efficiency is greater than CPT. Figure 7 show that multimodel ensemble is better than CPT especially in central Patagonia, as maximum negative (positive) difference values of POD and H (FAR) were observed. Multimodel ensemble tends to better detect the "wet" events rather than the "dry" or "normal" cases. Southern Comahue (the Limay River basin) is the only area where the efficiency of multimodel ensemble and CPT is similar.

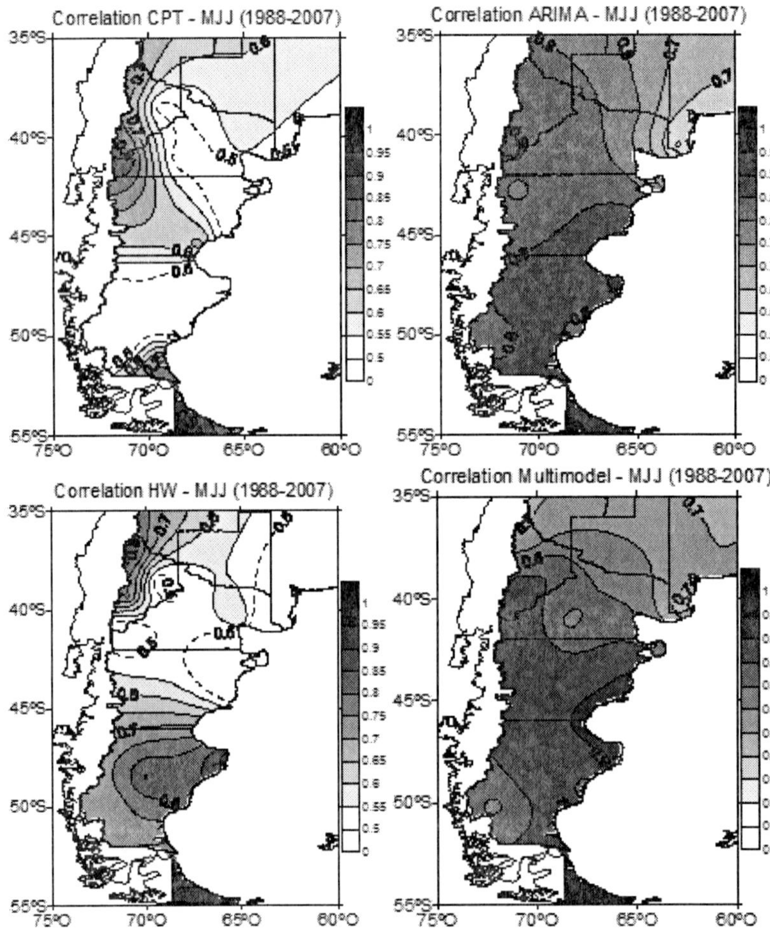

Figure 6. Correlation between MJJ rainfall and CPT (top left panel), ARIMA (top right panel), HW (bottom left panel) and Multimodel (bottom right panel) in 1988-2007. Correlations greater than 0.46 are significant (95% confidence level).

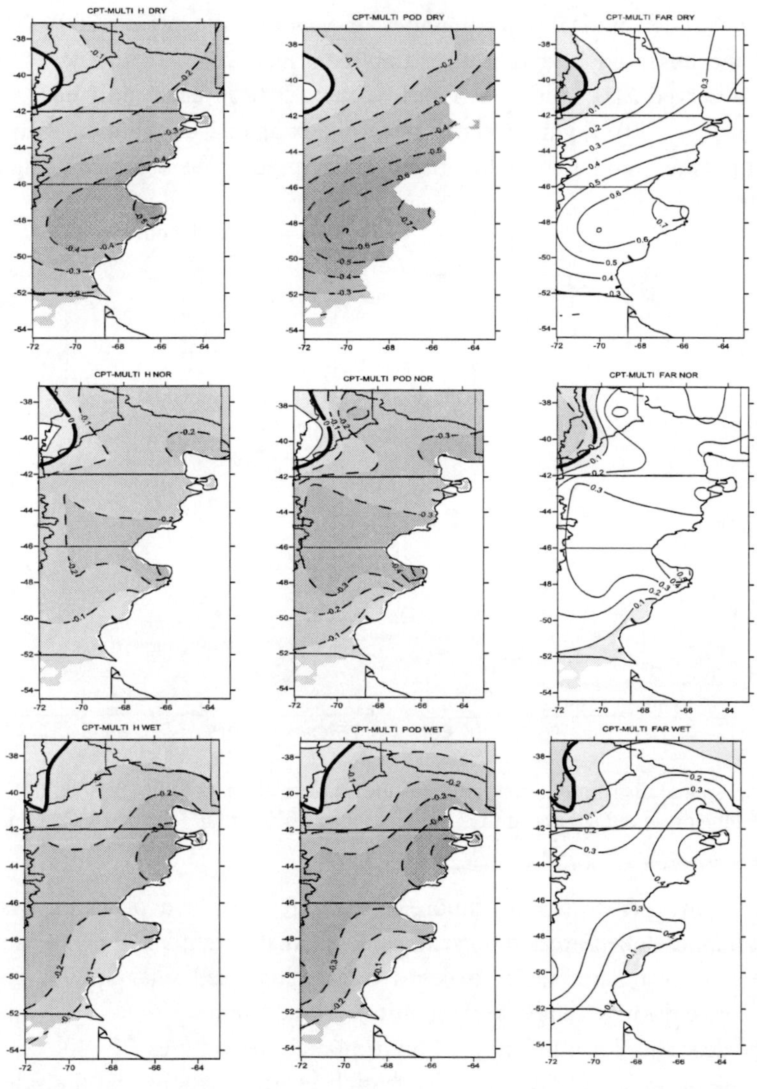

Figure 7. Difference between efficiency statistics (H left panel, POD central panel, FAR right panel) calculated for CPT and for multimodel ensemble for "DRY" (top panel) , "NORMAL" (central panel) and " "WET" (bottom panel) (1988-2007).

The five years period (2008-2012) was used as verification period. The CPT model and the multimodel ensemble were applied to forecast MJJ rainfall. In order to quantify the efficiency of the models, the *I index* was defined as:

$$I = \sum_{1}^{5} Ii$$

$$Ii = 1 \; if \; forecast \; and \; observed \; MJJ \; rainfall \; differs \; less \; than \; 20\%$$

$$Ii = 0 \; if \; not$$

Therefore, I index indicates the number of years of the period 2008-2012 when the forecast and the observed MJJ rainfall differs less than 20%. If all (none) the years the forecast and observed MJJ rainfall differs less than 20% in each individual station, so the index I will be I = 5 (0). Figure 8 shows the I index spatial distribution. It is important to point out that the multimodel ensamble improves the predictions obtained using only CPT all over the region.

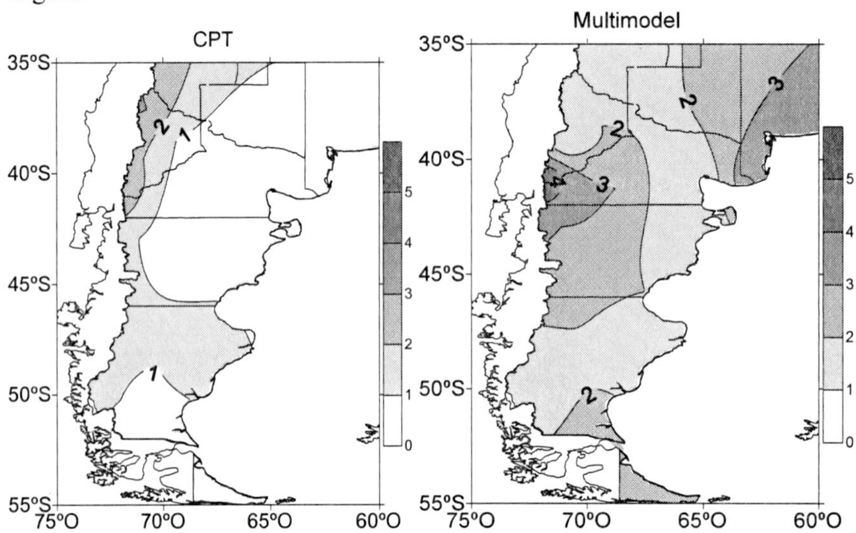

Figure 8. *I index* spatial distribution that indicates the number of years of the period 2008-2012 when the forecast and the observed MJJ rainfall differs less than 20% for CPT (left panel) and multimodel (right panel).

Figure 9 shows percentage deviation between forecast and observed MJJ rainfall in individual years, the dotted areas represents underestimated rainfall. It is noticeable that both models tend to overestimate rainfall except for some localized areas in 2008 and 2009. It is important to notice the high temporal variability of the deviation.

To better describe the efficiency of both models, percentiles 20, 40, 60 and 80 were calculate for the observed and the forecast rainfall in each stations and the rainfall could be classified in:

N (normal) if rainfall was between percentile 40 and percentile 60
SUB-N (sub-normal) if rainfall was between percentile 20 and percentile 40
SUB (extremely sub-normal) if rainfall was less than percentile 20
N-OVER (over-normal) if rainfall was between percentile 60 and percentile 80
OVER (extremely over-normal) if rainfall was greater than percentile 80

For each year in the period 2008-2012, a new index Ei was defined. It evaluates if the forecast and the observed rainfall differ in one or two categories as it is detailed in table 2. Ei is maximum (2) if forecast and observed categories are the same; Ei is 1 if they differ in one category, and so on.

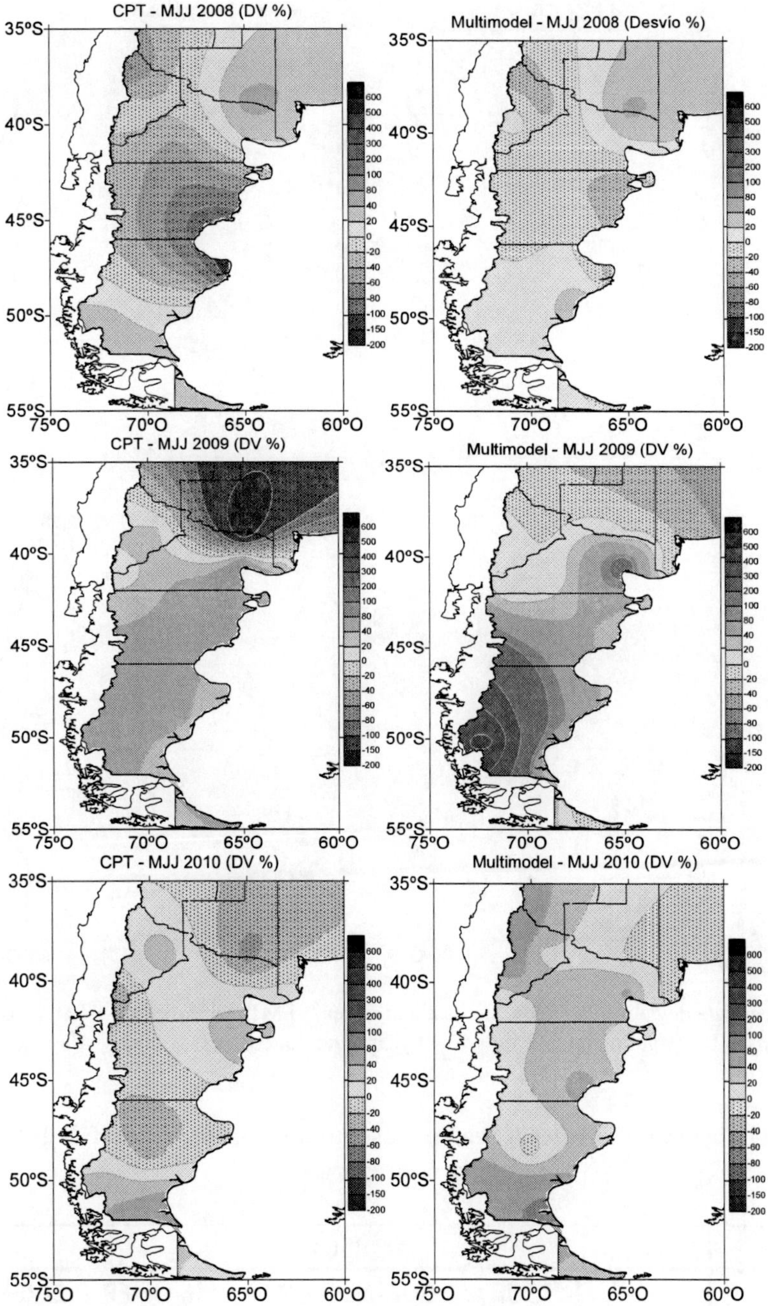

Figure 9. (Continued).

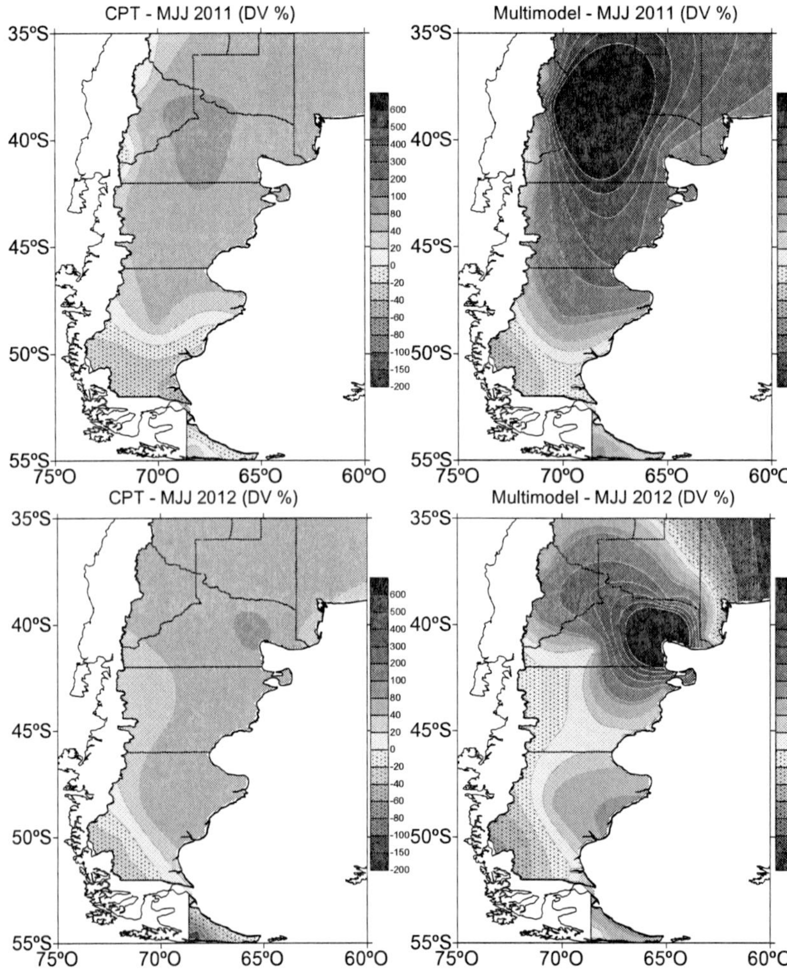

Figure 9. Percentage deviation between forecast and observed MJJ rainfall for CPT (left panels) and multimodel ensemble (right panels) in individual years, the dotted areas is underestimated rainfall.

Table 2. Ei index that indicates if the forecast and the observed rainfall differ in one or two categories

		FORECAST				
OBSERVED		SUB	SUB-N	N	N-OVER	OVER
	SUB	2	1	0	-1	-2
	SUB-N	1	2	1	0	-1
	N	0	1	2	1	0
	N-OVER	-1	0	1	2	1
	OVER	-2	-1	0	1	2

Then, an index E was defined as the sum of Ei for the 5 years in the period 2008-2012, expressed in percentage of the best performance over the 5 five year period (10):

$$E = \left(\sum_{1}^{5} Ei * 100\right)/10$$

Figure 10 shows the spatial distribution of E for CPT and for multimodel ensemble. This figure reflects the fact that the multimodel ensemble improves the CPT efficiency, especially in the northern and eastern Patagonia.

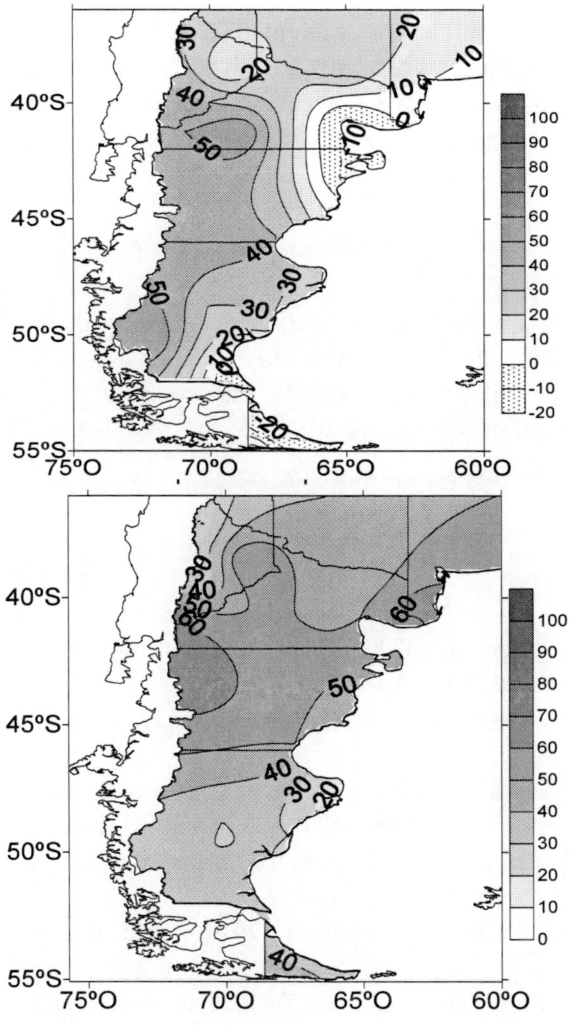

Figure 10. Spatial distribution of E, as defined in the text, for CPT (left panel) and for multimodel ensemble (right panel).

Conclusion

Different temporal scales variability in Patagonia was studied in this paper. A low frequency rainfall trend analysis showed a decrease of annual rainfall in central and east Patagonia meanwhile an increment is observed in southern part of Los Andes in 1981-2012. The percentage of the mean annual rainfall that these changes represent were less than 1%, therefore we can consider that the changes are not relevant. The percentage of change respect to the mean seasonal rainfall showed that the change was always less than 2,5% of the seasonal precipitation. Four regions were derived using a principal component analysis and they all showed an annual cycle with maximum rainfall in late autumn and winter. So, the three month period MJJ was selected to prove the efficiency of CPT model and an ensemble multimodel (CPT-ARIMA-HW) to forecast rainfall. They were designed with MJJ rainfall for the period 1988-2007 and proved for 2008-2012. The main result of this work is that the consideration of ARIMA and HW in the multimodel ensemble improves the efficiency obtained using only CPT, especially in northern and eastern Patagonia.

Acknowledgment

Rainfall data were provided by the Territorial Authority of the Limay, Neuquen and Negro rivers basins (AIC), the National Meteorological Service of Argentina (SMN) and the Secretariat of Hydrology of Argentina (SRH). This research was supported by UBACyT 2010-2012 CC02, UBACyT 2011-2014 01/Y128 and CONICET PIP 112-200801-00195, PICT 2010-2110.

References

Aravena, J.C. and Luckman, B.H. 2008, Spatio-temporal rainfall patters in Southern South America, *Int J Climatol DOI:* 10.1002/joc 1761.

Box, G. and Jenkins, Gwilym M. *Time series analysis: forecasting and control;* San Francisco, Holden-Day, 1976.

Castañeda, M.E. and González, M.H. 2008. Some aspects related to precipitation variability in the Patagonia region in Southern South America. *Atmósfera,* 21(3), 303-317.

González, M.H. and Vera, C.S. 2010. On the interannual winter rainfall variability in Southern Andes. *Int J Climatol,* 30, 643-657.

González, M.H.; Skansi, M.M. and Losano, F. 2010. A statistical study of seasonal winter rainfall prediction in the Comahue region (Argentine). *ATMOSFERA,* 23 (3), 277-294.

González, M.H. and Cariaga, M.L. 2011. Estimating winter and spring rainfall in the Comahue Region (Argentine) using statistical techniques. *Advances in Environmental Research,* 11, 103-118.

González, M.H. Some indicators ofinterannual rainfall variability in Patagonia (Argentine). In *Climate Variability: Regional and Thematic Patterns;* Editor, Dr. AondoverTarhule; INTECH; London, United Kingdom, 2013, in press.

Green, P. *Analysing Multivariate Data;* Dryden Press, 1978, pp 341-387.

Holt, C.C.1957. Forecasting Trends and Seasonal by Exponentially Weighted Averages. *International Journal of Forecasting,* 20 (1), doi:10.1016/ j.ijforecast.2003. 09.015

Kalnay, E.; Kanamitsu, M.; Kistler, R.; Collins, W.;Deaven, D.; Gandin, L.; Iredell, M.; Saha, S.; White, G.; Woollen, J.; Zhu, I.; Chelliah, M.; Ebisuzaki, W.; Higgings, W.; Janowiak, J.; Mo, K. C.; Ropelewski, C.; Wang, J.; Leetmaa, A.; Reynolds, R.; Jenne, R.; Joseph, D. 1996. The NCEP/NCAR Reanalysis 40 years- project. *Bull Amer Meteor Soc,* 77, 437-471.

Minetti, J.L.; Vargas, W.M.;Poblete, A.G.;Acuña, L.R. and Casagrande, G. 2003. *Non-linear trends and low frequency oscillations in annual precipitation over Argentina and Chile,* 1931-1999.Atmósfera, 16, 119-135.

Panofsky, H. and Brier, G. *Some aplications of Statistics to Meteorology; College of Mineral Industries, The* Pennsylvania State University, 1958; 224 pp.

Pasquini, A.I. and Depetris, P.J. 2007. Discharge trends and flow dynamics of South American rivers draining the southern Atlantic seaboard: an overview. *Journal of Hydrology,* 333, 385-399.

Prohaska, F.J. Climates of Central and South America. In: *World Survey of Climatology*; Elsevier Cientific Publishing Company; Amsterdam, Holland, 1976, pp 57-69.

Richman, M. 1986. Rotation of Principal Components. *J Climatol,* 6, 293-335.

Russian, G.; Agosta, E. and Compagnucci, R.H. 2010.Variabilidad interanual e interdecádica de la precipitación en la Patagonia norte. *Geoacta,* 35, 27-43.

Russian, G.; Agosta, E. and Compagnucci, R.H. 2012. Circulación troposférica de gran escala asociada a la precipitacion en Patagonia norte. In *Proceedings of CONGREMET XI,* 28May-1June 2012, Mendoza, Argentina.

Walsh, J. E.; Richman, M.B. and Allen, D.W. 1982. Spatial coherence of monthly precipitation in the United States. *Mon Weather Rev,* 110, 272-286.

Wilks D. S. *Statistical methods in the atmospheric sciences* (An introduction); International Geophysics series; Academic Press, San Diego, California, USA, 1995, 467 pp.

In: Horizons in Earth Science Research. Volume 11
Editors: Benjamin Veress and Jozsi Szigethy
ISBN: 978-1-63321-296-1

Chapter 8

PECULIARITY OF RESEARCH OF SEISMICITY AT AN OCEANIC AND SEA BOTTOM

Dmitry G. Levchenko, [*] ***Ivan P. Kuzin,***
Leopold I. Lobkovsky and Konstantin A. Roginsky
Laboratory of Seismology and Geodynamics Shirshov Institute of Oceanology RAS

ABSTRACT

The modern world seismological network consists mainly of stations on land while most earthquakes occur on seas and oceans. To create a single tectonic scheme of the Earth it is necessary to develop global network of bottom stations. However, there are many problems. In this paper the features of seismic signals in the bottom waters are discussed, which still has not been sufficiently studied. These problems are including: coupling a seismograph housing, resonance phenomena in recording signals on the bottom, noises from near bottom currents, and others. Describes broadband bottom seismographs and bottom observatory, developed in the Shirshov Institute of Oceanology Russian Academy of Science, taking into account the specific conditions recording at the sea bottom. There are present some results of studies with bottom seismographs. It is considered the possibility of using the bottom observatory for the study hydro-physic precursors of strong earthquakes in the water areas.

1. INTRODUCTION

The majority of the earthquakes (above 80%) occur under the ocean and sea bottom; however, the stationary marine seismological network is absent now. The land seismographs record signals from the remote marine earthquakes with substantial distortions. The signals

[*] Corresponding author: Email: levch35@mail.ru.

from the moderate and weak marine earthquakes they are not recorded at all. For this circumstance, the geotectonic processes occurring in the Earth's interior cannot be adequately estimated in their unity and variety. At present the deep structure of the ocean crust and the upper mantle is known very approximately. The prediction of the strong marine earthquakes and excitation of tsunami are highly important, but warning of them is very complex problem. Therefore, the formation of the integrated seismological network on land and at sea is extremely important (Gutenberg, 1959; Lobkovsky et al., 2005; Montagner, et al., 1992; Purdy and Dzievonski, 1995; Romanovicz et al., 2006; Soloviev, 1986).

The bottom earthquakes and the related phenomena become extremely hazardous for the marine structures and ecology of the region because of intense extraction of hydrocarbons in oil and gas fields at the shelves. It seems important to study the effect of the near and remote earthquakes on well yield in process of exploitation. Seismicity of seas has also potential danger for shore structures and populated areas. Absence of marine seismological network does not allows to estimate this danger properly. Therefore, seismological securing the marine oil- and gas-production complexes and other big shore and underwater structures and population is very actual problem (Levchenko et al., 2013; Lobkovsky, et al., 2005).

Recording of the seismic signals at the sea bottom considerably differs from that on land. The main difference are the influence of a water layer and a water saturated sedimentary layer, the interaction between the seismograph housing and the soft bottom, and excitation of noise by near-bottom currents (Levchenko, 2006; Sutton et al., 1981).Unfortunately, solid rocks, which are the most suitable for recording seismic signals, occupy very small areas of the bottom of seas and oceans. The most common are the soft sediments, mainly silts of different composition with fine-grained silty and clayey oozes (Figure 1) (Lisitsin, 1974).

Besides recording seismic signals on the bottom is accompanied by specific distortions caused by resonance phenomena. In the known publications such distortions are explained by the natural oscillations of the heavy housing of a seismometer on a soft elastic sediment layer (Godin and Chapman, 1999; Chapman et al., 1993; Sutton et al., 1981a, 1987). Meanwhile, there are experimental results that contradict this model (Bradner et al., 1970; Levchenko, 2006). In the present paper, we consider the rheological properties of the bottom sediments, which in fact were not taken into account earlier. The model of a plastic viscosity medium was used (the Bingham model), and parameters of the model were experimentally determined. Research have shown that at the big deformations of a bottom, commensurable with the sizes of a seismograph case, deposits viscosity was extremely large. Therefore the seismograph case is "pasted in" a bottom silt and oscillates together with it.

Estimations show that, in the frequency range from 0.003 to 30 Hz used in broadband bottom seismology, the effect of the mass of the seismometer on the results of recording on a soft bottom is negligible. Large errors can be introduced only when a seismometer is placed on rubberlike media such as peat soil, algae aggregations, etc. Resonance phenomena in recording signals on the bottom can occur when seismic waves propagate through a layer of water-saturated sediments. These phenomena are more pronounced for shear waves, whereas the distortions of the longitudinal waves in this case are relatively weak (Chapman, et al., 1993; Levchenko, 2006).

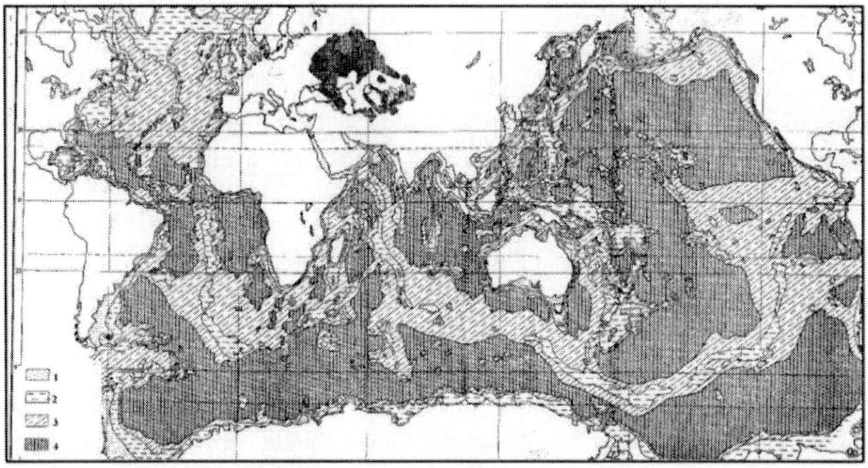

1 – gravel- pubble sediments and sands, 2 – silts (aleurite) and silting oozes, 3 – siltic-pelitic oozes, 4 – pelitic oozes.

Figure 1. Granulometriccomposicion of bottom sediments World ocean (Lisitsin. 1974).

The formation of an integrated seismological system, which can unite the land and marine networks for recording and processing of the earthquake signals and simultaneously search possible precursors has vital importance. However, for some objective and subjective reasons the marine seismological network not exist so far (Araki and Suyehiro, 2001; Levchenko and Matsievsky, 2000; Montagner et al., 1994; Romanowicz et al., 2006).

In our opinion, the main difficulties preventing the formation of the marine seismological network are the following. Since the bottom stations operate in the automatic regime, they should be highly reliable during long period of time. The stations should have durable housings for preventing their crash from depth pressure and shocks on a ship board and during installation on a firm bottom. At present practically there are not commercial seismic sensors for operation in the bottom stations, which could have the metrological parameters close to those for the land seismometers. Moreover it is difficult to provide long- term power supply of instrumentation and transmission of seismological information to the shore.

A principal solution of the problem could be use of submarine cables for feeding the bottom stations, control of their operation, and direct information transmission to the shore. However, the submarine cable installation is extremely expensive. The promising trend seemed to be connection to the existing overseas bottom communication cables, which currently are not used for their direct purpose (VENUS, POSEIDON, and GEO-TOK projects). However, such projects also require substantial investments. In addition, old submarine cables have low strength and water tightness on pulling out and reutilization. Therefore, there are few stationary marine seismographs with cable connection to the shore. All of them are located in USA (HUGO-2) and Japan and placed relatively not far (100–200 km) from the shore (Kasahara et al., 2003; Mikada et al., 2003).

The lack of the stationary marine seismological network leads to use mainly autonomous bottom seismic stations (ABSS). They do not solve all the existing problems; nevertheless, the ABSSs have some advantages over the cable stations. Their high mobility allows fast deployment of the station network in any World Ocean region of interest. Any polygon configuration in accordance with the specific problems and available instrumentation can be chosen. Instrumentation composition and parameters of seismometers can vary prior to the installation. Finally, the ABSSs are less expensive than the stationary cable stations by orders of magnitude. The autonomous bottom seismic stations are available in the USA, Russia, Japan, German, and France; however, the main metrological parameters of the stations are, as a rule, worse than those of the best land models (Jacobson et al., 1991; Levchenko, 2001; Lobkovsky et al., 2005; Shiobara et al., 2001; Wielandt and Steim, 1986; Zubko et al., 2003).

According to our opinion, the development of the marine network should begin within the so called hot points. First of all, it is the regions of marine oil and gas deposits. The hot points in Russia are the Barents, Pechora, and Kara Seas in the north, the Black and Caspian Seas in the south, and the Sea of Okhotsk in the east. The cable bottom seismographs can be integrated in the ecological monitoring network which should deploy around each drilling or producing platform. In addition to monitoring of seismicity, including that caused by redistribution of tectonic stresses during production, this network can solve other problems, such as supplementary exploration and estimation of reserves. A data collection post, which placed on a platform, would provide operative information transmission and the conditions required for information processing (Lobkovsky et al., 2005; Mikada et al., 2003; Shiobara et al., 2001).

Other areas of preferential deployment of local stations should be in the "seismic gaps," where the expected strong sea earthquakes and possible tsunami. One of these areas is Avacha Bay (Kamchatka), others are in the southern and central parts of the Kuril Islands. Because of significant removing seismic active zones from the shore it should be install the autonomous long-term bottom stations with operative transfer of express information. The main tools of such stations should be seismograph broadband and strong-motion seismograph, the device operative data analyzer, and communication system via satellite or radio channels.

Autonomous bottom seismographs available in above mentioned countries are intended, as a rule, for marine exploration seismology and solving specific seismological problems, such as, the study of the seismicity of some water areas. These seismographs have limited characteristics than with land seismographs, in particular, narrower frequency band, lower channel sensitivity, and shorter operation time. Therefore, for long-term monitoring of water areas requires the qualitatively different bottom seismic stations, the parameters of which are close to land stations (Levchenko, 2002; Montagner et al., 1992; Ostrovsky, 1998; Soloviev, 1986).

In the Shirshov's Institute of Oceanology, Russian Academy of Sciences for about 30 years, developed and used in investigations broadband (0.003–20 Hz) SHADS bottom seismometers. They were used to obtain unique records of microseisms, local and distant earthquakes, vibrations from underwater slides, etc., on sea bottom (Levchenko, 2002, 2009). In order to test the reliability of seismograph SHADS for 4 years, carry out continuous seismological monitoring on the territory of the Experimental Design Bureau of Oceanological Equipment, Russian Academy of Sciences (EDBOE). The device was installed in a specially equipped underground bunker. In the course of observations were recorded regularly the level of local noise, and oscillations of distant seismic events. In particular, there was recorded a catastrophic earthquake March 11, 2011, off the coast of Japan (Levchenko et al., 2012).

Based on operational experience seismographs SHADS designed in the Shirshov Institute of Oceanology, Russian Academy and the EDBOE, we can assume that their technical characteristics allow us to recommend them as a basis for the development of marine seismic networks in the areas of expected strong earthquakes.

Other promising direction of short-time forecasting of marine earthquakes may be monitoring of bottom water in areas expected strong marine earthquakes. This consideration is based on the idea of the escape of gases and fluids with an increase stresses before strong earthquake and as a result of changes in the physical and chemical parameters of the near-bottom water. Such experiment have been carried out by the "Kamchatnedra" company in Avachinsky Bay in 1987-1995 and yielded good results (Gavrilov et al., 2000; Lobkovsky et al., 2005).

2. The Coupling of the Seismometer Housing with a Soft Bottom

The coupling of the seismometer housing with a soft bottom is an important problem for marine experimental seismology. Under unfavorable conditions, specific interference and distortions of signals are possible, which makes it difficult to record and interpret them reliably. A number of theoretical and experimental studies are devoted to this problem (Neprochnov et al., 1983; Ostrovsky, 1998; Prothero, 1984; Soloviev, 1986; Sutton et al., 1981a; Trehu, 1985; Zelikovvitz, 1981, and others). Most of the studies are predominantly devoted to the consideration of the distortions associated with the free oscillations of a heavy seismometer on an elastic water-saturated bottom. This model was based on the resonance phenomena observed in recording the seismic signals by of bottom seismometers. By the late 1980s, a concept was proposed according to which recommendations on the optimal design of bottom seismometers were formulated (Sutton, and Duennebier, 1987).

It should be noted that the number of papers published experimental results earlier as well as recently have no explanation within the framework accepted theory. In particular, in recording seismic signals in the bottom wells, and elastic oscillations in water column there are observed resonance phenomena analogous to those described for installation of instruments at the bottom, although in these cases there is no contact between the soft bottom and seismometer (Araki, and Suehiro, 2001; Chapman et al., 1993; Duennebier et al., 1985; Godin, and Chapman, 1999). Moreover there were found out resonance phenomena in recording seismic-acoustic noise at the bottom which were not related to seismological observations. This raises doubts about the adequacy of the theory and developed on the basis of its recommendations for the design of bottom seismographs.

The ambiguity of the previous results requires more complete to consideration the problem of coupling of seismometer housing with soft bottom. In addition it is necessary take into account both density and elastic properties of the bottom sediments and other characteristics (viscosity, adhesion, and internal friction), which previously practically were not taken into account. It is also necessary to distinguish between the supporting power of the bottom sediments during the installation of the seismometer and the transfer of seismic pulses from sediments to its housing since in these cases quite different properties of bottom sediments are observed. From standpoint of supporting power and characteristic features of the propagation of seismic signals, bottom sediments, in the first approximation, can be divided into solid and soft types. Solid bottom can be subclassified into rocky and porous kinds. Rocky beds have either a magmatic origin (basalt, granite), or represent consolidated sediments. Porous rocks are pebble-gravel-stones, and sandstones (Ilnitskaia et al., 1969; Geoph. Refer. book, 1966; Handboock, 1969; Frolov and Korotkih, 1983).

The main difference between the supporting power solid and soft bottom sediments consists of different reaction to load due to weight of the seismograph. In thus sinking device on solid rock is slight and short-lived (of a second). In soft sediments, the seismometer sinks to a considerable depth commensurable with its size, and the process, due to viscosity of soil, lasts for several hours. Volume of a solid porous bottom does not increase significantly during the saturation of pores with water and does not decrease with water removal. At the same time a soft bottom rocks subject to considerably swelling when it contacts with water because occurs connection solid skeleton with filling at a molecular level. The material of a solid bottom has weak adhesion (stickiness). In porous solid rocks the particles are bonded due to internal friction and in hard rocks due to internal cementation. The material of a soft bottom has large stickiness as to the sediment particles within the rock and in relation to external bodies (e.g., the seismometer housing).

Propagation longitudinal and transverse seismic waves through a solid bottom within the area of elastic deformations take place without distortions and slightly depend on the filling water in the pores. In this case the faces of the particles of a solid material are tightly adjoin with each other and form a rigid frame. Therefore porosity, properties of filling, and character of the contacts between particles have weak effect on propagation seismic waves in porous media. The propagation of seismic waves through soft sediments depends significantly on their water saturation and on the particles sizes of solid fractions. The conditions of propagation of longitudinal and transverse waves can differ considerably (Geoph. Refer. book, 1966; Hempton, 1977).

Some organic deposits (peat soils, benthic algae agglomerations and others), have specific properties. They are natural shock absorbers, and are characterized by significant sizes of the area of reversible elastic deformations. Their mechanical properties are close to those of rubberlike synthetic materials. Process subsidence of seismograph housing in these materials occurs quickly, with virtually no adhesion.. The recording seismic signals on this bottom essentially depends on the porosity and water saturation. This may result in large signal distortions due to the resonance and attenuation. Theoretical investigations and laboratory modeling of recording elastic oscillations using rubberlike materials were carried out by number American scientists and engineers (Sutton et al., 1981a; Sutton et al., 1981b, Zelikovitz and Prothero, 1981). It should be noted that these deposits on the seafloor are rare, mainly in shallow waters, where they can be easily detected.

Unfortunately, the share of solid sediments, most suitable for recording seismic signals, account for a very small area of the bottom of the seas and oceans. The most common are the sediments that are represented mainly by soft silts ($0.01 \leq d \leq 0.1$ mm) and pelitic muds ($d \leq 0.01$ mm) of different composition. The largest part of the bottom of oceans and seas is represented by silty and clayey ($d < 0.01$ mm) oozes (Lisitsin, 1974).

Let us consider some of the properties of soft bottom deposits, defining features of the coupling with the seismometer housing. These properties substantially depend on the sizes of the particles of the material and extent of water saturation. Table 1 shows the classification of fine-grained fractions bottom sediments and sandy–clayey grounds prevailing in the composition of a soft bottom. This classification is adopted in marine geology (Lisitsin, 1974) and in engineering geology (Frolov and Korotkih, 1983). Below, we refer to publications where both of the classifications are used.

Depending on the water content, the material of a silty bottom can be in solid, plastic, or fluid state. Solid material when the load increases to the limit of its strength breaks, plastic material deformed without destruction. The boundary of plasticity ϖP is the percentage content of water, usually close to the maximum molecular moisture capacity, at which water is in bonded state (Frolov and Korotkih, 1983). The bottom deposits, which are in plastic state, have increased stickiness and forms aggregate bonds between particles. During deformation of this material microfractures of aggregate bonds occur, while maintaining its overall integrity. With further increase in water content of the sediment it moves in a fluid state and acquire the properties of a viscous fluid. A significant portion of water passes into free state, and the solid fraction in the absence of mixing precipitate and consolidated. The boundary between the plastic and fluid state is determined by the adopted percentage of water content ϖF. The width of zone of the plasticity is evaluated by number of the plasticity, which is equal to the difference $\varpi_F - \varpi_P$.

Table 1. Grain-size composition of the fractions of sandy-clayey grounds and bottom sediments

Type of ground fraction	Particle size, mm	Type of bottom sediment fractions	Particle size, mm
Clayey	<0.005	Clay	<0.01
Silty	0.005 – 0.05	Silt	– 0.1
Sandy	0.05 – 2	Sand	0.1 – 1.0
Gravel	>2	Gravel	1.0 - 10

Table 2. Classification and properties of sandy – clayey ground

Ground type	Concentration of clayey fractions ($d <$ 0.005 mm), %	Moisture content at the plasticity limit ϖ_P, %	Plasticity number ϖ_1-ϖ_P, %	Extent of sticky
Clay	>30	>25	>17	Strongly sticky
Loam	10 – 30	15 – 25	7– 17	Tenacious
Sandy loam	3 – 10	7 – 15	1– 7	Weakly sticky
Sand	<3	<7	<1	Nonsticky

As a rule, there is a mixture of various fractions at the bottom, and the characteristics of the bottom sediments depend on their proportions. Table 2 presents a classification and lists the characteristics of sandy–clayey grounds. Depending on the percentage of clayey and sandy fractions, the plasticity limit, plasticity number, and extent of tenacity decrease with the increase in the concentration of large particles.

A number of studies show that fine-grained bottom sediments are mainly in a plastic condition, at least in the upper layer, whose thickness ranges from a few centimeters to a few meters. This is also evident from the fact that the bottom relief is stable in time, and heavy objects (anchors, sunken ships, bottom instruments, and others) lie on the bottom for a long time and do not submerge down to greater depths. On the other hand, the material of the bottom is pronouncedly deformed under the action of external loading, but the deformations are slow, which indicates the high viscosity of the material. As follows from the experiments, when placed on a soft bottom, the seismometer housing submerges for several hours, generating specific seismic interferences that represent short pulses (Neprochnov et al., 1983; Ostrovsky, 1998; Soloviev, 1986). These noises can be explained by breaking bonds in the plastic–viscous material of the bottom.

A series of physical models and appropriate theories have been developed for the estimation of the physical and mechanical properties of plastic–viscous materials (Ilnitskaia et al., 1969; Landau and Lifshits, 1987, 1988;Geoph. Refer. Book, 1966). Such theories are based on the concepts of internal friction, viscosity, relaxation time, etc. According to the theory, for any solid body affected by a moderate loading, there is a region of reversible elastic deformations. For sedimentary rocks, this region is rather small and is no more than 10^{-4} of the value of the residual deformation. The presence of such a region allows the seismic oscillations to propagate over large distances with a low attenuation (Gutenberg, 1959; Savarensky, 1972; Geoph. Refer. book, 1966).

With the loading increase, the plastic deformations of solid bodies become irreversible. During the plastic deformation, energy equivalent to the internal friction in the viscous fluid is spent on the breaking of internal bonds between the particles (the Boltzmann model). In contrast to the fluids, in solid plastic bodies there is an initial shear stress τ_o, which is similar to the friction of rest between surfaces. According to the Bingham law, the shear stress for plastic–viscous bodies is

$$\tau = \tau_o + \eta(dv/dl), \qquad (1)$$

where τ_o is the initial shear stress, η is the dynamic viscosity, and dv/dl is the derivative of the velocity of motion across the section (Jiang and Le Blond, 1993; Liu and Mei, 1989; Mei and Liu, 1987).

It should be noted that the strict determination of viscosity for the assessment of the internal friction exists only for the so-called ideal Newtonian fluid, which has a constant viscosity and is incompressible. The actual fluids differ from the ideal one in their properties, but the concept of viscosity is also used for them. The concept of viscosity is also used for solid plastic objects (Ilnitskaia et al., 1969; Landau and Lifshits, 1987).

As is known, viscosity depends on a number of factors (composition of the material, temperature, pressure, etc.) and may change within a wide range for various fluids. For most of the fluids, under normal conditions, the scattering of the values of viscosity is approximately four orders, from 10^{-3} to 10. For solids, it ranges from 10^{7} to 10^{8} (the region of elastic– viscous deformations) and to 10^{15} (the region of plastic deformations) (Kariakin et al., 1962; Geoph. Refer. book, 1966). The relaxation time (time of disappearance of stresses after applying a load or a deformation) is directly dependent on viscosity (Savarensky, 1972):

$$\beta = \eta/\mu, \tag{2}$$

where the viscosity η and the shear modulus μ are expressed in Pa·s and Pa, respectively, and the relaxation time β is expressed in seconds.

Table 3. Comparative mass-size characteristics of bottom seismographs

Type, developer	**Mass in air, kg**	**Housing sizes, m**	**Total volume, m^3**	**Base area, m^2**	**Weight in water, N**	**Specific weight, N/m^3**	**Pressure on the bottom, Pa**
ADSS-1, IO RAS (Russia)	40 [sph] +40[cup]	[R] 0.22 [r] 0.25 [H] 0.30	0.066	0.20	140	2120	700
ShADS, IO RAS, (Russia)	53 [sph]	[R] 0.22	0.048	0.07	50	1040	714
DSS, IO RAS, (Russia)	80 [cyl]	[r] 0.1 [L] 1.0	0.03	0.30	500	16666	1666
SEDIS-4, GEO-Pro, (Germany)	57 [sph]	[R] 0.21	0.045	0.20	120	2666	600
SIOL, SIO, (USA)	230	[R] 0.30	0.110	0.21	1200	10909	5714
UWFL, WSU, (USA)	45 [sph]+ 80 [cup]	[R] 0.30 [r] 0.33 [H] 0.50	0.083	0.126	420	5060	3360

Note: [sf] is the spherical housing, [cup] is cup-shaping ballast, [cyl] is cylindrical housing, [R] is the radius of the sphere, [r] is the radius of the housing cylinder or the ballast radius, [L] is the cylinder length, and [H] is the ballast height.

IO RAS – Institute of Oceanology Russian Academy of Science, GEO-Pro.Co (Hamburg, Germany), SIO – Scripps Institute of Oceanography, WSU – Washington State University.

Table 4. Grain size composition of the experimental material

Size, mm	1-0.5	0.5-0.25	0.25-0.1	0.1-0.05	0.05-0.01	0.01-0.005	0.005-0.001	<0.001
Content, %	0.5	1.03	5.61	7.67	21.6	25.15	25.02	13.4

The process of the seismometer subsidence essentially depends on its pressure on the bottom and on the ratio of the specific weight of the instrument to that of the bottom sediment. The averaged specific weight of the upper layer of fine-grained bottom sediments of natural humidity, containing 50–70% of clayey fraction, ranges from 12000 to 18000 N/m^3. On the average, it is14000 N/m^3, 15000 N/m^3, and 13000 N/m^{3}for terrigenous, carbonate, and siliceous sediments, respectively (Lisitsin, 1974).Table 3 presents masses and sizes of selected types of bottom seismometers and the values of their pressure on the bottom.

For ADSS-1 and SEDIS-4 single-module pop-up seismometers, the specific weight was calculated for the instrument itself and for the ballast secured to it. For a two-module SHADS, the specific weight was calculated only for the housing with the seismograph. Table 3 shows that the specific weights of the instruments in the air are approximately equal to those of the sediments, while, in the water, their equivalent specific weight is several times lower. The specific weight of a DSS seismometer with the surface buoy exceeds on the average the specific weight of the sediments by a factor of 1.5 in the air and is approximately equal to it on the bottom. The specific weight in the air of heavy pop-up seismometers developed in the USA slightly exceeds the specific weight of the sediments, and in the water it is several times lower.

The data presented indicate that the excess specific weight of most of the bottom seismometers submerging into a soft bottom is low or completely absent. The excess specific weight may be balanced by the static viscosity of the bottom sediment. Therefore, the bottom seismometers are incapable of penetrating into a soft bottom down to great depths. In order to assess the supporting power of a soft bottom and to qualitatively verify the theoretical formulae presented, a series of experiments was carried out. The experiments were carried out with clay, whose grain size composition (Table 4) is close to that of the silty–clayey variety, the most abundant material in the surface bottom sediments (Lisitsin, 1974).

The density of the dry material is 1660 kg/m^3. The density of the humid material is close to the maximum, and is equal to 1980 kg/m^3. The humidity is about 19%of the dry residual and is approximately equal to the plasticity limit ϖ_P. The yield point ϖ_F is about 45%, and the plasticity number equals 26%. In the course of the experiments, the humidity changed if necessary to enhance the plasticity or to reach the fluid state. It is of interest to consider the process of formation of a soft bottom and to estimate its bearing strength. For this purpose, we prepared a suspension from the clayey material mentioned above in which the water concentration was enhanced up to 40%. A 125-mm layer of this mixture was placed in a glass vessel. The results of the measurements of the thickness of the layers of water and sediment during the settling are presented in Table 5.

The data presented show that the process of sedimentation virtually comes to a halt within two days. The level of the sediment layer continued to drop, probably due to the compaction of the mass. The boundary between the layers is clearly distinguished after 1.5 h. During this process, the fractions were slightly stratified: the upper sedimentary layer 10 mm in thickness had a lighter color. The mean density of the sediment was equal to 1980 kg/m^3, which corresponds to the density where humidity is equal to the plasticity limit.

Table 5. Time dependence of settling of the experimental material

Time, hours- minutes	00-00	00-15	01-30	20-30	44-30	70-00
Water layer, mm	0	15	20	22	25	25
Sediment layer, mm	125	110	104	100	98	97

To estimate the supporting power of the sedimentary layer, 21 h after the beginning of the experiment, a rod 20 mm in diameter and 56 mm in length was placed on the surface of the layer (the density is 7800 kg/m^3). The rod submerged down to a depth of 25 mm. For the next 24 h, the rod further sank by 3 mm. 49 h later, the experiment was repeated and the rod submerged down to a depth of only 10 mm, probably due to the compaction of the upper layer of the sediment. Thus, we can conclude that, for a given grain-size composition of the sediment (that is close to that in the most common bottom samples) and free settling over two days, its supporting power is sufficient to bear a heavy steel-made cylinder.

Another experiment was carried out to assess selected the mechanical characteristics of the sedimentary material (viscosity and static internal friction, which are not discussed in the known literature). The experiment was based on the modified Bingham formula (1):

$$\tau = \tau_o + \eta V/d, \quad (3)$$

where τ is the shear stress, τ_o is the initial (static) shear stress, η is the dynamic viscosity, V is the velocity of the body, and d is the thickness of the shear layer.

A perforated plate was mounted inside or on the surface of the medium under study and was set into a uniform motion by a force F acting parallel to its plane. The shear stress was defined as the ratio F/S, where S is the surface area of the plate. The velocity of the motion was defined as the mean ratio of distance to time. A perforation was made for improving the coupling of the plate with the media. The layer thickness d was visually estimated from the value of the layer stuck to the plate when it moves perpendicular to the surface of the medium and by placing elastic threads in the media under the moving plate.

To enhance the plasticity of the material, its humidity was increased up to 34%. The initial stress τ_o was defined as the force needed for shifting the plate from the static condition at the velocity $V \approx 0$. The viscosity was determined from Eq. (3). Humidity was then increased up to 50% (above the yield point), and the same values were determined. The data for five measuring runs are listed in Table 6.The experimental results show that, for a plastic medium of a given composition, the initial shear stress τ_o is comparable with the stress τ for low velocities *V*. With increase in the velocity, τ increases considerably, and the viscosity drops. For a fluid, the initial shear stress τ_o and the viscosity are relatively low.

Table 6. Results of determining viscosity and static internal friction

Experimental conditions	τ_0, Pa	D, mm	τ, Pa	V, mm/s	η, Pas
Plastic media, Plate inside	450	8	568	0.2	5714
Plastic media, Plate outside	570	8	1136	2.0	4545
Fluid, plate outside	70	8	1704	4.5	2020
Fluid, plate inside	56,8	2	284	3.0	94.6

The goal of one more experiment was to assess the process of submerging of the heavy housing of a seismometer into a soft soil. In the experiment, we used an aluminum thin-walled hemisphere 80 mm in radius and 0.1 kg in mass. The hemisphere was placed on the surface of the clayey sediment whose composition was described above and the humidity was 34% (plastic medium with a density of 1790 kg/m^3) and 58% (fluidith a density of 1600 kg/m^3). The hemisphere received different weights, and the depth of penetration into the sediment over a given time was measured.

We calculated the area of the contact with the sediment, the volume of the submerged portion, the buoyancy force, the pressure on the bottom (1), the pressure with regard to the buoyancy force caused by the water (2), and the pressure with regard to the buoyancy force caused by an elastic bottom (3). The data obtained for plastic (P) and fluid (F) media are listed in Table 7. It was found that the hemisphere submerges into the plastic medium rather slowly (within several hours), slowing gradually in line with an exponential model. For the fluid, the submergence was rather fast (a few minutes). Under a load of 6 N, the hemisphere was virtually floating on the surface; at a load of 11 N, the submergence lasted for 10 min; and at a load of 16 N, the hemisphere submerged completely.

The results of the experiment on submerging the hemisphere into a plastic medium for various loadings ranging from 6 to 21 N are presented in Figure 2. With the increase in the load and the relative stabilization of the hemisphere at a new site, the pressure on the bottom slightly increased in comparison with that in the previous state. This can be explained by an increase in the volume displaced to the surface of the material during the submergence of the hemisphere.

In the plastic condition, the medium exhibits viscoelastic properties, which are manifested by a gradual decrease in the value of the residual deformation of the trace when the hemisphere is removed. The reduction equaled a few percent and lasted for a long time (a few hours). Nevertheless, in the calculations of the equivalent buoyancy force, we took into account the conditions of the submergence of the hemisphere into both the water and elastic mass of the material (the results are presented in Table 7 in brackets).

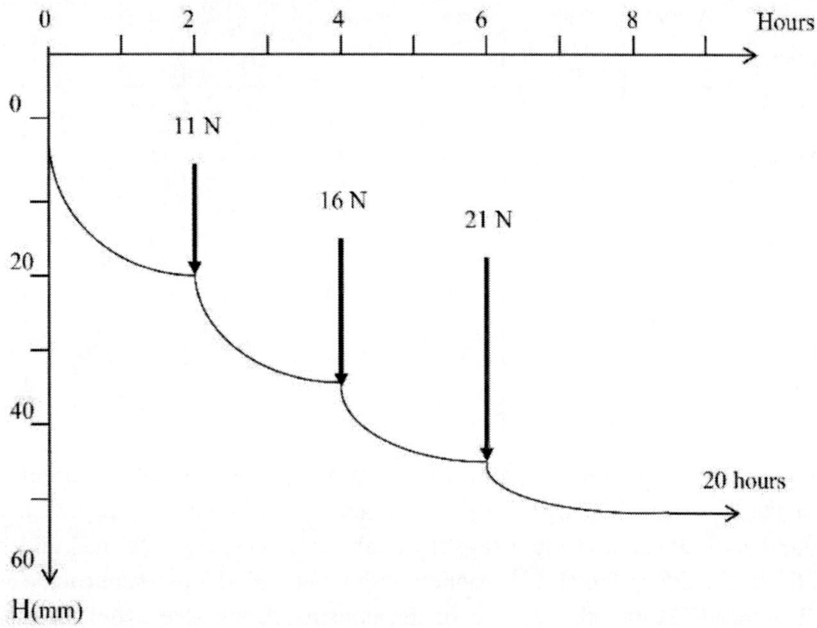

Figure 2. Depth of submerging of a spherical housing into a soft bottom as a function of time for different loadings (N – newtons).

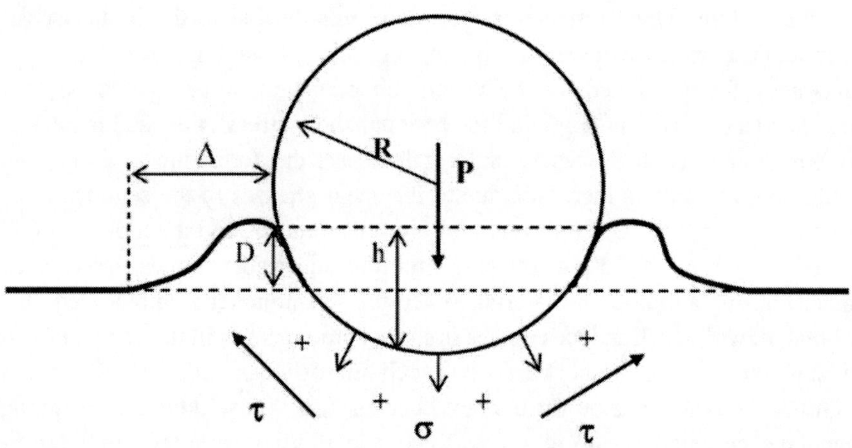

Figure 3. Schematic of coupling of the seismometer housing with a viscoplastic bottom. *R* and *P* are the radius and weight of the seismometer, *h* is the depth of submerging of the housing, *D* and Δ are the values of the near-wall deformations of the sediment, σ is the normal pressure, and τ is the tangential stress.

Table 7. Estimation of the carrying properties of the experimental material

Time, hours-minutes	Weight, N	Depth of sub-merging, mm	Area, dm^3	Volume, dm^3	Buoyancy force, N	Pressure		
						1	2	3
P 00-00	6	5	0.25	0.12	1.2 (2.2)	2400		
P 02-00	6	20	1.0	0.26	2.6 (3.9)	600	480	380
P 02-00	11	20	1.0	0.36	3.6 (6.5)	1100		
P 04-00	11	35	1,76	0.5	5.0 (9.0)	625	477	311
P 04-00	16	35	1.76	0.5	7.5	909		
P 06-00	16	43	2.16	0.58	9.3	740	574	440
P 06-00	21	45	2.26	0.26	4.16	929		
P 20-00	21	50	2.5	0.33	5.3	840	640	480
T 00-00	11	50	2.5			440		120
T 00-10	16	55	276			578		244
T 00-30	6	35	1.76			341		104
T 01-45	6	38	1.91			314		36.6

Let us consider with regard to the experimental results a schematic representation of the coupling of the spherical housing of a seismometer to a soft bottom (Figure3). Under the action of the weight of the housing a region of enhanced pressure is formed under it. In the vicinity of the walls, due to the elastic reaction of the medium, the pressure can be assumed to be directed perpendicular to the surface of the housing. Under the action of pressure, the material of the bottom is plastically deformed.

Due to the low compressibility of the medium, a portion of the mass of the sediment moves sideways, over the housing surface, and upward, forming a ring projection on the surface of the bottom. This motion is rather slow, mainly due to the high viscosity of the medium. In addition to its own weight, the housing of the seismometer undergoes the action of the buoyancy force induced by the water, the buoyancy force induced by the bottom sediments, the viscous friction drag, and the residual shear stress in the sedimentary layer. As the housing is submerged, the pressure under it decreases due to the increase in the area of the base. In addition, the drag is increased due to the shear stresses in the sediment. As a result, the counteracting forces balance the weight of the seismometer, and the motion stops.

Due to the strong adhesion of silt, multiple aggregate bonds arise between the seismometer housing and bottom material. When the seismometer is submerged, these bonds abruptly break down, which induces noises in the seismic records in the form of microshocks. It should be noted that some authors associate such microshocks with biological activity at the bottom. This is a possible reason for the rare “background” pulse interference throughout the records and the occasionally observed seasonal and diurnal variations of their frequency. However, the “biological” theory cannot explain the above-mentioned gradual decrease in the frequency of the microshocks after the installation of the instrument on the bottom. The nature of the microshocks is most likely associated with a combination of these reasons.

It should be noted that, during the submerging of the seismometer into a soft bottom, irreversible plastic deformations occur. The spatial scale of such deformations is comparable with the sizes of the seismometer housing; i.e., the relative deformations are close to unity.

The scale of reversible elastic deformations on which the seismic vibrations propagate is a small fraction of the wavelength (less than 10^{-4}); i.e., it differs by four orders of magnitude (Aki and Richards, 1983; Gutenberg, 1959; Geoph. Refer. book, 1966). The time constants of the processes that take place in the course of the submerging of the seismometer into a soft bottom range from tens to a few hours, and the periods of vibration of the seismic signal used range from a few tens of seconds to a few tenths of second. The above-mentioned differences in the spatial and temporal scales result in the fact that the bottom material demonstrates completely different properties; therefore, the phenomena that take place during the submerging of the seismometer housing into a soft bottom and while recording the seismic signals are independent.

Let us describe the motion of the seismometer housing with the use of the Maxwell model for a viscoelastic medium. The soft bottom generally represents precisely such a medium. The characteristics of such a medium can be described in the following manner. When deformations cease, the residual stresses continue to decay with time (relaxation time). Depending on the ratio of the rate of the processes to the relaxation time β, the same media can exhibit the properties of fluids or solids. A theoretical criterion for determining the properties of a viscoelastic medium in the case of harmonic vibrations with a frequency ω is presented in (Landau and Lifshits, 1987). The relationship between the stress τ_{ik} and the shear deformation γ_{ik} for such a medium is

$$\tau_{ik} = 2\mu\gamma_{ik}(1 + i/\omega\beta)^{-1}. \tag{4}$$

If $\omega\beta < 1$, then the medium behaves like a fluid with a viscosity $\eta = \mu\beta$. If $\omega\beta > 1$, then the properties of a solid with an elasticity μ prevail.

$$\mu = \tau_{ik}/\gamma_{ik} \tag{5}$$

In the former case, the seismometer housing moves with respect to the sediment like a float. In the second case, it oscillates together with the sediment. The relaxation time β is characterized by the time of submerging of the seismometer housing into the layer of the sediments and is generally equal to a few hours. Therefore, for the entire frequency range (0.003–30 Hz) usually used for the bottom seismometers, the value of ωβ ranges from 68to 68 × 10^4 (for β equal to 1 h). Thus, in accordance with the criterion presented, the seismometer should move together with the sediment layer within its elastic deformations.

Let us estimate the elastic deformations of the bottom under the action of the seismometer weight. The weight of the seismometer is assumed to be a vertical force concentrated at the origin of the coordinate system. The x and y axes define the sea bottom, and the z axis points downward. The shift of the points of a semi infinite axisymmetric media ($x = y$) is defined as (Landau and Lifshits, 1987)

$$u_x = u_y = [xz/r^3 - (1 - 2\nu)x/r(r + z)] \times (1 + \nu)P_z/2\pi E,$$
$$u_z = [z^2/r^3 = [-2(1 - \nu)/r](1 + \nu)P_z/2\pi\, E, \tag{6}$$

where P_z is the weight of the seismometer, ν is the Poisson coefficient, E is the Yong modulus, and $r = \sqrt{x^2 + y^2 + z^2}$ is the distance from the coordinate origin. Assuming for humid clayey sediments that $E = 2 \times 10^{10}$ N/m^2, ν= 0.47 [17, Geof, 1976], and the seismometer weight in the water P_z= 100 N (see Table 3), we obtain $u_z(x) = 1.2 \times 10^{-9} x^{-1}$ for the vertical deformation at the surface ($z = 0$, $x = y$) and $u_z(z) = 2.4 \times 10^{-9} z^{-1}$ for the vertical deformation over the vertical axis ($x = y = 0$). For the horizontal deformation at the surface ($z = 0$, $x = y$),$u_x(x) = -0.7 \times 10^{-10} x^{-1}$ and the horizontal deformation along the vertical line is equal to zero. It follows that static elastic deformations of the bottom under the action of the seismometer weight are negligibly small.

Let us consider the coupling of a seismometer to the bottom in recording the seismic signals. The seismic signals can propagate in the earth without a pronounced attenuation only at very small relative elastic deformations of the media, no more than 10^{-4} (Aki and Richards, 1983; Gutenberg, 1959; Geoph. Refer. book, 1966). Let us assess the mass and sizes of the seismometer–elastic bottom system with regard to the mass of the oscillating bottom. The previous models used the concept of a vertical pendulum composed of a massless spring and a concentrated mass of the seismometer with apparent masses of the water and the overlying sediments (Soloviev, 1986; Sutton et al., 1981a, 1981b, Sutton and Duennebier, 1987; Trehu, 1981). Such models represent a standing wave whose loop is at the top and node under the bottom. The wavelength of seismic oscillation is defined by the parameters of the media and the frequency. The mean wavelength in the layer of the sediments at a frequency of about 10 Hz for longitudinal oscillations is approximately equal to 200 m, and for transverse oscillations it equals 100 m. The length of the "spring" in the model is one fourth of a wavelength and is equal to 50 m and 25 m for longitudinal and transverse oscillations, respectively.

The elastic oscillation of the bottom occurs not only under the base of the seismometer but, according to the concept of continuity of the media, occupies an adjacent domain that is approximately equal to half the wavelength (100 m and 50 m for longitudinal and transverse waves, respectively). Thus, a part of the massive whose volume and mass significantly exceed those of the seismometer should oscillate together with it. For longitudinal waves, the ratio of such masses is approximately equal to eight orders of magnitude, and for transverse waves, from four to seven orders of magnitude. Thus, it can be concluded that there is no pronounced effect of the seismometer mass on the results of recording seismic signals at the bottom within the region of elastic vibration of the media.

3. Resonance Phenomena in the Water-Saturated Sediment

Let's consider the reasons of resonance phenomena at recording signals by bottom seismographs. It is assumed with some degree of approximation that the marine environment consist of a set of horizontal layers whose properties are constant or vary smoothly enough, but at the layer boundaries change unevenly. If boundary layers sufficiently clear, and their thickness commensurable with the length of the seismic wave, it may occurrence a constructive resonance (Ewing et al., 1957; Kaufman et al., 2002; Savarensky, 1972).

Layer thickness of bottom sediments varies from 100 m in the central Pacific and along the mid-ocean ridges up to 1000 m or more near the continental slopes and in inland seas. Water-saturated loose sediments typically have a thickness of several tens to hundreds of meters and are characterized by low shear wave velocity (see below, Table 7) (Lisitsin, 1974; Sutton et al. 1981a). However at depths greater than 100 m the layer of sediments is largely consolidated, as determined by an increase in seismic velocities. thickness up Frequency range of signals in marine seismology makes up from 0,003 Hz up to tens Hz. In the area of high-frequency oscillations (1-30 Hz) seismic wave length commensurable with thickness of the layer of unconsolidated sediments. Therefore, it can essentially influence on short-period seismic waves registration. At low frequencies (0,003 - 0,1 Hz) wavelengths reach tens kilometers, so the influence of loose sediment layer can be neglected of a short period of seismic waves can be essential. In low-frequency area of registration lengths of waves make tens kilometers, and influence of a friable layer of deposits in this case can be neglected.

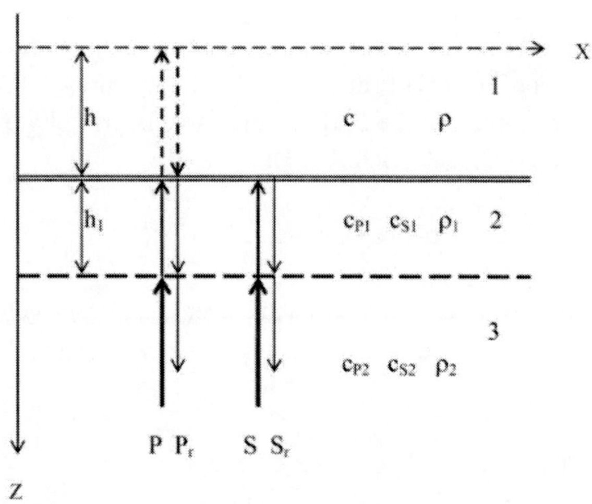

Figure 4. Model of oceanic medium: *1*—water layer, *2*—layer of water-saturated sediments, and *3*—consolidated basement; c, c_{P1}, c_{S1}, c_{P2}, and c_{S2} are the velocities of longitudinal and transverse waves in the water, sediment layer, and basement, respectively; ρ, ρ_1, and ρ_2 are the densities of the water, sediment layer, and basement; *P*, P_r, *S*, and S_r are the directions of arrivals of direct and reflected longitudinal and transverse waves.

To illustrate the resonance phenomena occurring in a watersaturated sediment layer, consider a simple model, which consists of the water layer with thickness h, sediment layer with thickness h_1, and consolidated halfspace (Figure 4).

In these media the following parameters are indicated: densities and phase velocities P and S waves. It is assumed that the plane harmonic P and-S waves propagate perpendicular to the layers from below upwards, as determined by a sharp decrease in characteristic impedance for unconsolidated sediments. At the lower boundary of the sediments more likely to occur only transmitted and partly reflected P and S-waves. On the border of sediments-water shear waves are reflected completely, as in liquid medium, they do not propagate. Assume also that the sea depth is much larger then the wavelength in the water and transmitted wave is dispersed in water, i.e. there is absent coherence between the transmitted through and the wave reflected from the water surface.

For the analysis, it can be used the method of coherent summation of the waves reflected from the layer boundaries sediments (Savarensky, 1972; Chapman N. and Chapman D., 1993), extending it to the case of medium with attenuation. It should be noted that the presence of attenuation the velocity of propagation of the body waves depends on the damping coefficient and the frequency (Savarensky, 1972). However, for low frequency ($f < 30$ Hz) is usually used in seismology, for small distances commensurable to wavelength, these dependencies can be neglected.

Consider the propagation of longitudinal wave (the displacement velocity of the particles of the medium), coming from lower halfspace

$$v_P = V_{P0} \exp j(-\omega t - kz + j\delta_P fz), \tag{7}$$

where V_{P0} - amplitude of a incident wave, k - wave number, δ_P – attenuation coefficient, f - frequency, $\omega = 2\pi f$. The coefficients of reflection r_∂ and transmission ϕ_∂ of waves at the upper boundary sediment layer (bottom of the sea) for this wave is given by expressions (Ewing, et al., 1957; Kaufman et al., 2002; Isakovich, 1973):

$$r_\partial = (1 - \alpha_\partial)/(1 + \alpha_\partial), \phi_\partial = 2/(1 + \alpha_\partial), \alpha_\partial = \rho c/\rho_1 c_{P1}, \tag{8}$$

where ρ and ρ_1 - densities of water and a sediment layer, c and c_{P1} - velocities of longitudinal waves. Coefficients reflection r_o and transmission ϕ_o of waves at lower border of the sediment layer (basement):

$$r_o = (\alpha_o - 1)/(1 + \alpha_o),\ \phi_o = 2/(1 + \alpha_o),\ \alpha_o = \rho_1 c_{P1}/\rho_2 c_{P2}, \tag{9}$$

where ρ_2, c_{P2} - density of medium and velocity longitudinal waves in lower halfspace. Parameters of wave, incident at the top border of a sediment layer, are defined by the sum of members of an infinite geometrical progression with the common member

$$R_n = v_P \phi_o (r_\partial r_o)^{n-1} \exp[-j(2n-1)\beta], \tag{10}$$

$$\beta = \beta_1 - j\beta_2 = kh_1 - j\delta_P fh_1 \tag{11}$$

The sum Σ of the progression members is

$$\sum = v_o \phi_o \exp(-j\beta)[1 - r_\partial r_o \exp(-j2\beta)]^{-1}. \tag{12}$$

Speed motion of particles at of the sea bottom is determined by transmitted wave (or the sum of the incident and reflected waves in a layer of sediments). Let us determine the coefficient transformation wave in the sediment layer as the ratio of the amplitudes speed motion of bottom and the wave arriving to the lower boundary layer of sediments (7):

$$K_P = \phi_\partial \sum /V_{Po} = 2[(1+\alpha_\partial \alpha_o)\cos\beta + 2j(\alpha_\partial + \alpha_o)\sin\beta]^{-1} \quad (13)$$

Substitute β from (11) and find the modulus and argument (phase shift) for K_P:

$$|K_P| = 2[\{(1+\alpha_\partial\alpha_o)ch\beta_2 + (\alpha_\partial + \alpha_o)sh\beta_2\}^2\cos^2\beta_1 + \{(\alpha_\partial+\alpha_o)ch\beta_2 + (1+\alpha_\partial\alpha_o)sh\beta_2\}^2\sin^2\beta_1]^{-1/2}$$

$$tg\theta = [(1+\alpha_\partial\alpha_o)ch\beta_2 + (\alpha_\partial+\alpha_o)sh\beta_2]^{-1}[(\alpha_\partial+\alpha_o)ch\beta_2 + (1+\alpha_\partial\alpha_o)sh\beta_2]tg\beta_1 \;. \quad (14)$$

In these expressions parameters α_∂ and α_o define influence wave impedances of water environment, sediment layer, and the bottom; parameters β_1 and β_2, respectively, the time delay and attenuation of wave in the sediment layer. Consider these questions in detail.

For small differences between the wave impedances sediment layer and bottom $\alpha_o \to 1$, then

$$|K_P| \to 2[(1+\alpha_\partial)(ch\beta_2 + sh\beta_2)]^{-1} \text{ and } \theta \to \beta_1, \quad (15)$$

i.e. modulus of transformation coefficient will depend only on the wave transmission coefficient $2/(1 + \alpha_\partial)$ and losses β_2, and the phase shift is determined by the time delay in a layer.

For very low frequencies oscillations thickness sediment layer $h_1 << \lambda$, then

$\beta_1 = (2\pi h_1/\lambda) \to 0$ and $\beta_2 = (\delta_P c_{P1} h_1/\lambda) \to 0$.

In this case $|K_P| \to 2/(1+\alpha_\partial\alpha_o)$ and $\theta \to 0$. With an essential difference wave impedances of the water environment and the bottom product $\alpha_\partial\alpha_o = (\rho c/\rho_2 c_{P2}) \to 0$ and $|K_P| \to 2$. This expression corresponds to the known phenomenon doubling amplitudes of the speed and of particles displacement at the free surface.

For all $\beta_1 = \pi(2n - 1)/2$, (n = 1, 2 …) relations (14) are transformed in form:

$$|K_P| = 2[(\alpha_\partial + \alpha_o)ch\beta_2 + (1+\alpha_\partial\alpha_o)sh\beta_2)]^{-1}, \; \theta = \pi/2. \quad (16)$$

At small losses in a sediment layer $\beta_2 \to 0$ and coefficient Kp in expression (16) $|K_P| \to 2/(\alpha_\partial + \alpha_o)$. As α_∂ and α_o usually less than 1, the transformation K_P coefficient may be greater than 2, i.e. resonant amplification of the amplitudes of P-waves in the sediment layer is possible. Consider the transit of shear waves through the sediment. Since shear waves do not propagate in water, speed motion of the sea bottom is equal to the sum amplitudes of the incident and reflected waves, i.e. it doubles. In this case $\alpha_{\partial s} = 0$, $r_{\partial s} = 1$, $\phi_{\partial s} = 0$. Expressions for the coefficient modulus of transformation and phase shift take the form:

$$|K_S| = 2[(ch\beta_{2S} + \alpha_{0S} sh\beta_{2S})^2 \cos^2\beta_{1S} + (\alpha_{oS} ch\beta_{2S} + sh\beta_{2S})^2 \sin^2\beta_{1S}]^{-1/2} \quad (17)$$

$$tg\theta_S = (ch\beta_{2S} + \alpha_{oS} sh\beta_{2S})^{-1}(\alpha_{oS} ch\beta_{2S} + sh\beta_S) tg\beta_{1S}. \quad (18)$$

Compare these expressions with the above mentioned for longitudinal oscillations (14). At $\alpha_{oS} \to 1$,

$$|K_S| \to 2(ch\beta_{2S} + sh\beta_{2S})^{-1} \text{ and } \theta_S \to \beta_{1S}. \quad (19)$$

For very low frequencies $\beta_{1S} \to 0$ and $\beta_{2S} \to 0$. In this case $|K_P| \to 2$ and $\theta \to 0$.

For all $\beta_{1S} = \pi(2n-1)/2$, $(n = 1, 2 \ldots)$ expressions (17) and (18) have the following form:

$$|K_S| = 2(\alpha_{oS} ch\beta_{2S} + sh\beta_{2S})^{-1},\ \theta = \pi/2. \quad (20)$$

For small losses in the sediment layer $\beta_2 \to 0$ and $|K_S| \to 2/\alpha_{oS}$. Since $\alpha_{oS} < 1$, in the sediment layer may be resonant amplification of transverse oscillations. Expressions (16) and (20) can be used to estimate quality factor Q determined its resonant properties. It should be noted that quality factor Q depends on two factors:

a) the ratio wave impedances of water medium, sediment layer, and foundation (parameters α_∂, α_o, α_{oS});
b) attenuation in the sediment layer.

Since $1 > (\alpha_\partial, \alpha_o, \alpha_{oS}) > 0$ this quality factor is always limited, even in the absence of damping. From comparison of (16) and (20) is follows that the quality factor for sediment layer of longitudinal oscillations less then traverse, as $|K_P| = 2/(\alpha_\partial + \alpha_o)$ is always less then $|K_S| = 2/\alpha_{oS}$ due to the additional term α_∂ in the denominator for and $\alpha_{oS} = \rho_1 c_{S1}/\rho_2 c_{S2}$ it is usual less than $\alpha_o = \rho_1 c_{P1}/\rho_2 c_{P2}$. In the discussed examples there were considered the seismic waves transmitted through the sediment layer from its footing to the top. However, similar phenomena may occur when the hydroacoustic noises input from above out of water (Godin, Chapman, 1999). In this case, increasing the amplitude of oscillations will not occur.

Consider the numerical examples based on Table 8. The velocities of seismic waves are accepted according to the generalized velocity sections for the bottom of the Atlantic and

Pacific oceans (Webb, 1992). Attenuation in a layer of unconsolidated sediments were taken from works (Hampton, 1977; Volovov, 1992). Attenuation in a layer of friable deposits was defined according to works

Table 8. Initial data for models of a layered bottom (Figure 4)

Region	ρ_1 kg/m^3	ρ_2 kg/m3	c_{1P} km/s	c_{2P} km/s	c_{S1} km/s	c_{S2} km/s	δ_P dB/mHz	δ_S dB/mHz
Atlantic Ocean1	1900	2500	1,8	5,5	0,45	3	0,5	1,6
Atlantic Ocean 2	2000	2500	2,0	5,5	0,53	3	0,2	0,9
Pacific Ocean	1600	2600	1,6	5,2	0,35	2,5	0,07	0,6

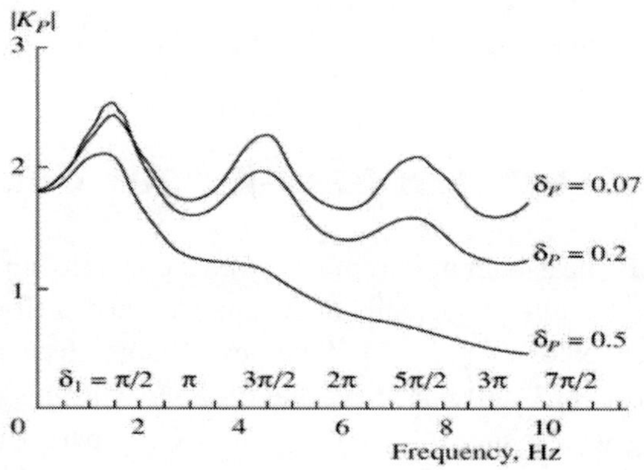

Figure 5. Transfer coefficient $|K_P|$ for compress waves as a function of frequency at various a ttenuation δ_P.

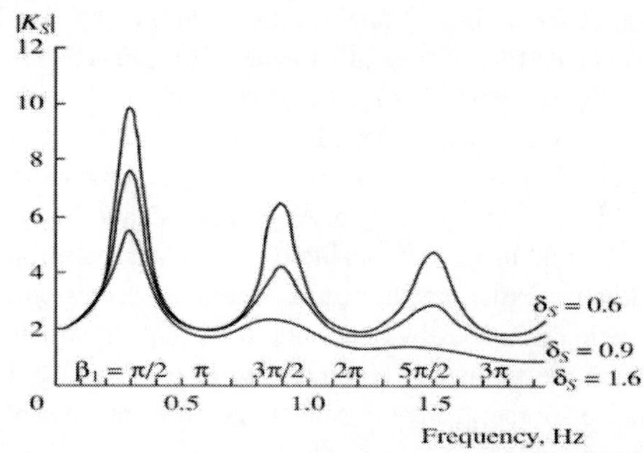

Figure 6. Transfer coefficient $|K_S|$ for sheer waves as a function of frequency at various attenuation δ_S.

Figures 5 and 6 show the results of calculations in accordance with the model velocity amplitudes botton (Figure 4) for the compressional $|K_P|$ and transversal $|K_S|$ waves depending on frequency for different damping coefficients δ in sediment layer. At all graphs are observed resonance maxima and minima, whose amplitudes decrease with increasing frequency and attenuation similar to those observed in experiments. At the same time despite of greater attenuation ($\delta_S > \delta_P$) resonance phenomena for transverse waves expressed more strongly than for longitudinal.

The reason is semitransparency for longitudinal waves, and Q-factor bottom sediments determines not only attenuation but losses during transit through its borders. In calculation of effective thickness of the water saturated sediment layer was adopted 300 m for longitudinal waves and 100 m for the traverse. This corresponds to the generalized velocity sections for Atlantic and Pacific oceans. And calculated resonant frequencies close to those observed in experiments (Bradner et al., 1970; Sutton et al., 1981a; 1981b; Godin and Chapman, 1999).

It should be noted that the obtained dependences $|K_P|$ and $|K_S|$ from β_1 and β_2 are fairly universal. Parameters β_1 and β_2 are directly proportional to product of $h_1 f$ (11), so the shape of the curves in Figure 5 and 6 does not change its form at change thickness of a sediment layer and the corresponding change of scale on the frequency axis.

4. Influence of Near-Bottom Currents

As practice shows, near-bottom currents may cause serious noises with seismic records bottom seismographs. Results of the study of such noises, and their qualitative analysis given in many works (Duennebier et al., 1981; Soloviev, 1986; Levchenko, 2001). Modeling of these processes (Kasahara, 1980) confirmed the basic conclusions of the analysis. It has be found, that beginning with a speed 3-5 m/s, the projecting parts of bottom stations causing swirls (Karman vortices) that excite vibrations affecting highly sensitive sensors these stations. However, known publications have neither quantitative estimates of the noise intensities nor the recommendations for their reduce. This is due, apparently, insufficient working out quantitative theory of turbulent fluid motion (Landau and Lifshits, 1988), as well as lack of experimental data for this area of knowledge (Trehu, 1985b).

As a result, the 20-year practice of seismological research with bottom stations in the Institute of Oceanology RAS some experience of struggle against noises from near-bottom currents. This paper attempts to analyze these phenomena and formulate general recommendations to significantly reduce noises in a broadband bottom seismographs. In this case there are used a combination of qualitative estimates of turbulence in the published materials with quantitative experimental results. There are discussed the following points: a) mechanisms of excitation of noises by near-bottom currents at structural elements of bottom stations; b) estimate of the frequency and intensity of noises; c) estimates attenuation with distance from the noise source; d) recommendations on distance separating the sensors unit from the container seismograph.

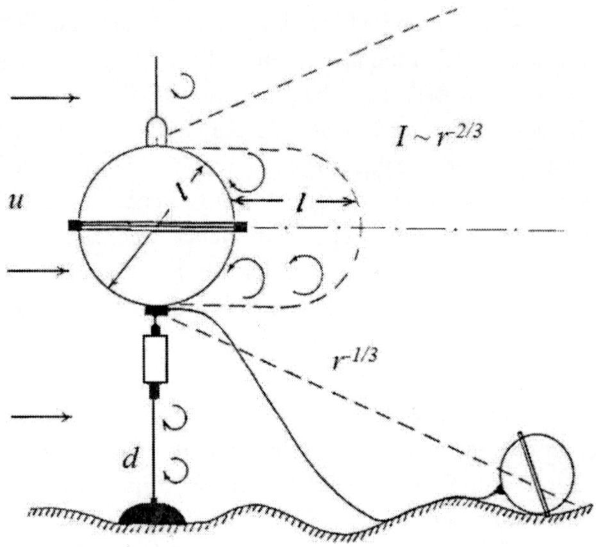

u – average speed of the current, *l* – typical dimensions of the seismograph case, *d*–diameter of the buoy-rope, *r*–distance from case, *I* – specific energy of the turbulence.

Figure 7. Effect of the near-bottom currents on bottom seismograph.

In the general case, the near-bottom currents may have both laminar and turbulent character (due to irregularities of the bottom). In both cases, the current speed decreases with proximity to water - bottom border. Hence obvious conclusion ensue to place elements of bottom station as close as possible to that border. However, in practice this is difficult to accomplish. For example, the housing bottom station with positive buoyancy to be installed in 2-3 m from the bottom to prevent them from sucking the soft silt, and all the thin-measuring elements can be placed inside the case (Figure 7). Therefore to reduce noises in recording for broadband bottom seismographs adopted design with block sensors removal in a separate container. However around the container may occure wortexes, whose frequency coinsides with an operating range of a broadband seismograph.

Let us consider the three main mechanisms of noise generation by near-bottom currents. First –this is is the direct swing seismograph housing under the influence of unsteady current. Second mechanism is excitation of vibrations in the water due to occurrence of turbulency around projecting parts bottom station. Third mechanism is excitation of acoustic oscillations due to vibration elastic elements under the turbulent flow. It should be noted that in special publications there are considered mainly only the second driving mechanism of noises generation and are given only estimations frequency oscillations (Duennebier et al., 1981; Soloviev, 1986). Therefore necessary to consider the mechanisms in greater detail. In general water is viscous compressible fluid. The properties of the flow characterized by the Reynolds number

$$R = \rho u l / \eta, \tag{21}$$

where ρ is the density, η - the dynamical viscosity, u -the mean velocity of the fluid, and l- the characteristic size (scale), close to the size source of the turbulence. At the critical Reynolds number ($R >> 2000$ for the water), flow becomes turbulent. Below, it should be considered development of turbulent or laminar flow at great Reynolds numbers $R < R_{cr}$.

According to (Landau and Lifshits, 1988), near the body placed in the flow directly behind the line of separation is formed large-scale turbulence characteristic size l. Then, with the extension of its area scale turbulence turbulence decreases and, finally, small-scale turbulence dissipates and turn into heat. It should be noted that at great Reynolds numbers fluid viscosity can be neglected for simplifying expressions for turbulence. Another simplification is the possibility of neglecting the compressibility of water when considering oscillations from a small source with dimensions $l << \lambda$, where λ is the wavelength of acoustic oscillations (Isakovitch, 1973). Let`s consider the question of the frequency range of pulsations, as it determines the frequency of the excited noise. In paper (Landau and Lifshits, 1988) are given lower and upper estimates frequencies of pulsation associated with the scale turbulence and Reynolds numbers :

$$u/l<\omega< (u/l)(R/R_{cr})=u^2\rho/\eta(R_{cr}). \qquad (22)$$

Consider the mechanism of the direct impact of flows on the container with seismic receivers. Behind the body placed in a flow, either a laminar or turbulent trail is formed with respect to the character and velocity of the running flow. In either case, a vortical domain is formed behind the body, which causes its oscillations. The laminar trail is characterized by the absence of layer mixing in the vortex-free domain of the trail (Landau and Lifshits, 1988). In the case of a turbulent flow, additional rocking of the container emerges caused by the pulsations of the flow itself. It should be noted that, in this case, the changes in the flow direction may significantly influence the noise level, since the container with seismic receivers may find itself in the zone of a turbulent trail from a major irregularity of the bottom topography of from the parent instrumental container. This may explain the occurrence of the so-called "near-bottom storms" (Duennebier et al., 1981, Levchenko, 1994). The radius of the turbulent trail grows with distance r as $r^{1/3}$, while the mean pulsation velocities decrease proportionally to $r^{2/3}$. This is the reason for the relative stability of the turbulent trail with distance. However, the Reynolds numbers decrease together with the pulsation velocity. At this, the turbulent trace transforms into a laminar one. The flow acts on a spherical container with a force proportional to the mean flow velocity, pulsation velocity Du and squared radius of the sphere a (Landau and Lifshits, 1988):

$$\Delta F \sim u\rho\Delta u a^2 \qquad (23)$$

Oscillation velocity at a spherical container with seismic sensors is directly proportional to the velocity of pulsations flow of fluid and depends on its mass and water resistance (added mass). In accordance with the work (Landau and Lifshits, 1988), this rate is conveniently expressed in terms of the average density of the container ρ_{con} and the density of liquid ρ :

$$V = 3\rho\Delta u/(\rho + 2\rho_{cont}). \tag{24}$$

From the relations (23) and (24) follows that for reduction the influence of currents it is necessary to increase the mass of the container with seismic sensors and reduce its cross-section and volume. It should be noted that this conclusion is contrary to the requirements of the theory adhesion of seismic sensors unit with the bottom (Sutton et al., 1981a), therefore in this case necessary a compromise solution.

Consider the problem of acoustic noise excitation turbulent flow. To estimate the energy relations due to changes in linear scale and speed we introduce the concept of specific energy per unit mass of fluid in the unit time (Landau and Lifshits, 1988). For example the specific energy dissipation

$$\varepsilon_{dis} \sim (\Delta u)^3 \eta/(lR_{cr}). \tag{25}$$

Evaluations in (25) and below ar given in the form of proportionality, not equality, so sign ~ is used. This is due to the fact that quantitative estimates of the ratio u and Δu for the turbulent motion is generally unknown.

The main source of acoustic noise is located directly behind the streamlined body at distances ~ l from the line of separation, since within this region pulsations are correlated, which determines the maximum radiation. In the works (Landau and Lifshits, 1988) ratios for specific sound energy caused by turbulence are of the form:

$$\varepsilon_{sound} \sim (\Delta u)^8/(lc^5) = (\Delta u/c)^5(\Delta u/l)^3 l^2 \tag{26}$$

where c is sound speed in the water. Here, the ratio $(\Delta u/l)$ has the dimension of frequency and corresponds to a certain frequency of the flow pulsations. The ratio $\varepsilon_{sound}/\varepsilon_{dis}$ allows to determine the energy dissipation, which in the first approximation is equal the initial energy of turbulence: $\varepsilon_{sound}/\varepsilon_{dis} \sim ((\Delta u/c)^5$. Since Δu is significantly smaller than c, the efficiency of sound radiation due to turbulence flow should be very low.

Since the frequency range of bottom broadband seismographs is characterized by the following relation between r-spaced seismic sensors, the radiator size l and the wavelength of sound λ : $\lambda >> r_{cont} >> l$, noise radiation region can be considered a point source of sound field close to a monopoly because of correlationary pulsations. Therefore, the acoustic wave velocity in the fluid in the near-field will decrease inversely proportional to the square of the distance r (Isakovitch, 1973; Landau and Lifshits, 1988), i.e.

$$v_{tur} \sim \Delta u/r^2 \tag{27}$$

Consider the nature of the radiation oscillating elastic elements bottom station (thin-measuring tubular frame, antenna, buoy rope, junction cables etc.). The noises radiation has a dipole character, their oscillations are excited of stream-lined flow and may have resonance maxima. However due to large resistance of water the quality factor of noises is small.

According to (Isakovitch, 1973) for the case of small size ($l << \lambda$) harmonically oscillating body in the near zone (($kr << \lambda$) may be wrote expression:

$$v_{osc} \sim (1/r^3)\Delta u e^{jkr}\cos\theta, \quad (28)$$

where v_{osc} - the speed of particles in the acoustic wave, k - the wave number, r - distance, θ-angle between the velocity vector of the body and the direction of wave propagation.

Compare levels of acoustic noise excited by turbulent vortex and oscillating elements station. Let`s define the ratio of the specific intensity (per unit length) of radiation in form relation of the oscillating (J_{osc}) and pulsing (J_{pul}) the same diameter cylinders d.

$$J_{osc}/J_{pul} = 2\pi^2 d^2/\lambda^2. \quad (29)$$

Since the diameter size substantially smaller than the acoustic wavelength ($d << \lambda$) , this value is small. In the near zone according to the relation (28) the radiation does not depend on the direction, while in the second according to the expression (29) it will dipole character and will depend on the direction. From (29) it also follows that the intensity noise due to oscillations of the elastic elements bottom seismograph is significantly less than due to vortices near the elements and damping rate of the excited acoustic wave with distance is much faster - respectively by the square (see (27) or cubic (see (28) law. It should be noted that due to the small size of the emitter, and the low speed of movement in comparison with the wavelength and the velocity of sound efficiency radiation in both cases quite low. It increases with the pulsation frequency and size of the radiator is about the third degree (Landau and Lifshits, 1988). However, despite the fact that only a small fraction of the energy is converted into radiation of turbulent sound noises, due to the high sensitivity of the sensors (the threshold is 10^{-9} m/sec), this noise may significant affect their operation in the immediate proximity to the source of turbulence (Duennebier et al., 1985; Levchenko, 2001; Wieland and Steim, 1986).

Based on these ratios, consider some specific questions choice of parameters construction broadband bottom seismographs. First, we define the frequency range noise exciting turbulent phenomena by the structure of seismographs. In accordance with a scale ratio of turbulence and Reynolds number (22). Lower limit frequency turbulence depends on the average flow rate and size of the elements and the upper - of Reynolds numbers. Characteristic dimensions of these elements are as follows: body seismograph - 1 m, the external elements (hydroacoustic antenna, flashing beacon, etc.) - 0.1 m, buoy rape, connecting cables, frame, radio antenna - 0.01 m. Typical velocity bottom currents causing noise from 0.03 to 0.5 m/s. Considering the conditions of fully developed turbulence (R / Rcr > 10) at the coincidence frequency band noise with a range of pulsations turbulence obtain frequency 0.03 - 500 Hz. Since the frequency range of the broadband seismograph is 0.003-20 Hz, it largely overlaps with the range of noises. In the low frequency band of the seismograph may interfere due to turbulent phenomena in large uneven bottom (10 m). Therefore, these seismographs desirable to install as far as possible enough flat bottom.

Very important is the question of the spatial separation of the sensor unit and the instrumental container. In several papers have shown the expediency of such a decision, however, is not based significance of this separation (Soloviev, 1985, 1986; Voronina et al., 1995 ; Duennebier et al., 1981). The necessity of moving away the sensor unit from the container determined by the excitement of various types noises:

a) from the mechanisms in the instrumental container;
b) from near-bottom currents exciting various sources,
c) mechanical connection between the horizontal and vertical seismic sensors due to swinging of large container at soft ground.

In view of these reasons, since the 80s. the last century, in installations bottom seismographs are used primarily remote blocks seismic sensors. At the same time remained uncertain choice of the distance, at which must be installed this unit. However, in experiments to measure seismic noise level excited currents in seismographs combined with sensors, it was found that its average rate is of 1000 nm / s (Voronina et al., 1995; Kontar et al., 1991; Duennebier etc., 1981). For approach the level of acoustic noise to the level of background microseisms (about 1 nm / s) it is necessary to reduce it in 1000. Taking into account in the worst case noise attenuation with distance is inversely proportional to the square of the distance (see relation (28), (29) at the maximum characteristic scale of turbulence l = 1 m spacing value $r \geq 30$ m.

When the turbulent trace of the housing seismograph affects on the block seismic sensors it is necessary to remove it to a distance at which the turbulent trace becomes laminar , which corresponds to a decrease in the Reynolds number of about one order of magnitude. Given that the speed in the turbulent layer decreases inversely to the cube of the distance ($r^{2/3}$), with a characteristic scale of turbulence 1 m obtain spacing around 33m . Therefore, the length of the connecting cable instrumental unit with the sensor unit in the bottom of broadband seismographs design IO RAS is 50m, which eliminates the effect of noise from the main unit to the sensor unit.

5. Instruments for Research of the Seismicity of the Bottom

The autonomous bottom seismographs available in a number of developed countries, such as the USA, Japan, Russia, Germany, and France are intended, as a rule, for marine exploration seismology and solving specific seismological problems, for example, determination of seismicity of some water areas. For this reason, such seismographs have limited characteristics as compared to those of land seismographs, in particular, lower frequency band, lower channel sensitivity, and shorter operation time (usually not longer than two–three weeks). Therefore, long-term monitoring of water areas requires the qualitatively different bottom seismic stations whose parameters would be close to those of the land stations (Jacobson et al., 1991; Levchenko, 1994, 2001; Montagner et al., 1992; Prothero, 1984; Romanovicz et al., 2006; Shiobara et al., 2001).

A limited recording frequency band of the marine seismographs is related, first of all, to the absence of reliable compact economical broadband seismic sensors. In the highquality land instrumentation, pendular seismic sensors are currently used (Streckeisen, Switzerland and Guralp, England) with a frequency band from 0.01 to 20 Hz (Usher et al., 1979; , Wielandt, 1986). They are big and heavy, require manual installation, and fear of shocks and moisture. In the autonomous bottom seismographs, compact sensors with a frequency bands from 2–3 to 20 Hz are mainly used. They do not record surface low-frequency Rayleigh and Love waves and low-frequency bottom vibrations caused by tsunamis.

For more than twenty years, in the broadband bottom seismographs of the Institute of Oceanology of the Russian Academy of Sciences (IO RAS), EKhP17 electrochemical seismic sensors have been used developed at the Institute of Electrochemistry of the Russian Academy of Sciences (Abramov and Graphov, 1978; Levchenko et al., 1994, 1996, 2010b). The main advantages of the electrochemical seismic sensors are low shock sensitivity (up to 30 g), possibility of operation at any slope, small dimensions and weight, and economical feed. The known shortcomings of the electrochemical seismic sensors, such as some temperature and external pressure dependence of the parameters, are inessential for the bottom seismographs and now successfully overcome constructively.

The intrinsic noise level of the electrochemical seismic sensors was estimated during processing the data on microseism recording at the bottom of the Atlantic Ocean in October 1991. The microseism spectrum maximum at a frequency of 0.17 Hz was $3.6 \cdot 10^{-7}$ m/s^2 (Hz)$^{1/2}$. At the frequencies from 0.03 to 0.12 Hz, the spectrum minimum is located which is lower than the maximum by 30 dBs. This is the estimate of the sum noise consisting of microseisms and intrinsic noise of the seismic sensor and input noise of an amplifier. Consequently, the level of spectral density of OBSS the intrinsic noise of seismic sensor in this region should be lower than 10^{-8} m/s^2(Hz)$^{1/2}$.

Recently, novel EP105OBS and CME-4111-3X broadband molecular kinetic (a kind of electrochemical) seismic sensors have been developed by an EENTEC company (USA) and the Center of Molecular Electronics of the Moscow Institute of Physics and Technology. At present, these detectors are being experimentally tested in the Experimental Design Bureau of Oceanologic Engineering of the Russian Academy of Sciences (EDBOE RAS).

Construction of the bottom seismograph determines, to a large extent, the level of noise caused by the near-bottom currents, reliability of installation at the bottom and emersion, and usability. Installation of the modern broadband autonomous seismograph of the IO RAS is schematically illustrated in Figure 7; the appearance of the seismograph is shown in Figure 8. The seismograph consists of a remote container of the seismic sensors 450 mm in diameter and an instrumental container 650 mm in diameter made of titanium and connected by a multicore cable with a length of 50 m. The total mass of the equipped seismograph is of about 80 kg. A bottom container holds a vertical and two horizontal electrochemical seismic sensors, a three channel low noise amplifier, low-pass filters, AD converters, economic microcontroller with a buffer memory device, and an orientation block.

Figure 8. Appearance of the instrumental container (big sphere) and seismic sensor block (small sphere) of the IO RAS bottom seismograph.

In general, the construction of the broadband bottom seismograph of the IO RAS is sufficiently reliable and usable for operation in the autonomous regime. At the same time, the instrumentation of the container of the seismic sensors represents a complete complex which can be connected to the communication and power supply cable for operation in the stationary regime.

For twenty years of recording of the earthquakes and microseisms at the bottom of water areas within a wide frequency band (0.003–20 Hz), the researchers of the IO RAS have collected great experimental experience. Below, some examples are presented.

In October 1991, during the fifth trip of the scientific and research vessel "Akademik Ioffe", bottom seismic recording was performed in the center of the Atlantic Ocean southward of the Azores.

The seismograph was installed at a depth of 1660 m by about 10 km eastward of the axis of the rift valley of the Mid Atlantic Ridge and about 23 km northeastward of crossing the valley by the transform fault Oceanographer. A remote (9800 km) earthquake with the magnitude M = 6.7 was recorded which happened on October 19 in the Himalayas. Figure 9 depicts a time compressed three component record of the earthquake. Earthquake duration was 10 s, but the signal was recorded for about two hours because of wave dispersion.

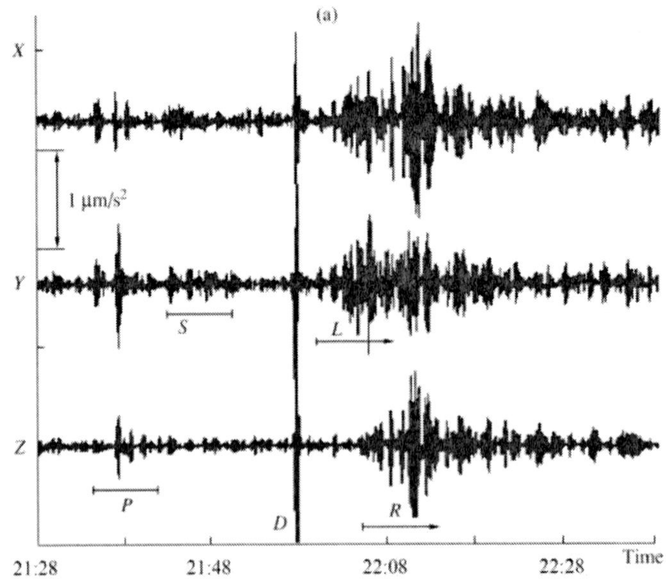

Figure 9. Compressed record of the Himalaya earthquake (October 19, 1991, M = 6.7) obtained by the IO RAS bottom seismograph in the central part of the Atlantic Ocean at a distance of 9800 km from the epicenter; the characters denote P and S groups of the body waves and the Love (L) and Rayleigh (R) waves.

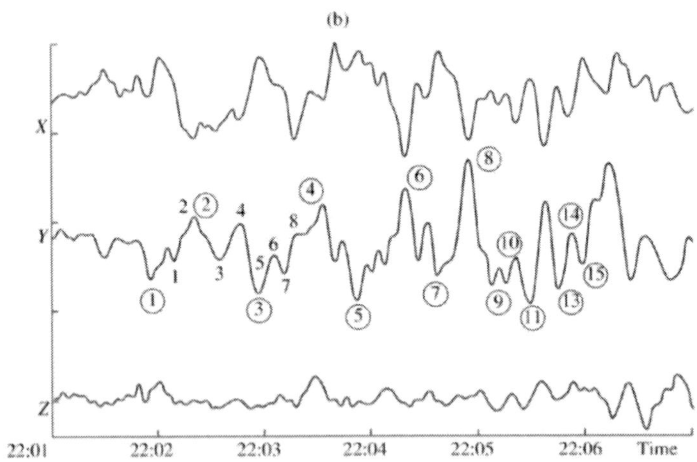

Figure10. Record of the Love wave from the Himalaya earthquake (see Figure 4), separated and filtered in the 0.015- 0.08 Hz frequency band; the numerals denote maxima of the first and second harmonics.

A separated record of the Love wave this earthquake with the marked maxima of the first and second harmonics is shown in Figure10. The analysis of dispersion curves of the group velocities of these waves revealed waveguides at depths from 60 to 80 km and from 160 to 200 km which were previously unknown in this region (Levchenko et al., 1994).

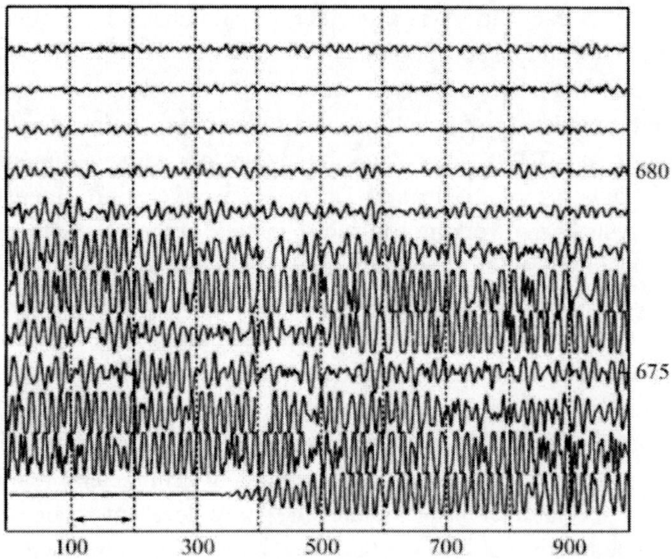

Figure 11. Record of the long-term seismic event (hypothetically, from an underwater landslip), obtained August 25, 1999, by IO RAS bottom seismograph in the Caucasian continental slope of the Black Sea (the arrows point an interval of 2 s).

From 1999 to 2001, researchers of the IO RAS fulfilled geological and geophysical investigations in the northeast part of the Black Sea continental slope in the regions of the constructed Russian-Turkish undersea gas line and oil loading terminal near Novorossiysk. As a result of the bottom seismologic observations in August and September 1999 and September 2001, over 1200 seismic events were recorded. The overwhelming majority of them (90%) were local micro-earthquakes and weak shocks with a magnitude of less than 1 or 2 which were not recorded by shore seismic stations.

Figure 11 shows a record of a long local seismic event ($t > 200$ s) which has no features of the tectonic phenomena and may be caused by an submarine landslip. During the expedition, several tens of such long records (about 80–250 s) were recorded by only one seismograph. It should be noted that high seismicity and extraordinary steepness (sometimes up to 30 deg) of this region favor the descent of a critical mass of sediments from the upper part of the continental slope. Results of registration of earthquakes on this expedition are described in more details in the collection "The Black Sea: Dynamics, Ecology and Conservation" (Levchenko and Kuzin, 2011).

The challenge of the autonomous bottom stations is the operative communication with the shore or a providing vessel. This communication is necessary when a seismograph is intended, for example, for tsunami danger warning. Ordinary hydro-acoustic communication operates at small distances, requires much energy, is subjected to reverberation noise, etc. It should be noted that recently developed communication systems with the use of complex coding and digital processing are much more noise immune and have lower signal powers (Catipovic, 1997). Also, in the newly developed techniques on the basis of floatable modules an express signal is transmitted via a satellite or radio channel. However, these methods have not been used in the bottom seismographs.

In the current autonomous tsunami warning systems, surface retranslation radio buoys are used which are connected to the bottom station via a hydro-acoustic channel (DART system, USA) (Tang et al., 2009). However, the buoys of average dimensions (overwater and underwater heights are of 2.5 m and 1.8 m, respectively) are intensively affected by sea waves, which reduces the working life and reliability of the communication channels. Loss of such buoys is mainly resulted from buoy rope breaks due to fatigue loading. At the same time, there exist other constructions of long-term buoys that have been used in marine practice.

In 70-s and 80-s of the last century, in the USA and USSR the so called stabilized buoys were developed and used in research. Simplistically, such a buoy is an anchored vertical pipe with a length of 50 m submerged in water on three quarters of its length. Because of high stability and small windage, such a buoy has a small slope (units of degrees) even during strong storms and small vertical oscillations (units of centimeters) relative to the average sea level. Due to a special anchoring system, the buoy does not rotate around the axis, which makes it possible to connect it to the bottom stations by cables. Such a buoy was installed in the South Branch of the IO RAS and used for ten years (1974–1984) (Lobkovskii et al., 2005).

There are long term buoys with the use of an inter mediate subsurface floatage (Froude spar buoy). Such an anchored floatage is located at a depth of 50–100 m where the surface waves cannot penetrate. Therefore, practically no dynamic loads affect the buoy rope. A stationary or submerging surface buoy is connected to the floatage. As the dimensions of the buoy are relatively small, in this case it does not considerably load a coupling element, which provides long (about 10 years) working life of the construction.

In the Institute of Oceanology according to the contract with the Ministry of Emergency Situations of the Russian Federation in 1996-1997 has been developed bottom observatory for the study of short-term precursors of strong earthquakes. It allows to record a number of parameters of water environment: temperature, pressure, current speeds, water acidity (pH), the electrical conductivity, the sound speed in water and the spectrum of transparency in the visible optical range. The principle of operation of observatory is based on known phenomena of change parameters and the composition of near-bottom water in pressure variations within the earth`s crust during preparation of a strong earthquake (King, 1980).

Figure 12 shows of the observatory on the sea bottom. Set of equipment of the observatory consists of the bottom instrumental complex, underwater buoy with replaceable power supply and a pop-up radio buoy. Points for data acquisition and control observatory are located on the shore. Figure 13 shows appearance of the instrumental container of the bottom observatory.

To install and recovery observatory and replace the power supply it is used auxiliary vessel. In 1997, the observatory has been tested in Avacha Bay (Kamchatka) and then was given into trial operation Institute of Volcanology and Seismology RAS.

During testing the observatory in 1999 in Avacha Gulf in 50 km from it occurred earthquake with M = 4.4 and focal depth of 37 km. Shakes near observatory installation at a depth 30 m was impossible to determine, but in Petropavlovsk- Kamchatsky they reached 2-3 points macroseismic scale MSK-64. Figure 14 shows graphs of changes in temperature, pressure and speed of sound, which is marked by arrow at moment of earthquake.

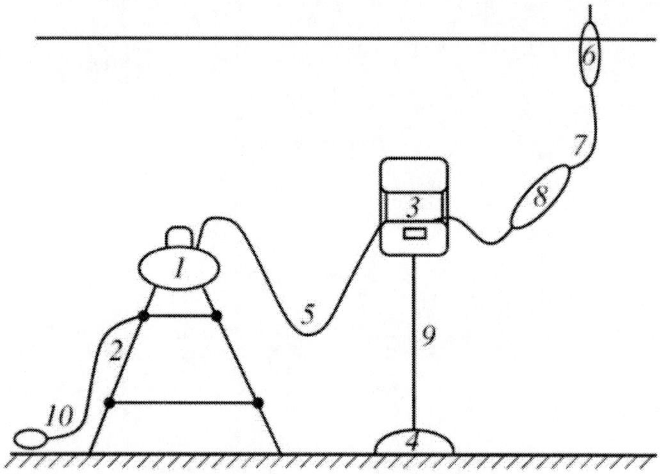

1 – instrumental container, 2 – tripod basement, 3 – buoy with power supply unit, 4 – anchor, 5, 7 – cables-rope, 6 – radio buoy, 8 – reel, buoy rope, 10 – emergency rope.

Figure 12. Scheme of the observatory installation on the bottom.

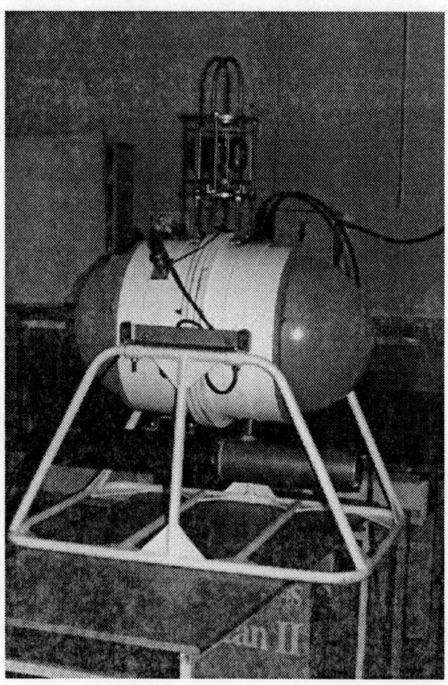

Figure 13. Appearance of the instrumental container of the bottom observatory IO RAS.

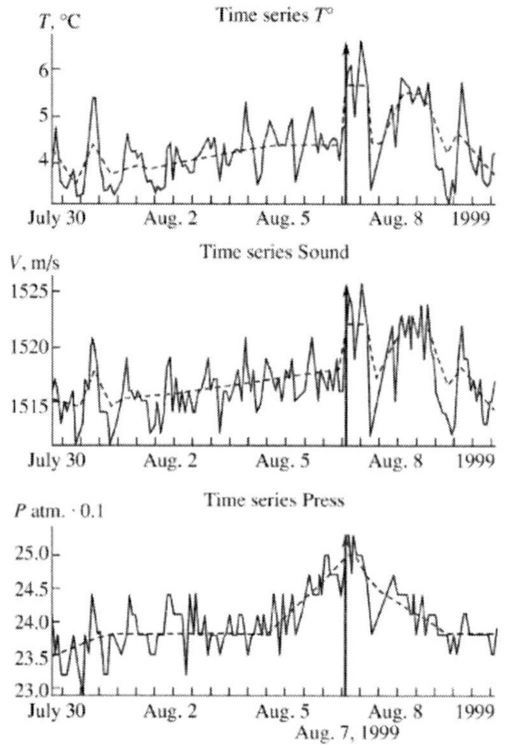

Figure 14. Records of the time series (temperature, sound velocity, and bottom pressure), obtained by bottom observatory IO RAS. Vertical arrow marks the moment of the Avacha bay earthquake on August 7, 1999 (M = 4.4, distance 50 km).

The course graphs affected storm noises, but on it background there are noticeable changes in temperature and speed of sound. The graph of pressure showed changes three days before earthquake. Unfortunately, this observatory is not use now by the Institute of Volcanology and Oceanology because of the lack of funds in the Administration of Kamchatka region (Gavrilov et al., 2000; Lobkovsky, et al., 2005).

Sea gas and oil platforms can be a good basis for placement centers data collection and transmission system of environmental monitoring, including data from seismic activity bottom. With seismological equipment can monitor changes in the state of the bottom area in the extraction process large volumes of hydrocarbons. Connection with measuring devices can be wire, cable, radio or hydroacoustic.

Figure 15 shows the layout of a number of measuring devices for geo-ecological monitoring in the area of extracting platform. They include: bottom and moored buoy autonomous station, cable bottom seismographs, manned and unmanned underwater moving vehicles, etc. With the development of infrastructure in the area of industrial development of the oilfield will be expanded network of marine monitoring stations.

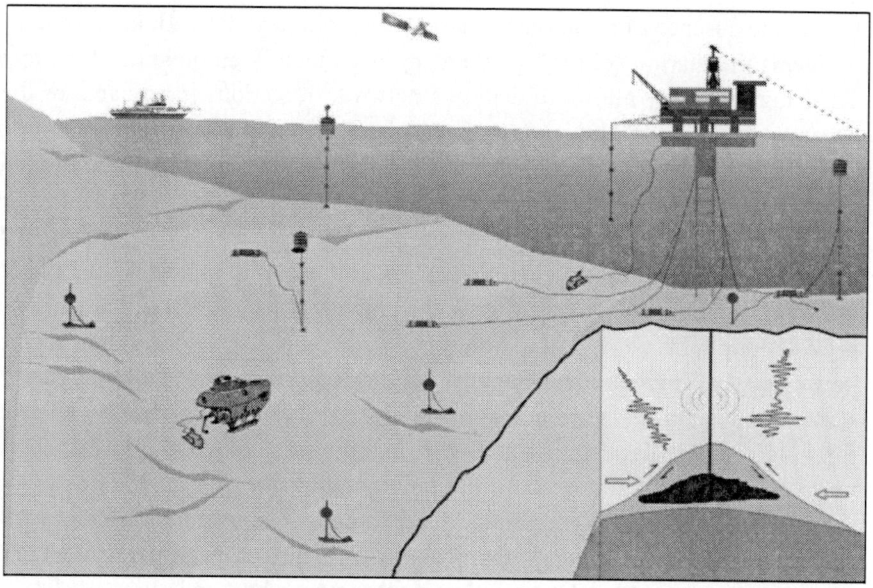

Figure 15. Scheme of the geo-ecological monitoring in the region near drilling platform.

Conclusion

1. The majority earthquakes occur under the bottom of the oceans and seas, but stationary marine seismological network still does not exist. Seismographs network of ground stations registrated records strong distant earthquakes with large distortions and weak earthquakes are never recorded. For this reason, it is impossible to adequately study the deep geotectonic processes under the seas and oceans. Insufficiently investigated also the structure of the oceanic crust and upper mantle, excitation mechanisms of tsunamis and other seismic events. Actual problems remain strong sea earthquake forecast, especially the registration of their short-term precursors and warning tsunami hazard. Therefore there is an urgent need for a unified global network of land and marine stations.
2. Differences seismic recording at the bottom waters consist in the influence of the water layer and water saturated layer of unconsolidated sediments , features coupling housing seismograph with the soft bottom and excitation noises near- bottom currents. There are technical difficulties due to autonomous operation station, which requires them to ensure high reliability. Furthermore, the station housing should be durable to protect the pressure at great depths and from impacts during installation. Substantial difficulties arise also providing duration power autonomous apparatus and transfer of seismological information to land stations. Radical solution to this problem is possible by using submarine cables, but the cost of their laying is high and reuse cables intercontinental communication is impossible due to their low reliability.

3. Due to the absence of a stationary marine seismological network at present in solving problems the marine seismology there are used mainly autonomous bottom seismic. Their high mobility allows to deploy a network of stations in any area of the World ocean. Configuration of this network may be selected in accordance with the specific objectives and having equipment and the cost of autonomous bottom stations are many times less than the cost of stationary cable stations. The achievements of modern science and technology in the field of marine instrumentation overcomes many difficulties in the development and operation of bottom seismographs. Modern electronic components measuring and computing technique and memory devices provide long-term work on the bottom of autonomous devices without replacing the power supply. There are known constructions long- term buoys allowing to provide a stable connection with the autonomous station via radio or satellite channels. The foregoing shows that creating stationary marine seismological network approaching to its metrological parameters to the land network, it is advisable to use long-term autonomous bottom seismographs. The foregoing shows that creating stationary marine seismological network approaching to its metrological parameters to the land network, it is advisable to use long-term autonomous bottom seismographs.

ACKNOWLEDGMENTS

The authors express their appreciation to the Special Design Bureau of Oceanological Technology RAS for active participation in the development and modernization of broadband autonomous bottom seismographs, Laboratory Seismology and Geodynamics Institute of Oceanology RAS, crews research vessels, which helped in a unique experimental work on recording seismic signals at the bottom, as well as Labortory analytical chemistry Institute of Oceanology for help in determining the composition and physical characteristics of the sediments.

REFERENCES

Abramov, O. K. & Grafov, B. M. (1978). Electrochemical receiver of mechanical oscillations, and possibility iof their use in seismology. Moscow: *Nauka*, *11*, 203-208. (in Russ.).

Aкi, K. & Richards, P. (1983). Quantitative seismology. *Theory and Methods*, V-*1*, S-Fr. Freeman & Co. 515 p.

Araki, E. & Suyehiro, K. (2001). Long period seismic noise in deep ocean boreholes. *Proc. OHP/ION Joint Symp*, Japan, P. 112-114.

Bradner, H., de Jerphanion, L. G. & Langlois, R. (1970). Ocean microseism measurements with a neutral buoyancy free-floating midwater seismometer. *Bull. Seism. Soc. Am.*, *60*(4), 1139-1150.

Catipovic, J. (1997). Acoustic Communication and Networking – State of the Art and Current Applications, *Int. Worksh. Sci entific Use of Submarine Cables*, Okinava, Japan, 1–6.

Chapman, N. R. & Chapman, D. M. (1993). A cogerent ray model of plane-wave reflection from a thin sediment layer. *JASA*, *94*(5), 2731-2738.

Duennebier, F. K., Blackinton, G. & Sutton, G. N. (1981). Current generated noise recorded on ocean bottom seismometers. *Mar. Geophys. Res*., *5*(1), 109 - 115.

Duennebier, F. K., McCreery, Ch. S., Harris, D., Cessaro, R. K., Fisher, C. & Anderson, P. (1985). *OSS IV: Noise levels, signal - to noise ratios, and noise sources* // Report, Part 7, 89-103.

Ewing, W. M., Jardetzky, W. S. & Press, F. (1957). *Elastic waves in layered media*. N-Y.: Mc-Gr. Hill Book Co.380 p.

Frolov, A. F. & Korotkih, I. V. (1983). *Engineering geology*. M.: Nedra,. 300 p. (in Russ.).

Gavrilov, V. A., Levchenko, D. G., Utyakov, L. L. & Shekhvatov, B. V. (2000). Hydrochemical bottom station for detection of short term predictors of the marine earthquakes. *Oceanology*, *40*(3), 456–467.

Geophysical Reference Book. (1966). Moscow: *Nedra*, 750 p.

Godin, O. A. & Chapman, D. M. F. (1999). Shear - speed gradients and ocean seismoacoustic noise resonances. *JASA*, *106*(5), 2367-2382.

Gutenberg, B. (1959). *Physics of the Earth's interior*, N-Y. Academic Press, 260 p.

Hempton, L. (1977). *Acoustic of the sea deposits*, M.: Mir, 550 p. (In Russ.).

Ilnitskaia, E. I., Teder, R. I., Vatolin, E. S. & Kuntish, M. F. (1969). *Property of rocks and methods of their definition*, M.: Nedra., 390 p. (in Russ.).

Isakovich, M. A. (1973). General acoustic. M.: *Nauka*. 495 p. (in Russ.).

Jacobson, R. S., Dorman, L. M., Purdy, G. M., Shultz, A. & Solomon, S. C. (1991). Ocean bottom seismometer facilities available, *Eos Trans. Am*., *12*, 506–515.

Jiang, L. & LeBlond, P. H. (1993). Numerical modeling of an underwater Bingham plastic mudslide and the wave which it generates. *J. Geophys. Res*., *98*(C6.), 10303-10317.

Kariakin, N. I., Bystrov, K. N. & Kireev, P. S. (1962). Brief handbook on physics. M.: *Vishaya Shkola*, 560 p. (in Russ.).

Kasahara, J., Koresava, S. & Nagumo, S. (1980). Experimental results of vortex generation around ocean bottom seismograph due to bottom current. *Bull. Earth. Res. Inst. Univ*., Tokyo, *55*, 169–182.

Kasahara, J., Iwase, R., Nakatsura, T., et al. (2003). Multidisciplinary VENUS observation at the Rynkyu Trench using Guam-Okinawa geophysical submarine cable. The 3rd Int. Worksh. *Scientific Use of Submarine Cables and Related Technologies*, Tokyo, Japan, 25–30.

Kaufman, A. A., Levshin, A. L. & Larner, R. L. (2002). *Acoustic and elastic wave fields in geophysics*, N-Y. Els. 661 p.

King, C. Y. (1980). Geochemical measurements pertinent to earthquake prediction. *J Geph. Res*., *85*(B6), 3051-3121.

Landau, L. D. & Lifshits E. M. (1988). Theoretical physics. Hydrodynamics. M.: *Nauka*, 720 p. (in Russ.).

Landau, L. D. & Lifshits E. M. (1987). Theoretical physics. The theory of elasticity. M.: *Nauka*, 245 p. (in Russ.).

Levchenko, D. G., Soloviev S. L., Sonkin A. V. & Voronina E. V. (1994). Recording of ocean-bottom seismic noise and of a strong earthquake in the Himalayas by broadband digital OBS installed on the Mid-Atlantic ridge. *Physics Earth and Planet. Int.*, *84*, 305-320.

Levchenko, D. G., Timoshuk, E. P., Grafov, B. M., Sirotinskiy, Yu. V., Rykov, A. V., Ulomov, I. V. & Marchenkov, A. Yu. (1996). The comparative analysis various seismic sensors in structure digital broadband seismic station. *Proc. 4th Intern. Worksh. on Electrochemical Flow Measurements - Fundamentals and Applications*. Lanstein, Germany, (*17*), 1-10.

Levchenko, D. G. & Matsievsky, S. A. (2000). Broadband digital bottom seismic stations. *Seism.Instruments*, *33*, 52–68.

Levchenko, D. G. (2001). Peculiarities of designing broadband bottom seismographs. *Oceanology*, *41*(4), 586–599.

Levchenko, D. G. (2002). Results of the registration of broadband (0.003–10 Hz) seismic signals on the sea bottom. *Oceanology*, *42*(4), 594–605.

Levchenko, D. G. (2006). Effect of the water-saturated sediment layer on recording seismic signals with a bottom seismometer. ISSN 0001-4370, *Oceanology*, *46*(5), 739–750. Pleiades Publishing, Inc. DOI: 10.1134/S0001437006605050146

Levchenko, D. G. (2009). Propagation of storm microseisms at the ocean–continent boundary. ISSN 0001-4370, *Oceanology*, *49*(2), 257–264. Pleiades Publishing, Inc. DOI: 10.1134/S0001437009020106.

Levchenko, D. G., Ledenev, V. V., Il'in, I. A. & Paramonov A. A. (2010a). Long term seismological sea bottom monitoring using autonomous bottom stations. ISSN 0747_9239, *Seismic Instruments*, *46*(1), 1–12. © Allerton Press, Inc. DOI: 10.3103/S0747923910010019.

Levchenko, D. G., Kuzin, I. P., Safonov, M. V., Sichikov, V. N., Ulomov, I. V. & Kholopov, B. V. (2010b). Experience in seismic signal recording using broadband electrochemical seismic sensors. *Seismic Instruments*, *46*(3), 236-250. DOI: 10.3103/S074 79239 10030072.

Levchenko, D. G. & Kuzin, I. P. (2011). General seismicity of Black Sea and recent seismic investigation in this region. In A.L. Ryann and N.J. Perkins (Eds). *The Black Sea: Dynamics, Ecology and Conservation. P.*, 207-225. N-Y, U.S.: Nova Science Publishers.

Levchenko, D. G., Kuzin, I. P., Zubko, Yu. N. & Chervinchuk, S. Yu. (2012). The Japan earthquake: problems of forecast. ISSN 0747_9239, *Seismic Instruments*, *48*(1), 10–15. © Allerton Press, Inc. DOI: 10.3103/S0747923912010070

Levchenko, D. G. & Levchenko, V. D. (2013). Nature of "Transparency windows" in the broadband spectrum of seafloor microseisms. ISSN 0001_4370, *Oceanology*, *53*(3), 294–307. © Pleiades Publishing, Inc., DOI: 10.1134/S0001437013020100

Levchenko, D. G., Lobkovski, L. I. & Rimski-Korsakov, N. A. (2013). Problems and prospects of creation of a Global seismological network. Rep. Confer. MSOI-2013. P.

201-212. ISBN 978-5-904761-36-3. Moscow, RF.: Shirshov Institute of Oceanology RAS.

Lisitsin, A. P. (1974). Formation of deposits at oceans. Moscow, Russia: *Nauka*, 438. (in Russ.).

Liu, K. F. & Mei, C. C. (1989). Slow spreading of a sheet of Bingham fluid on an inclined plane. *J. Fluid Mech*, *207*, 505-529.

Lobkovsky, L. I., Levchenko, D. G., Leonov, A. V. & Ambrosimov, A. K. (2005). Geoekologicheskii monitoring morskikh neftegasonosnykh akvatorii (Geoecological monitoring of oil and gas bearing water areas). Moscow, Russia: *Nauka*. 326 p. (in Russ.)

Mei, C. C. & Liu, K. F. (1987). Bingham plastic model for a muddy seabed under long waves. *J. Geophys. Res*, *92*, 14581-14594.

Mikada, H., Hirata, K., Matsumoto, H., et al. (2003). Scientific results from underwater earthquake monitoring using cabled observatories. *The 3rd Int. Worksh.: Scientific use of submarine cables and related technol.*, Tokyo, Japan, 3–7.

Montagner, J. P., Karczewski, J. F., Romanowicz, B. & the Geoscope Group. (1992). The GEOSCOPE network: the state of the art and perspectives// *Abst. Book. Intern. Simp. Ten Years of GEOSCOPE Broadband Seismology*, Paris, France, 6 – 15.

Neprochnov, Y. P., Sedov, V. V. & Ostrovsky, A. A. (1983). Experimental research of bottom seismic noise at ocean // *Oceanology*, *23*(2), 276-283.

Ostrovsky, A. A. (1998). Bottom seismic experiments. Moscow, Russia: *Nauka*, 255 p. (in Russ.).

Prothero, W. A. (1984). Ocean bottom seismometer technology. EOS. *Trans. Amer. Geoph. Un.*, *65*(13), 113-116.

Purdy, G. M. & Dzievonski, A. M. (1995). Towards a permanent network of ocean floor seismological observatories. *Abstr. Book, Intern. Worksh. Multidisciplinary Observatories on the deep sealfoor*. Marseille, France, 165 - 168.

Romanovicz, B., Ctakes, D., Dolenc, D., Neuhauser, D., MeGill, P., Uhrhammer, R. & Ramirez, T. (2006). The Monterey Bay broadband ocean bottom seismic observatory // *Ann. Geoph*., V. *49*. № 2/3. P.607-622.

Savarensky, E. F. (1972). Seismicheskie volny (Seismic waves), Moscow, russia: *Nedra*, 290 p. (in Russ.).

Soloviev, S. L. (1986). Seismological bottom supervision in the USSR and abroad. Moscow, Russia: *Nauka*, 118 p. iIn Russ.).

Shiobara, H., Moshisuki, M., Shinohara, M., et al. (2001). Long term OBS array observations – development and preliminary results. *Proc. OHP/ION Joint Symp*, Japan, 106–108.

Sutton, G. H., Duennebier, F. K. & Iwatake, B. (1981a). Coupling of ocean bottom seismometers to soft bottoms. *Mar. Geo phys. Res*., *5*(1), 35–51.

Sutton, G. H., Levis, B. T. R., Ewing, J., Duennebier, F. K., Iwatake, B. & Tuthill, D. (1981b). An overview and general results of the Lopez Island OBS Experiment. *Mar. Geophys. Res*., *5*(1), 3-34.

Sutton, G. H. & Duennebier, F. K. (1987). Optimum design of ocean bottom seismometers. *Mar. Geophys. Res*., *9*, 47-65.

Tang, L., Titov, V. V. & Chamberlin, C. D. (2009). Development, testing, and applications of site-specific tsunami inundation models for real-time forecasting. *J. Geophys. Res.*, *114*(C12025), DOI:10.1029/ 2009JC005476

Trehu, A. & Solomon, S. (1981). Coupling parameters of the MIT OBS at two near-shore sites. *Mar. Geophys. Res.*, *5*(1), 69-78.

Trehu, A. (1985). Coupling of ocean bottom seismometers to sediment: results of tests with U.S. Geological Survey ocean bottom seismometer. *Bull. Seism. Soc.*, *75*(1), 271-289.

Trehu, A. (1985). A Note on the effect of bottom currents on an ocean bottom seismometers. *Bull. Seism. Soc. Amer.*, *75*(4), 1195–2004.

Usher, M. J., Burch, R. F. & Gurlap, C. (1979). Wide-band feedback seismometers. *Phys. Earth Planet. Interiors*, V-*18*. P.38-50.

Volovov, V. I. (1997). Reflection of a sound from a bottom at ocean. Moscow, russia: *Nauka*, *300* (in Russ.).

Webb, S. C. (1992). The equilibrium oceanic microseism spectrum. *JASA*, *92*(4P1), 2141-2157.

Wielandt, E. & Steim, J. M. (1986). A digital very broadband seismograph. *Annales Geophys.*, *4*, 227–232.

Zelikovitz, S. J. & Prothero, W. A. (1981). The vertical response of an ocean bottom seismometer: Analysis of the Lopez Island vertical transient tests. *Mar. Geoph. Res.*, *5*(10), 53-68.

Zubko, Yu. N., Levchenko, D. G., Ledenev, V. V. & Paramonov, A. A. (2003). Modern bottom stations for seismic survey and seismic monitoring. *Nauchnoe Priborostroenie*, *13*(4), 70–82. (in Russ.).

INDEX

C

D

E

F

G

H

I

J

K

L

M

N

O

P

Q

R

S

T

X

Y

Z